Advances in

Heterocyclic Chemistry

Volume 74

Editorial Advisory Board

R. A. Abramovitch, *Clemson, South Carolina*
A. T. Balaban, *Bucharest, Romania*
A. J. Boulton, *Norwich, England*
H. Dorn, *Berlin-Bohnsdorf, Germany*
J. Elguero, *Madrid, Spain*
S. Gronowitz, *Lund, Sweden*
E. Lukevics, *Riga, Latvia*
O. Meth-Cohn, *Sunderland, England*
V. I. Minkin, *Rostov-on-Don, Russia*
C. W. Rees, FRS, *London, England*
E. F. V. Scriven, *Indianapolis, Indiana*
D. StC. Black, *Kensington, Australia*
E. C. Taylor, *Princeton, New Jersey*
M. Tišler, *Ljubljana, Slovenia*
J. A. Zoltewicz, *Gainesville, Florida*

Degenerate Ring Transformations of Heterocyclic Compounds

Henk C. van der Plas
Laboratory of Organic Chemistry
Wageningen University
Wageningen, The Netherlands

Advances in Heterocyclic Chemistry
Volume 74

Edited by
ALAN R. KATRITZKY, FRS

ACADEMIC PRESS
San Diego London Boston New York
Sydney Tokyo Toronto

This book is printed on acid-free paper.

Copyright © 1999 by ACADEMIC PRESS

All Rights Reserved.
No part of this publication may be reproduced or transmitted in any form or by any means, electronic or mechanical, including photocopy, recording, or any information storage and retrieval system, without permission in writing from the Publisher.
The appearance of the code at the bottom of the first page of a chapter in this book indicates the Publisher's consent that copies of the chapter may be made for personal or internal use of specific clients. This consent is given on the condition, however, that the copier pay the stated per copy fee through the Copyright Clearance Center, Inc. (222 Rosewood Drive, Danvers, Massachusetts 01923), for copying beyond that permitted by Sections 107 or 108 of the U.S. Copyright Law. This consent does not extend to other kinds of copying, such as copying for general distribution, for advertising or promotional purposes, for creating new collective works, or for resale. Copy fees for pre-1999 chapters are as shown on the title pages. If no fee code appears on the title page, the copy fee is the same as for current chapters.
0065-2725/99 $30.00

Explicit permission from Academic Press is not required to reproduce a maximum of two figures or tables from an Academic Press chapter in another scientific or research publication provided that the material has not been credited to another source and that full credit to the Academic Press chapter is given.

Academic Press
A Harcourt Science and Technology Company
525 B Street, Suite 1900, San Diego, California 92101-4495, USA
http://www.apnet.com

Academic Press
24-28 Oval Road, London NW1 7DX, UK
http://www.hbuk.co.uk/ap/

International Standard Book Number: 0-12-020774-5

PRINTED IN THE UNITED STATES OF AMERICA
99 00 01 02 03 04 BB 9 8 7 6 5 4 3 2 1

Contents

Editor's Preface ix
About the Author xi

Chapter I

A. Introduction 1
B. Classification of Degenerate Ring Transformations of Heterocyclic Systems 1
C. Ring-Bond-Redistribution Graphs 5

Chapter II S_N(ANRORC) Reactions in Azines Containing an "Outside" Leaving Group

A. The Discovery of the S_N(ANRORC) Concept 9
B. S_N(ANRORC) Reactions in Monoazines 14
    1. Pyridines 14
    2. Isoquinolines 19
C. S_N(ANRORC) Substitutions in Diazines 21
    1. Pyrimidines 21
        a. Aminodehalogenation of 6-Halogeno-4-Substituted Pyrimidines 21
        b. Aminodehalogenation in 4-Halogeno-2-Substituted Pyrimidines 31
        c. Aminodehalogenation of 2-Halogeno-4-Substituted Pyrimidines 34
        d. Aminolysis of Pyrimidines Containing a Leaving Group at C-2 Different from Halogen 39
        e. Aminodebromination of 5-Bromopyrimidines 41
        f. Aminodemethoxylation of Dimethoxypyrimidines 45
        g. Aminodehydrogenation of Pyrimidines 46
    2. Quinazolines, Purines, and Pteridines 53
        a. Aminodechlorination of 4-Chloroquinazolines 53
        b. Aminodehalogenation of 2-Halogenoquinazolines 55
        c. Aminodeoxogenation of Quinazolin-4-one 56
        d. Aminodehydrogenation of Quinazoline(s) 58
        e. Aminodehalogenation of Halogenopurines 58

f. Aminodehalogenation of 2-Halogenopteridines and Aminodethiomethylation of 2-Methylthiopteridines... 62
3. Pyrazines ... 65
4. Pyridazines and Phthalzines ... 67
D. S_N(ANRORC) Substitutions in Triazines and Tetrazines ... 69
1. 1,2,4-Triazines and Benzo-1,2,4-Triazines... 69
   a. Aminodemethylthiolation of 3-Methylthio-1,2,4-Triazines... 69
   b. Aminodehalogenation of 3-Halogeno-1,2,4-Triazines ... 71
   c. Aminolysis of 1,2,4-Triazines Containing at C-3 a Leaving Group Different from Halogen ... 74
   d. Aminodeoxogenation ... 75
   e. Aminodechlorination of Chlorobenzo-1,2,4-Triazines... 75
2. 1,3,5-Triazines ... 77
   a. Aminodehydrogenation (Chichibabin Amination) of (Di)phenyl-1,3,5-Triazines ... 77
   b. Aminolysis of 2-X-4,6-diphenyl-1,3,5-Triazines ... 79
3. 1,2,4,5-Tetrazines... 81
   a. Hydrazinodehydrogenation of 1,2,4,5-Tetrazines ... 81
   b. Hydrazinodeamination and Hydrazinodehalogenation of Amino- and Halogeno-1,2,4,5-Tetrazines ... 85

Chapter III S_N(ANRORC) Reactions in Azaheterocycles Containing an "Inside" Leaving Group

A. Degenerate Ring Transformations Involving the Replacement of the Nitrogen of the Azaheterocycle by Nitrogen of Reagent ... 87
1. Pyridines ... 87
2. Pyrimidines... 94
   a. N-Alkylpyrimidinium Salts... 94
   b. N-Aminopyrimidinium Salts... 104
   c. N-Arylthiopyrimidones, N-Arylpyrimidinium Salts, and Quinazoline (Di)ones... 108
   d. N-Nitropyrimidones ... 112
   e. Photostimulated Degenerate Ring Transformations of Thymines... 117
3. N-Aryl-1,2,4-Triazinones... 121
4. Benzodiazaborines ... 122
5. Imidazoles... 123
B. Degenerate Ring Transformations Involving the Replacement of the C–N Fragment of the Ring by a C–N Moiety of a Reagent ... 130
C. Degenerate Ring Transformations Involving Replacement of a Three-Atom Fragment of the Heterocyclic Ring by Three-Atom Reagent Moiety ... 131
1. CNN and CCC Fragment Replacement... 131
   a. Pyridines ... 131
   b. Pyrimidines... 135
2. Replacement of a CCC Fragment ... 136
3. Replacement of the NCN or CNC Fragment ... 137
   a. Pyrimidines... 137
   b. 1,3,5- and 1,2,4-Triazines... 148

D. Degenerate Ring Transformation Involving the Replacement of a Carbon Atom of the Heterocyclic Ring by a Carbon Atom of a Nucleophilic Reagent 149

Chapter IV Degenerate Ring Transformations Involving Side-Chain Participation

A. Introduction 153
B. Degenerate Ring Transformations Involving Participation of One Atom of a Side Chain 155
    1. Five-Membered Heterocycles 155
        a. 1,2,3-Triazoles and Tetrazoles 155
        b. 1,2,4-Triazoles 158
        c. 1,2,4-Thiadiazolines and 1,2,4-Thiadiazolidines 159
        d. 1,2,4- and 1,3,4-Dithiazolidines 161
        e. Thiazolines and Imidazolidines 162
    2. Six-Membered Heterocycles 163
        a. Pyridines 163
        b. Pyrimidines 165
        c. Triazines 172
        d. Pteridines, Purines, Quinazolines, and Azolopyrimidines 174
C. Degenerate Ring Transformations Involving the Participation of Two Atoms of a Side Chain 189
    1. Five-Membered Heterocycles 190
        a. Oxazoles and Isoxazoles 190
        b. 1,2,3-Triazoles 191
    2. Six-Membered Heterocycles 195
        a. Pyridines 195
D. Degenerate Ring Transformations Involving Participation of Three Atoms of a Side Chain 199
    1. Degenerate Ring Transformations Involving Nitrogen as Pivotal Atom 200
        a. 1,2,4-Oxadiazoles (RA = CN, DCB = NCO, Scheme IV.3) 200
        b. 1,2,5-Oxadiazoles (RA = CN, DCB = CNO, Scheme IV.3) 205
    2. Degenerate Ring Transformations Involving Sulfur as Pivotal Atom 207
        a. 1,2,4-Thiadiazoles (RA = CS; DCB = NCN, Scheme IV.3) 207
        b. 1,2,3-Thiadiazoles (RA = CS, DCB = CNN, Scheme IV.3) 213
        c. Isothiazoles (RA = CS; DCB = CCN, Scheme IV.3) and Thiazoles (RA = CS; DCB = CNC, Scheme IV.3) 214
        d. 1,2 Dithioles (RA = CS, DCB = CCS, Scheme IV.3) 217
        e. Thiophenes (RA = CS, DCB = CCC, Scheme IV.3) 218
    3. Degenerate Ring Transformations Involving Carbon as Pivotal Atom 218
        a. Pyrroles (RA = CN, DCB = CNN, Scheme IV.3) 218
        b. 1,2,4-Triazoles (RA = CN, DCB = CNN, Scheme IV.3) 220
        c. 1,2,3-Triazoles (RA = CN, DCB = CNN, Scheme IV.3) 220

REFERENCES 223
INDEX 241

Editor's Preface

Volume 74 of *Advances in Heterocyclic Chemistry* is a monograph dedicated to degenerate heterocyclic ring transformations and authored by Professor H. C. van der Plas of Wageningen University, The Netherlands.

Professor van der Plas provided the first comprehensive review of ring transformations of heterocycles in his authoritative monograph *Ring Transformations of Heterocycles,* which was published by Academic Press in 1973. Degenerate ring transformations form a subclass of heterocyclic ring transformations in which the final product has a heterocyclic system identical with the starting material, but one or more of the ring atoms have been interchanged with the same atoms from the reagent or starting material.

The ANRORC (addition of nucleophile, ring opening, ring closure) class of rearrangements was originally discovered by Professors van der Plas and den Hertog at Wageningen. Degenerate heterocyclic rearrangements in general, and ANRORC reactions in particular, have since been extensively studied at Wageningen and elsewhere, and this volume is the first comprehensive overview of this interesting field. Of equal fascination are degenerate ring transformations involving side-chain participation, which form the other main topic of this volume.

ALAN R. KATRITZKY

About the Author

Professor Dr. H. C. van der Plas received the Ph.D. from the University of Amsterdam. He served the Agricultural University at Wageningen (The Netherlands) from 1966 as a reader, from 1970 as Professor of Organic Chemistry, and from 1978–1982 and 1989–1995 as Rector Magnificus. His research interest is heterocyclic chemistry, mainly in the area of nucleophilic substitution and ring transformations. The results of his scientific reseach are set down in nearly 400 research papers, 20 review articles, and *Ring Transformations of Heterocycles* (Academic Press, 1973) and (together with O. Chupakhin and V. Charushin) *Nucleophilic Substitution of Aromatic Hydrogen* (Academic Press, 1994). Professor van der Plas was president of the Royal Netherlands Chemical Society, the International Society of Heterocyclic Chemistry, and the European Agricultural Network (NATURA). He received the first Award of the International Society of Heterocyclic Chemistry and was honored with honorary degrees from the University of Wrozlaw, University of Leuven, Agricultural University of Prague, Technical University of Cracow, and Agricultural University of Moscow. He is a foreign member of the Russian Academy of Sciences.

Chapter I
Degenerate Ring Transformations

A. Introduction

The ability of heterocyclic compounds to undergo ring transformation reactions (the first examples were discovered more than a century ago) is a fascinating feature of their chemistry. Great experimental developments in this area have been achieved and many examples of heterocyclic ring transformations have been found for almost all heterocycles of any size, with any type, number, and distribution of hetero atoms. An essential feature in all ring transformations being described is that at least one (hetero) atom of the ring in the starting material is incorporated in the ring of the final product.

The first classical review on the ring transformations of three-, four-, five-, and six-membered heterocycles appeared in 1973 (73MI1); it was followed by excellent reviews in specific topics such as the monocyclic rearrangements of five-membered heterocycles [74MI1; 77AG(E)572; 81AHC141; 82T3537; 84JHC627, 84MI1; 92AHC49], the transformation of pyridines into benzenes (81T3423; 88KGS1570), rearrangement reactions of pyrylium salts (82MI1), ring transformations of pyrimidines [74MI2; 78ACR462, 78KGS867; 84H289; 85T237; 94KGS1649; 95H(40)441], and pyrimidinum salts (78H33; 80WCH491).

A specific type of heterocyclic rearrangement is the degenerate heterocyclic ring transformation, which refers to reactions in which, after the rearrangement, the heterocyclic system in the final product is still the same as in the starting material, but with the important difference that one or more atoms of the starting material are "interchanged" with the *same* atoms present in reagent, the side chain, or even in the starting material itself. These rearrangements are often discovered by isotopic labeling methods or by low-temperature controlled NMR studies.

B. Classification of Degenerate Ring Transformations of Heterocyclic Systems

Degenerate ring transformations of heterocyclic systems that have been discovered to date can be classified in two main groups: (1) Nucleophilic

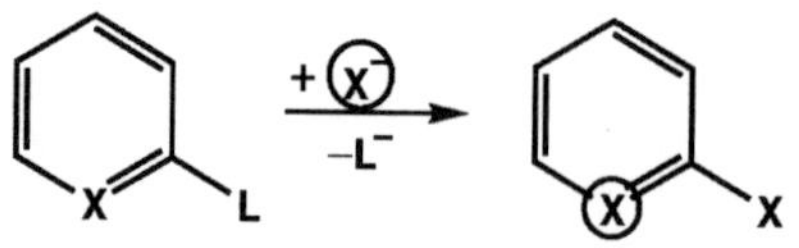

2 CLASSIFICATION OF DEGENERATE RING TRANSFORMATIONS

SCHEME I.1

substitution reactions in which the displacement of a leaving group takes place with a *simultaneous* replacement of one (or more) of the ring atom(s) by one (or more) identical atom(s) present in the nucleophilic reagent (78ACR462).

These reactions can be divided in two subgroups: reactions in which the leaving group is present as a substituent on the heterocyclic ring ("outside" leaving group) (Scheme I.1) and those in which the leaving group forms an integral part of the heterocyclic system ("inside" leaving group) (Scheme I.2). This "inside" leaving group can be one atom, but there are many examples known in which the "inside" leaving group consists of more than one atom (78H33).

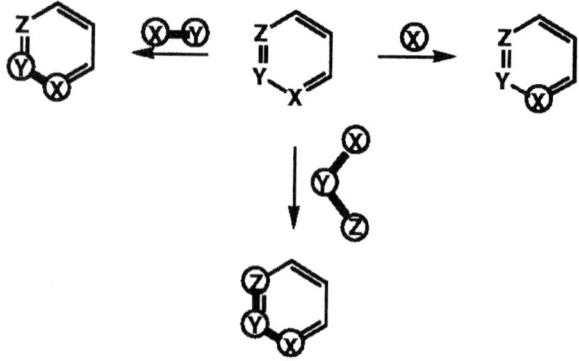

SCHEME I.2

These substitution reactions have been studied in detail, mainly in the laboratory of Organic Chemistry of the Agricultural University of Wageningen (the Netherlands). They are found to proceed in many heterocyclic ring systems.

An illustrative example of a reaction as represented in the general Scheme I.1 is the amino-debromination of 6-bromo-4-phenylpyrimidine (Scheme I.3). This nucleophilic substitution is described with the acronym S_N(ANRORC), indicating that it occurs according to a reaction sequence involving the initial *A*ddition of a *N*ucleophile, *R*ing *O*pening, and *R*ing *C*losure (71RTC1239; 72RTC1414; 73RTC145, 73RTC442). This type of reac-

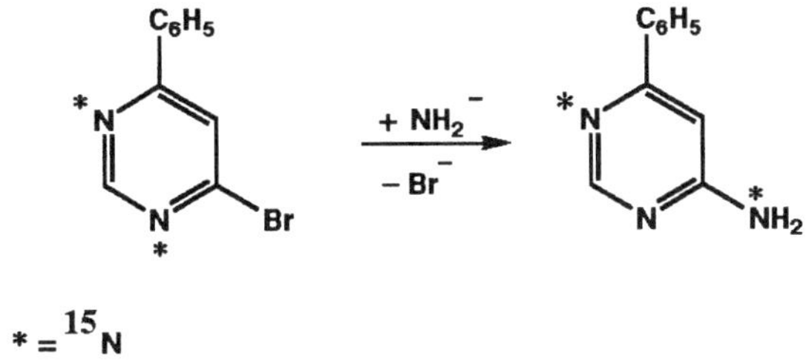

DEGENERATE RING TRANSFORMATIONS

SCHEME I.3

tion will be extensively discussed in this monograph in Chapters II and III. A reaction illustrating the role of an "internal" leaving group is the apparent demethylation reaction of *N*-methylpyrimidinium salts by liquid ammonia (Scheme I.4) (74RTC114).

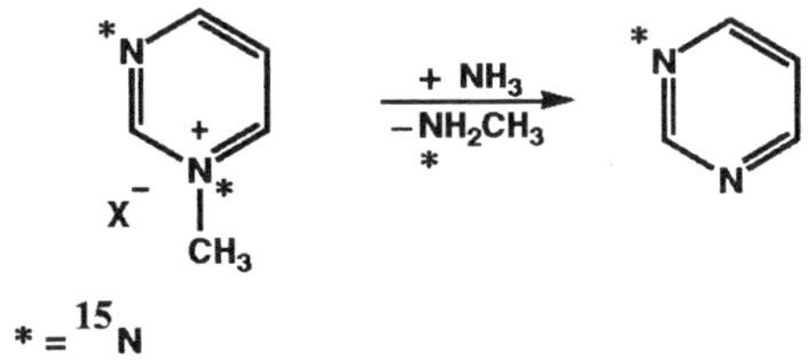

SCHEME I.4

In some reactions, the ring rearrangement takes place with side-chain participation, replacing one or more atoms of the heterocyclic ring by one or more atoms of the side chain (Scheme I.5). These reactions may be triggered by an initial reaction with a basic reagent, but in contrast to the reactions mentioned in Schemes I.1 and I.2, *no* incorporation takes place of the atom(s) of this reagent into the heterocyclic ring.

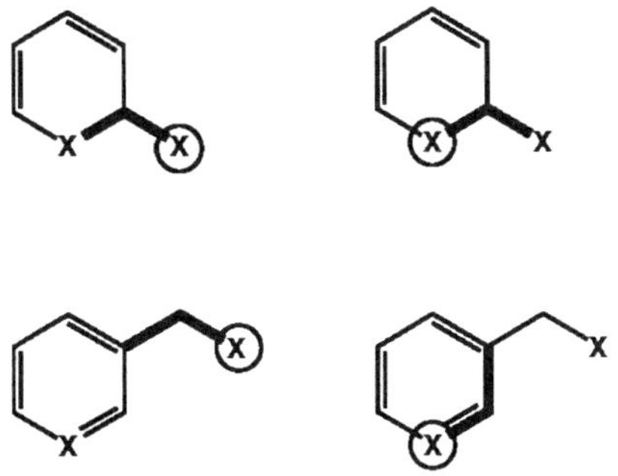

SCHEME I.5

4 CLASSIFICATION OF DEGENERATE RING TRANSFORMATIONS

These rearrangements can in general be classified as ANRORC reactions, since they are characterized by the "troika" process, involving the three steps of *A*ddition of a *N*ucleophile, *R*ing *O*pening, and *R*ing *C*losure. However, it needs to be emphasized that these reactions cannot be classified as S_N(ANRORC) substitutions, since in the rearrangement no nucleophilic substitution is involved. An example of this type of rearrangement is the well-known base-induced Dimroth reaction of 1-alkyl-2-iminopyridines into 2-alkylaminopyridines (Scheme I.6) (68MI1; 69ZC241; 84JHC627). These base-induced reactions with side-chain participation are extensively discussed in Chapter IV.

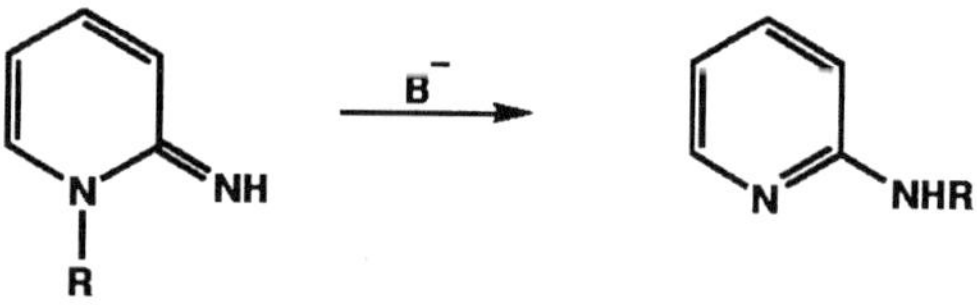

SCHEME I.6

Numerous degenerate ring transfomation reactions with side-chain participation can occur without base catalysis, but may proceed by thermolysis or by photostimulation. The degenerate reactions form a specific branch of the more generally occurring and widely studied ring transformations, known as the Boulton–Katritzky rearrangement reactions (72UK1788; 74MI1; 81AHC141; 84JHC627, 84MI1). They are schematically presented in Scheme I.7.

SCHEME I.7

A few examples, illustrating the principle of this type of degenerate rearrangement, are the thermo-induced equilibrium shift of 3-benzoylamino-5-methyl-1,2,4-oxadiazole into 3-acetylamino-5-phenyl-1,2,4-oxadiazole and the isomerization of 5-benzoyl-methylfuroxan oxime into 4-[α-nitroethyl]-3-phenylfurazan (82G181) (Scheme I.8). They are extensively discussed in Chapter IV.

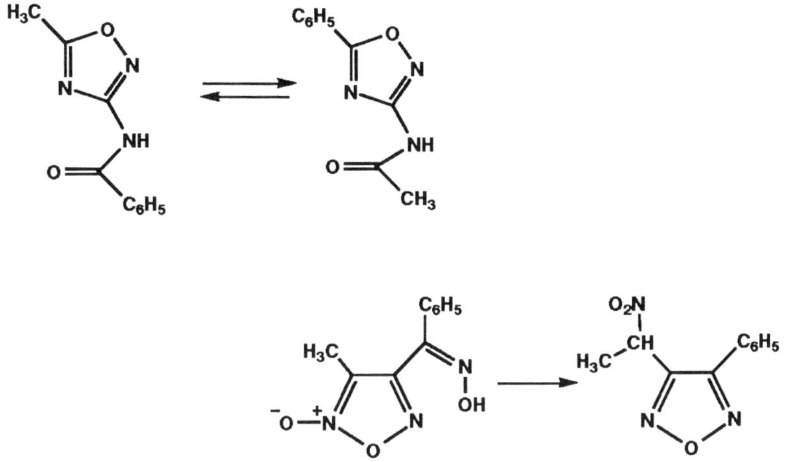

SCHEME I.8

C. Ring-Bond-Redistribution Graphs

The great diversity of the many heterocyclic ring transformations induced efforts to classify them. Different types of mathematical models have been suggested to describe and classify rearrangement reactions. Interesting papers have been published in which heterocyclic rearrangements were classified on the basis of ring-bond-redistribution graphs (RBR graphs). These graphs reflect the topological changes of the heterocyclic nucleus in the course of the ring transformations and offer the possibility of determining the degree of similarity (or dissimilarity) for heterocyclic recyclizations and rearrangements (92BSB67, 92KGS808, 92MI3; 93JA2416; 97MI2).

Although for more detailed information it is necessary to consult the original articles, in principle the RBR graph of a reaction consists of a set of solid, dashed, and bold lines, in which the solid lines represent the bonds that are involved in the redistribution process but are not broken, the dashed lines represent the bonds that are broken or formed during the rearrangement, and the bold lines represent the chain of atoms that are unchanged during the conversion and are thus common in both initial and final product.

To illustrate the principle of the construction of or RBR graph, a few examples are chosen from the different categories of degenearate ring transformations, as classified in Section I.B.

The amidine rearrangement of N-alkyl-2-iminopyridines into 2-alkylaminopyridines (Scheme I.9). In this rearrangement bond breakage occurs

SCHEME I.9

between positions 1,6 and bond formation between positions 6,7. Both bond-breaking and bond-forming are indicated by dashed lines. The bonds between positions 1,2 and 2,7 in the starting material are involved in the redistribution, but are not broken. They are represented in the graph by solid lines. It is also evident that the chain of atoms consisting of atoms 2, 3, 4, 5, and 6 is common in the starting material and in the final product; these atoms are represented as bold lines. So the Dimroth rearrangement can graphically be represented as graph G_1 in Scheme I.10. A more simplified molecular graph is the G_0 graph, which differs from the G_1 graph where all the dashed edges are changed to the solid ones, i.e., a graph with only two sort of edges: solid and bold ones.

In order to give more detailed information about the rearrangement, G_2 and G_3 graphs are developed from the G_1 graph. One constructs the G_2 graph from the G_1 graph by introduction of the electrophilic centers (indicated by empty circles) and nucleophilic centers (indicated by heavy dots) involved in the transformation. If one also marks the position of the hetero atoms involved in the rearrangements, one obtains the G_3 graph.

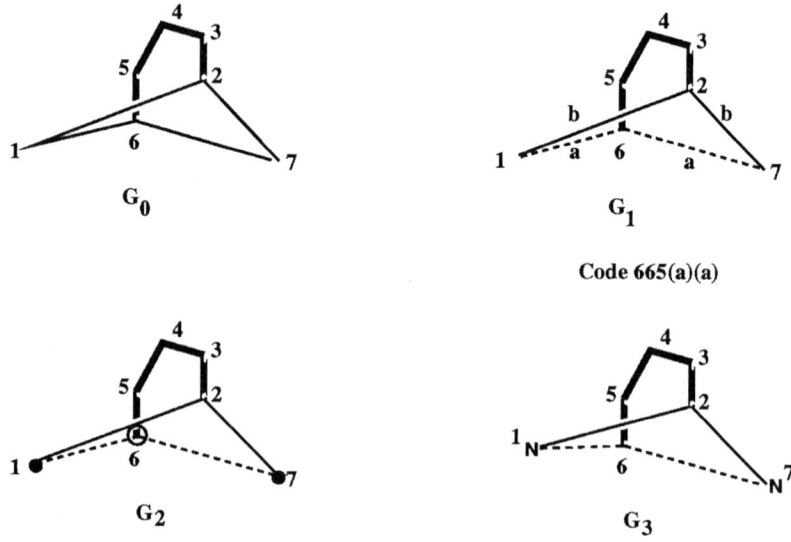

Code 665(a)(a)

SCHEME I.10

Thus, the G_3 graph can be considered as a G_1 graph in which the nature and position of the hetero atom(s) are indicated (Scheme I.10). It is evident that the geometric representation of the graphs describing this *degenerate* Dimroth ring rearrangement obeys visual symmetry (consider, for instance, *mirror plane* symmetry involving the chain C2–C3–C4–C5–C6). For a more accurate definition of symmetric properties of the RBR graphs the reader is referred to the literature (92MI1).

The amino-debromination of 6-bromo-4-phenylpyrimidine (Scheme I.3). The reaction has been proved to occur by the formation of an initial σ-adduct at C-2, which subsequently rearranges into the 6-amino product [S_N(ANRORC) mechanism].

Based on the rules outlined above, let us consider by what kind of G_0, G_1, G_2, and G_3 graph the amino-debromination can be described (72RTC1414; 73RTC145, 73RTC442). The reaction occurs by bond breaking between the atoms 1 and 2 and bond formation between atoms 6 and 7 (dashed lines). The bonds that are involved in the redistribution but are not broken are those between atoms 1 and 6 and those between 2 and 7 (solid lines). The G_1 graph and consequently the G_0, G_2, and G_3 graphs can thus be represented as pictured in Scheme I.11. It is evident that these G graphs have the

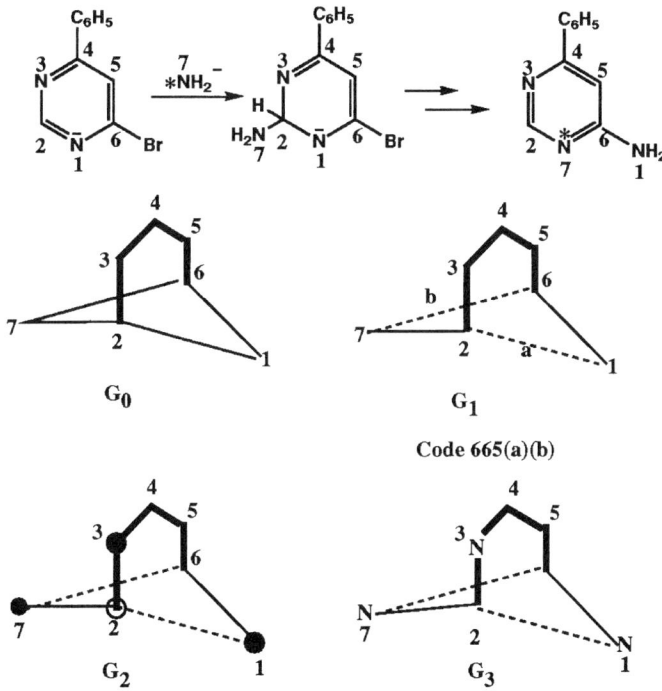

SCHEME I.11

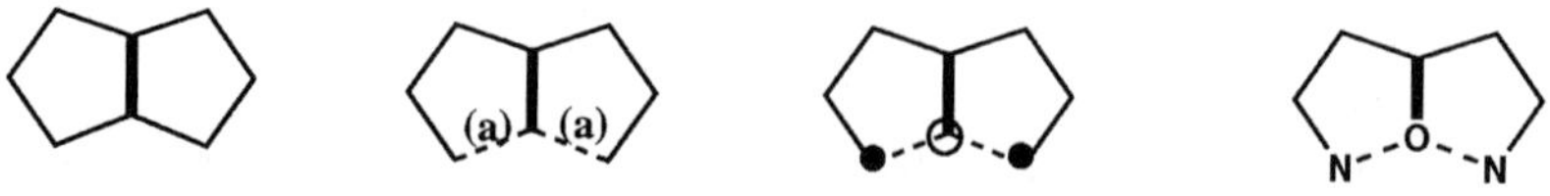

G_0 G_1 G_2 G_3

Code 552(a)(a)

SCHEME I.12

same topology as those of the Dimroth rearrangement (Scheme I.10). They also feature the characteristic symmetries: The G_0 graph has the same symmetry as the one describing the Dimroth rearrangement (symmetry plane); the G_3, however, has no plane, symmetry.

A third example to illustrate the construction of G graphs is *the furoxan–furazan rearrangement (see Scheme I.8)*. Based on the principles outlined previously it is evident that this rearrangement can be pictured by the G_0, G_1, G_2, and G_3, graphs as presented in Scheme I.12. They also show the mirror plane symmetry.

From the examples presented in Schemes I.10, I.11, and I.12, it is evident that the G_0 graphs in these *degenerate* ring transformations are all characterized by mirror plane symmetry.

A simple code has been developed to describe the different rearrangements. In Scheme I.10 the G graphs consist of two six-membered bicyclic rings that have in common a chain of five atoms. On the basis of the distribution of the *dashed* lines, the code for the Dimroth rearrangement (Scheme I.10) can be described as 665 (a)(a). In a similar way, for the amino-debromination (Scheme I.11), the code 665 (a) can be assigned and the code 552(a)(a) for the furoxan–furazan rearrangement (see Scheme I.12) [see for further details the original literature (93JA2416)].

Chapter II

S$_N$(ANRORC) Reactions in Azines, Containing an "Outside" Leaving Group

A. The Discovery of the S$_N$(ANRORC) Concept

The intermediacy of benzyne in reactions of halobenzenes with strong bases [potassium amide/liquid ammonia (53JA3290; 55JA4540; 56JA601; 60JA3629), lithium piperidide/piperidine (60AG91, 60MI1, 60T29)] induced great interest in studies on the possible occurrence of heteroarynes (didehydrohetarenes) in corresponding reactions of halohetarenes (67MI1). When 3-chloro-, 3-bromo-, and 3-iodopyridine as well as 4-chloro, 4-bromo-, and 4-iodopyridine were reacted with potassium amide at $-33°C$, it was found (1) that in all six reactions a mixture of 3- and 4-aminopyridine was obtained in the ratio 1 : 2 and (2) that this ratio was independent of the nature and position of the halogen substituent (61RTC1376; 65AHC121). Both facts were correctly considered as a strong indication for the occurrence of the intermediate 3,4-pyridyne. The case for its existence was further strengthened by trapping of this intermediate by cycloaddition with furan, the endoxide of 5,8-dihydroisoquinoline being isolated (Scheme II.1) (74RTC166).

Similar results were also obtained when 3-chloro- and 3-bromopyridine were treated with lithium piperidide/piperidine in boiling ether (61AG65; 62CB1528; 65AHC121). A mixture of 3- and 4-piperidinopyridine (ratio 1:1) was isolated, also strongly suggesting the occurrence of 3,4-pyridyne as intermediate. These reactions are found to occur according to the S$_N$(EA) mechanism, which means that in this overall nucleophilic substitution process, the first step is the *E*limination of hydrogen halide and the second step *A*ddition of the *N*ucleophile. In the reactions with the 3-halopyridines, mentioned earlier, no trace of a 2-(substituted amino)pyridine was found (61RTC1376; 62CB1528). This excludes the intermediacy of 2,3-pyridyne, since, based on mesomeric considerations (65AHC121) as well as molecular orbital calculations (64TL1577; 69JA2590), addition of the nucleophile only takes place to the carbon atom adjacent to the nitrogen in this ynamine structure, leading to nearly exclusive formation of a 2-substituted product (Scheme II.2) (69MI2). Moreover, deuterium–hydrogen exchange

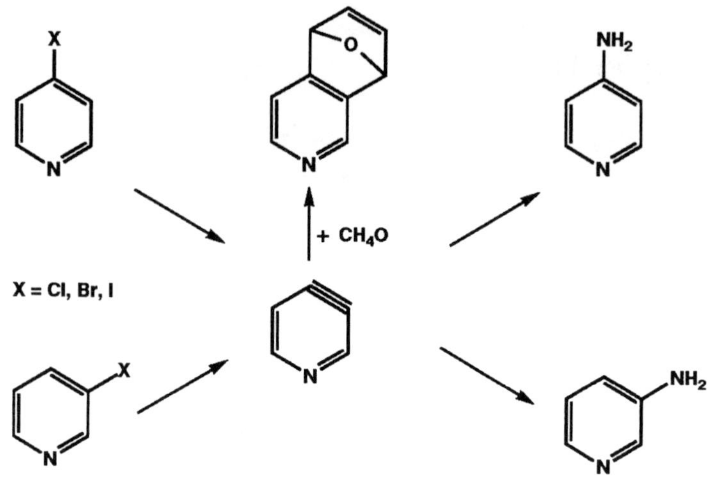

SCHEME II.1

studies with deuterated 3-chloropyridine have shown that the hydrogen on position 4 is more readily exchanged than the hydrogen on position 2 (4 > 2 and 6) (66JA4766; 67TL337), favoring the formation of 3,4-pyridyne over that of 2,3-pyridyne. It is of interest to note that there is convincing

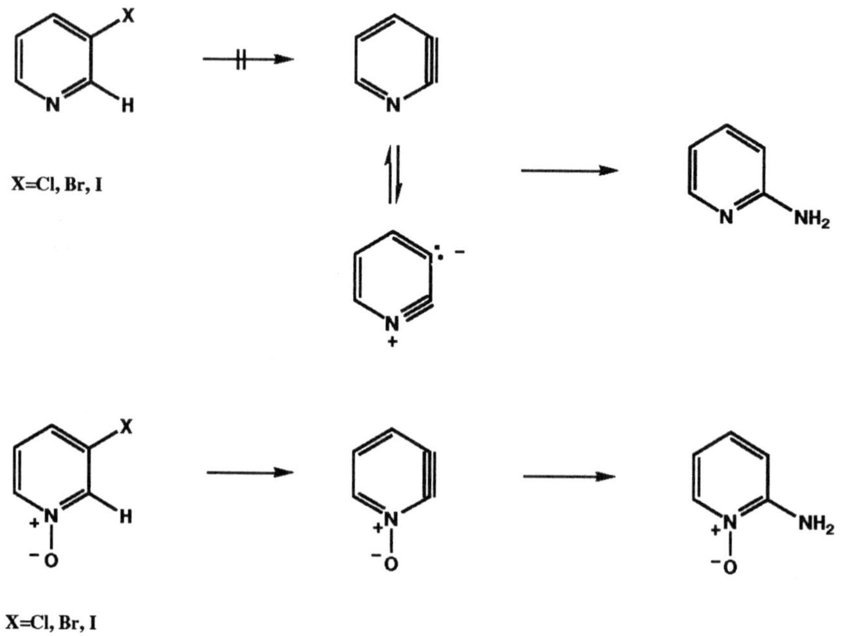

SCHEME II.2

evidence for the occurrence of 2,3-pyridyne *N*-oxide in the amination of 3-chloro-, 3-bromo-, and 3-iodopyridine *N*-oxide (Scheme II.2) (74RTC281). Deuterium–hydrogen exchange studies support these results [67JCS(CC)55, 67TL337; 69JOC1405].

Amination of 2-chloro-, 2-bromo-, and 2-iodopyridine with potassium amide in liquid ammonia gave exclusively 2-aminopyridine. In this case an *A*ddition–*E*limination mechanism ($S_N(AE)$ mechanism) has been proposed. The intermediacy of 2,3-pyridyne was considered to be highly unlikely (65AHC121).

In extension of this work, the amino-debromination reaction of 4-methoxy-, 4-phenyl-, 4-*t*-butyl-, and 4-(*N*-methylanilino)-5-bromopyrimidine was investigated. It was found that in all these reactions, good yields of the corresponding 6-aminopyrimidines are obtained; no indication for the formation of a 4-substituted 5-aminopyrimidine was observed (Scheme II.3) (64TL2093; 65TL555; 68TL9).

Experiments with 4-*t*-butyl-5-bromo-6-deuteriopyrimidine indeed showed the formation of a 6-amino compound that was *nondeuterated* (68TL9). In the liquid ammonia containing potassium amide, no deuterium–hydrogen exchange was found to take place in the starting material. Moreover, control experiments with 6-amino-4-*t*-butyl-5-deuteriopyrimidine clearly indicated that the deuterium at C-5 does not undergo deuterium–hydrogen exchange. All these results strongly suggested the intermediary existence of a 4-*t*-butyl-5,6-pyrimidyne. The occurrence of a *cine*-substitution process, involving the addition of the amide ion to C-6 and a subsequent 1,2-hydride shift from C-6 to C-5 with concomitant loss of the bromide ion (see structure **1**, is not in agreement with the results of the deuterium labeling studies and therefore can be rejected (Scheme II.4)(68TL9). Later experiments have shown that the hypothesis of a pyrimidyne as intermediate in aminations of 5-bromopyrimidines appeared to be incorrect (see Section II,C,1,e).

Similar experiments were carried out with 6-bromo-5-deuterio-4-phenylpyrimidine. When subjected to treatment of potassium amide/liquid ammonia, it was observed that the 6-aminopyrimidine did not contain deu-

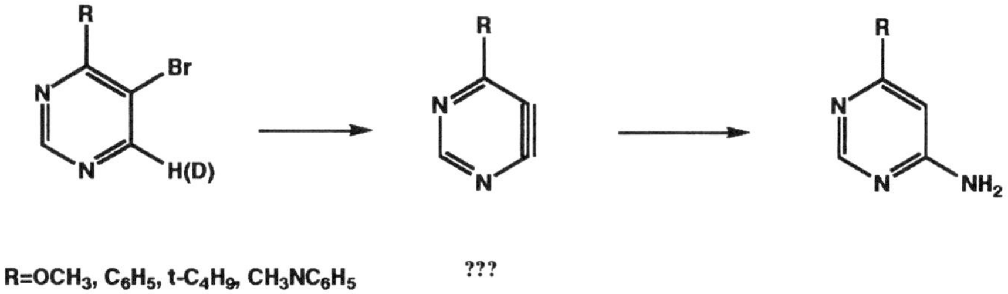

R=OCH$_3$, C$_6$H$_5$, t-C$_4$H$_9$, CH$_3$NC$_6$H$_5$

Scheme II.3

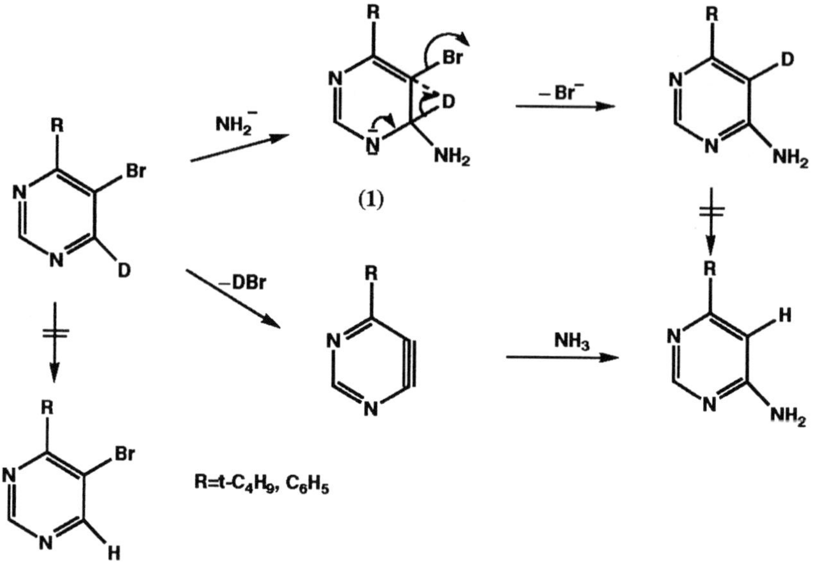

SCHEME II.4

terium and that the starting material has not undergone a deuterium–hydrogen exchange. These results seem to exclude the occurrence of an $S_N(AE)$ substitution and again to suggest a 5,6-pyrimidyne as intermediate.

However, when 6-bromo-4-phenylpyrimidine reacts with lithium piperidide/piperidine instead of potassium amide/liquid ammonia, it was surprisingly found that the corresponding 4-phenyl-6-piperidinopyrimidine was not obtained. Rather, the product was a compound whose structure was established to be a Z/E mixture of 2-aza-4-cyano-3-phenyl-1-piperidino-1,3-butadiene (**2**) (70RTC129). The presence of a cyano group in structure **2** strongly indicates a pyrimidine ring opening between N-1 and C-2, induced by initial addition of lithium piperidide across the bond between N-3 and C-2 (Scheme II.5).

This experimental finding induced reflection on the possibility that also in the reaction with potassium amide/liquid ammonia a process could take place involving an initial addition of the amide ion at C-2 and a subsequent ring opening between C-2 and N-1 to the ring-opened intermediate 1-amino-2-aza-4-cyano-3-phenyl-1,3-butadiene (**3**). This intermediate, however, in contrast to the azapiperidinobutadiene (**2**), may undergo a subsequent ring closure, leading to 6-amino-4-phenylpyrimidine. So, instead of forming the 6-amino compound by the classical $S_N(AE)$ mechanism or alternatively by an amide addition to a possibly formed 4-phenyl-5,6-pyrimidyne [$S_N(EA)$ mechanism], the nucleophilic substitution can now be

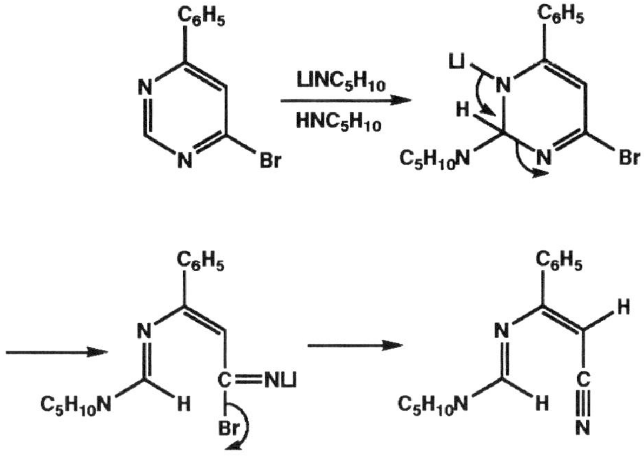

SCHEME II.5

described as involving a sequence of three steps: *A*ddition of the *Nu*-cleophile, *R*ing *O*pening, and *R*ing *C*losure [S_N(ANRORC) mechanism] (Scheme II.6). Unequivocal evidence for the occurrence of this S_N(ANRORC) mechanism was obtained when the amination was carried out with the *mono* ^{15}N-labeled 6-bromo-4-phenyl[1(3)^{15}N]pyrimidine. In the 6-amino compound obtained, about 50% of the ^{15}N-label was found to be present on the ring nitrogen and about 50% on the exocyclic amino nitrogen, proving the reality of this S_N(ANRORC) process. A new mechanism for heteroaromatic nucleophilic substitution was born (71RTC1239).

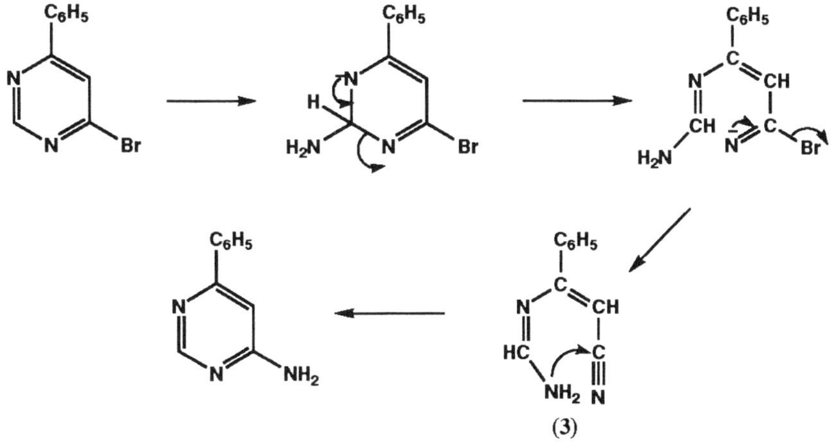

SCHEME II.6

It is of interest to mention that the addition of the amide ion to C-2 of the pyrimidine ring is in general not a favored reaction. There is sound NMR evidence that addition of the amide ion to the parent system pyrimidine occurs exclusively at C-4; there is almost no indication for addition at C-2 (72JA682). For a more detailed discussion on this topic, see Section II,C,1,a.

It is evident that in this substitution reaction the pyrimidine ring has undergone a dramatic change: *One of the ring nitrogens in the starting material is replaced by the nitrogen of the amide ion.* In fact, a ring transformation has taken place, although in starting material as well as in product the pyrimidine ring is still present. One refers to this type of ring transformation as *degenerate ring transformation* (85T237). Many degenerate transformations involving an $S_N(ANRORC)$ mechanism have since been discovered. It is one of the purposes of this book to discuss in detail this type of degenerate ring transformation in heterocyclic systems.

From the results mentioned in this section it is evident that substitution reactions that occur according to the $S_N(ANRORC)$ process and involve a degenerate ring transformation require (1) that the heterocyclic compound be able to undergo an initial addition reaction with nucleophiles (which means that the heterocycle is characterized by a considerable π-deficiency) and (2) that the nucleophilic displacement takes place with a reagent containing the *same* heteroatom(s) that are present in the heterocyclic ring. From the experimental results obtained thus far, it seems that the heteroatom in the reagent needs to carry al least one hydrogen atom. So, for example, ammonia and primary and secondary amines are found to be able to perform $S_N(ANRORC)$ substitutions with nitrogen-containing heterocycles; however, tertiary amines have not been observed to do so. In the following chapters these features will be extensively illustrated.

B. $S_N(ANRORC)$ Reactions in Monoazines

1. Pyridines

It has been reported that 2-bromopyridine on amination with potassium amide in liquid ammonia gives 2-aminopyridine. No indication for an $S_N(EA)$ process involving the intermediacy of 2,3-didehydropyridine was found (see Section II,A). The occurrence of an $S_N(ANRORC)$ mechanism was also excluded, based on the experimental result that amination of 2-bromo[^{15}N]pyridine gives 2-amino[^{15}N]pyridine, being exclusively labeled in the pyridine ring (69RTC1391; 74RTC195). A very classical $S_N(AE)$ process involving intermediate **4** seems to take place in this aminodebromination reaction (Scheme II.7). It was observed, however, that by

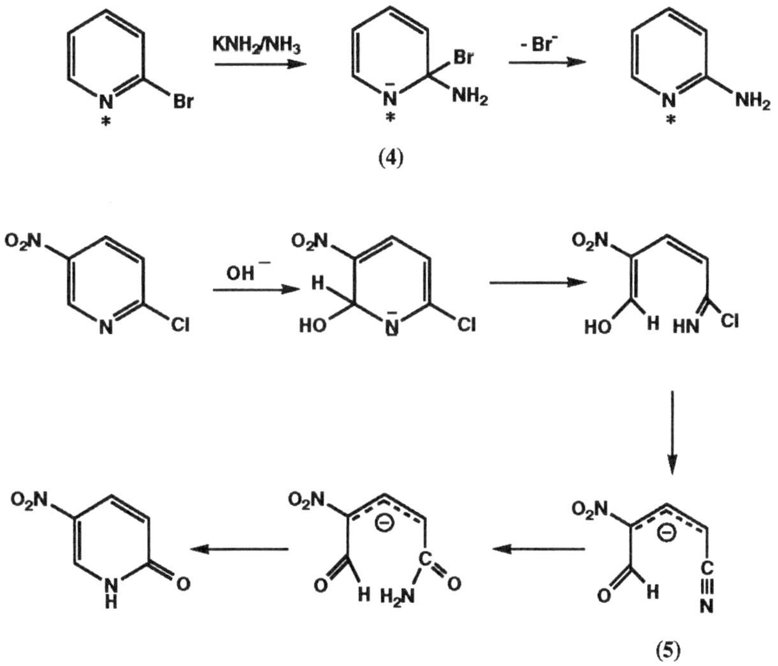

SCHEME II.7

introduction of a nitro group *para* to the halogen atom the reactivity of the molecule completely changed. Hydroxydechlorination of 2-chloro-5-nitropyridine, using an excess of base in the solvent dimethyl sulfoxide, yields 5-nitropyridin-2(1*H*)-one, whose formation, however, occurs by a ring opening, ring closure sequence (80JOC3097). The reaction involves the intermediacy of the (isolable) formylcyanonitropropenide salt (**5**), being formed by ring opening of the initially formed 1:1 hydroxy C-6 adduct. This ring-opened structure (**5**) was confirmed by negative ion fast-atom bombardment and negative ion chemical ionization tandem mass spectrometry when 2-chloro-5-nitropyridine reacts with the hydroxide ion (92MI2).

It became evident that the mechanism for the "simple" hydroxydechlorination replacement reaction decribed in Scheme II.7 occurs in a way very different from the aminodebromination of 2-bromopyridine, but very similar to that of 2-bromopyrimidine (see Section II,C,1,c). It is, however, good to stress the point that, although the aminodehydroxylation occurs according to an S_N(ANRORC) mechanism, *it does not involve a degenerate ring transformation.*

When 2-chloro-5-nitropyridine reacted with potassium amide in liquid ammonia at −33°C, a mixture of compounds was obtained: as main products, 2-amino-5-nitropyridine (40%) and 5-nitropyridin-2(1H)-one (12%), together with small quantities (<3%) of 6-amino-2-chloro-5-nitropyridine and 4-amino-2-chloro-5-nitropyridine (85JOC484). When the amination was carried out with ^{15}N-labeled potassium amide in liquid ammonia (^{15}N excess 9.4%), it was found that the 2-amino-5-nitropyridine also has a ^{15}N excess of 9.4%. After diazotation of the 2-amino compound into 5-nitropyridin-2(1H)-one (**8**), it was found that the ^{15}N excess in this pyridone amounted to 7.1%. From these data it was calculated that 100 × 7.1/9.4 = 75% of the ^{15}N label was incorporated into the ring nitrogen of 2-amino-5-nitropyridine during the aminodechlorination.

This result unequivocally proves that during the amination an S$_N$(ANRORC) process was operative involving as anionic intermediates the 1:1 covalent σ-adduct 6-amino-2-chloro-5-nitropyridinide (**6**) and the ring-opened intermediate 1-amino-4-cyano-2-nitrobuta-1,3-diene (**7**) (Scheme II.8). To substantiate more firmly the involvement of one or both of these intermediates, attempts were undertaken to measure NMR spectra of solutions of 2-chloro-5-nitropyridine in liquid ammonia, containing potassium amide, in different time intervals. Besides signals of the starting material and the product 2-amino-5-nitropyridine, peaks were observed that could be assigned to the aminocyanonitrobutadiene (**7**). These results are in excellent agreement with those obtained in the hydroxydechlorination (80JOC3097). However, NMR signals characteristic of the presence of the intermediary σ-adduct could not be detected. This suggests that the rate of the ring opening of the σ-adduct is relatively faster than the rate of its formation.

More *indirect* support for the occurrence of an intermediate σ-adduct is provided by the result that the yield of the 6-amino-2-chloro-5-nitropyridine is increased from less than 3% to about 10%, when the starting material is added to a solution of potassium amide in liquid ammonia, containing potassium permanganate. This threefold yield increase indicates the presence of this σ-adduct, since a solution of potassium permanganate in liquid ammonia has been found to be a highly effective and very specific reagent for oxidizing aminoazinides (82JHC1285; 85JHC353; 86IJ67; 88JHC831; 94MI1) and their conjugate acids the dihydroaminoazines (81JOC3805; 82JHC1527; 85S884; 86JHC477; 87JHC1657; 94MI1). Their respective nitro derivatives are also appropriate substrates for oxidation with liquid ammonia–potassium permanganate (81JOC3805; 83JHC9, 83JOC1354, 83RTC359; 85JHC353, 85S884; 87JOC5643; 88JHC831; 90LA653; 91LA875, 91PJC323; 92LA899; 93ACS95, 93LA7; 94MI1). Also, the introduction of an methylamino group at position 3 in 4-nitroquinoline has been reported using liquid methylamine–potassium permanganate (98KGS967).

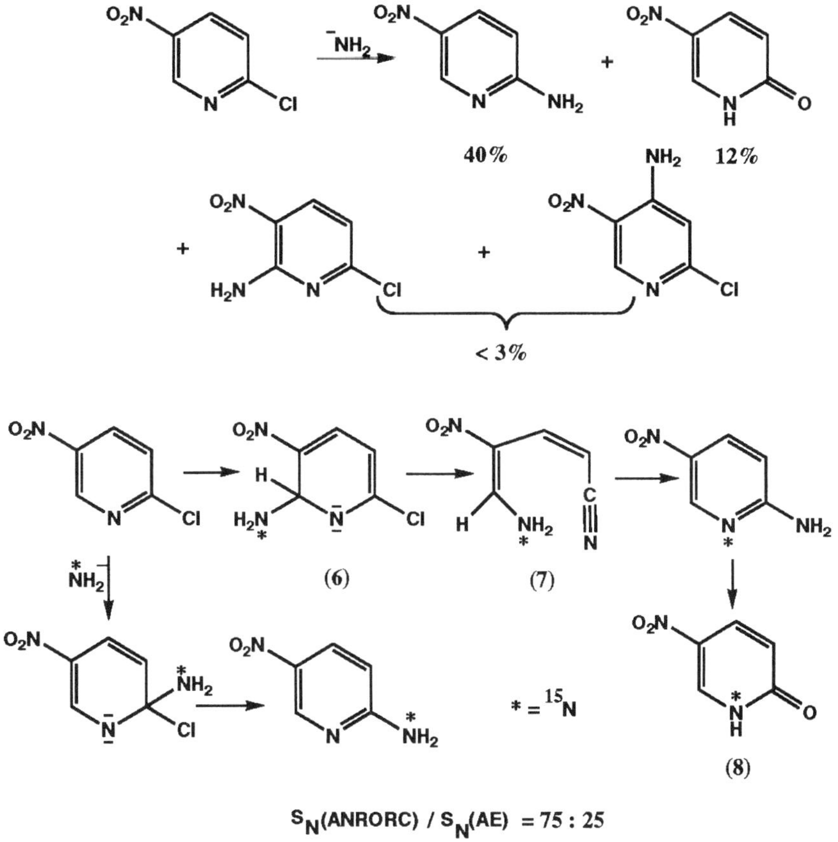

SCHEME II.8

In conclusion, the amination of 2-chloro-5-nitropyridine mainly occcurs according to the S_N(ANRORC) mechanism (about 75%) and to a smaller degree (about 25%) according to the S_N(AE) process.

This behavior is certainly very different from that found for 2-bromopyridine (see earlier discussion), where addition of the amide ion takes place exclusively at position C-2. Apparently, the nitro group at C-5 activates mainly its neighboring C-6 position (C-6 > C-2 > C-4). It remains remarkable that none of the intermediary σ-adducts are detected in the reaction mixture.

As an extension of this work, the aminodechlorination of 2-chloro-3,5-dinitropyridine was studied (85JOC484). Because of the high π-electron deficiency of this compound, it appears that the amination does not require the presence of a strong nucleophilic amide ion; liquid ammonia is already sufficiently reactive to perform the amination. In quantative yield, 2-amino-3,5-dinitropyridine is formed. However, surprisingly, hardly any incorporation of the ^{15}N-label in the pyridine ring was found when the amination was carried with with ^{15}N-labeled liquid ammonia. Apparently, the addition–elimination [S$_N$(AE)] substitution is the principal mechanism operative in the aminodechlorination.

The ^{1}H and ^{13}C spectroscopy of a solution of 2-chloro-3,5-dinitropyridine in liquid ammonia at −40°C showed the formation of the C-6 adduct (**10**). This adduct is rather stable, since after 1 hr standing, no change in the spectrum was observed. It is interesting that at a somewhat lower temperature (−60°C) the addition takes place at C-4, i.e., formation of (**9**). Apparently one deals with the interesting concept of kinetically and thermodynamically controlled covalent adduct formation. At −60°C the addition is kinetically controlled, and at −40°C the addition is thermodynamically favored. The higher stability of the C-6 adduct compared to the C-4 adduct is probably due to the more extended conjugate resonance system (Scheme II.9).

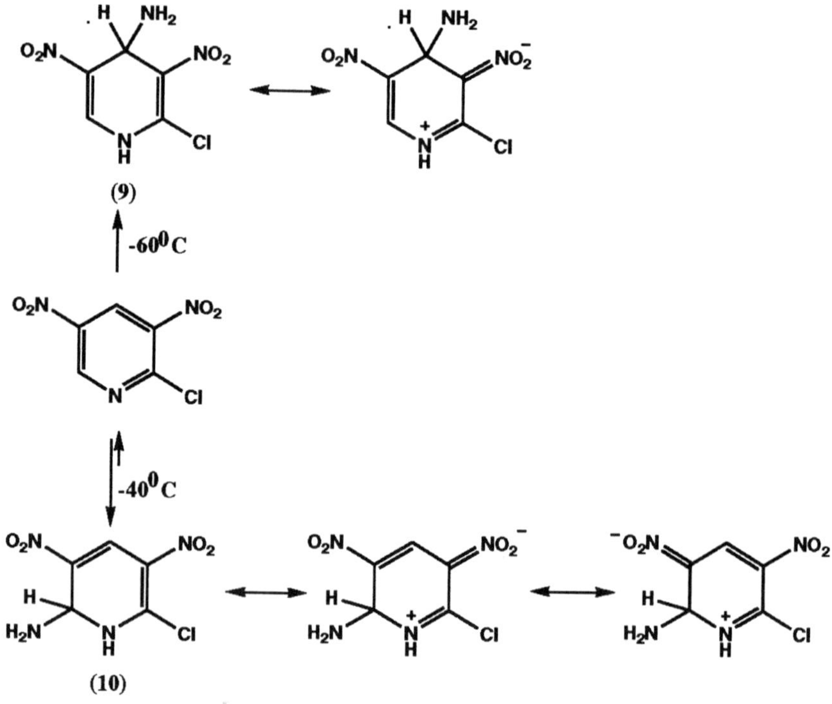

SCHEME II.9

2. Isoquinolines

Amination of 3-bromoisoquinoline with potassium amide/liquid ammonia involves (in contrast to the amination of 2-bromopyridine, see Section II,B,1) an ANRORC process. When 3-bromo[^{15}N]isoquinoline was used as substrate, the 3-aminoisoquinoline being obtained contains 55% of its ^{15}N enrichment on the exocyclic nitrogen, i.e., the formation of 3-[^{15}N-amino]isoquinoline (**14**), and 45% inside the heterocyclic ring, i.e., formation of 3-amino[^{15}N]isoquinoline (**13**) (74RTC198).

It is suggested that the covalent adduct 1-amino-3-bromoisoquinolinide (**11**) and the ring-opened benzylcyanide intermediate, i.e., *ortho*-iminomethylenebenzylcyanide (**12**), are involved in the formation of 3-aminoisoquinoline (**14**). It can be suggested that the ratio 55:45 for the distribution of the ^{15}N label over the exocyclic as well as ring nitrogen in 3-aminoisoquinoline may due to the occurrence of a scrambling process, taking place in the ring-labeled **13**. Control experiments show (74RTC198) that compound **13**, being obtained from 3-bromo[^{15}N]isoquinoline by an addition–elimination reaction with ammonia, is fully stable under the conditions of the amid-induced amination and undergoes no scrambling at all (Scheme II.10).

That 2-bromopyridine does not react according to an S_N(ANRORC) mechanism, and 3-bromoisoquinoline only partly, can be understood in the light of NMR-studies on covalent amination with the parent azines: Pyridine does not form a covalent σ-adduct with potassium amide, but isoquinoline gives σ-adduct formation at C-1, forming 1-aminoisoquinolinide (72JA682; 73JOC1947). It is evident that the isoquinoline ring is more inclined to a nucleophilic addition than the pyridine ring.

It is of interest to mention that 2-bromoquinoline, when treated with potassium amide in liquid ammonia, reacts very differently from 3-bromoisoquinoline; hardly any 2-aminoquinoline is obtained, but as main product 2-methylquinazoline has been isolated (65RTC1569; 67RTC187). This ring transformation involves breaking of the bond between C-3 and C-4 in the covalent σ-amino adduct at C-4 (Scheme II.10). Since this ring transformation cannot be considered degenerate, it is beyond the scope of this book to discuss this reaction in more detail.

It has been found that the cupric sulfate–mediated aminolysis of 3-bromo[^{15}N]isoquinoline with ethanolic ammonia (130°C, 7 days reaction time) yields a mixture of **13** and **14**, ratio 75:25 (74RTC198). It shows that also with the weaker nucleophile ammonia a part of 3-bromoisoquinoline can undergone a ring-opening, ring-closure process.

Interestingly, one of the first reports on ring-opening reactions during

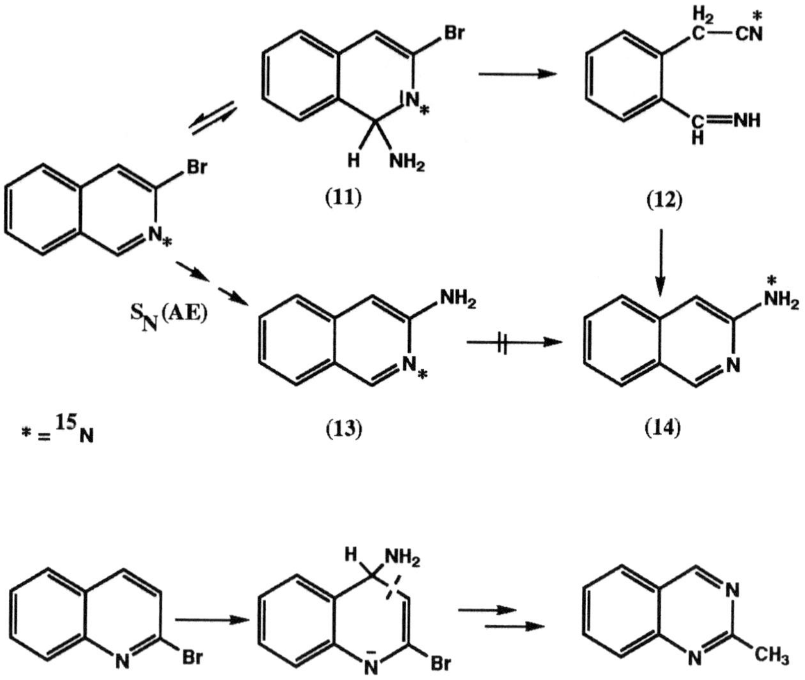

SCHEME II.10

aminolysis concerned the reaction of 3-chloroisoquinoline with ^{15}N-labeled aqueous ammonia (44 hr, 155–180°C) (68T441). In the 3-aminoisoquinoline obtained the ^{15}N label was equally distributed over the ring nitrogen and the exocyclic nitrogen. This result was explained by assuming that the initially formed product was 3-(^{15}N-amino)isoquinoline (**14**), and that this compound under the conditions of the reaction would undergo a covalent hydration into **15,** followed by a ring-opening into (*ortho*-formylphenyl) acetamidine (**16**). In the amidino group both nitrogens are scrambled, which leads to the result that after the ring closure the 3-aminoisoquinoline contains the ^{15}N label partly in the amino group and partly in the heterocyclic ring (Scheme II.11). Based on the results discussed earlier in this chapter, it seems not unlikely that a combination of an S_N(ANRORC) mechanism and an S_N(AE) process may also explain the results of these experiments.

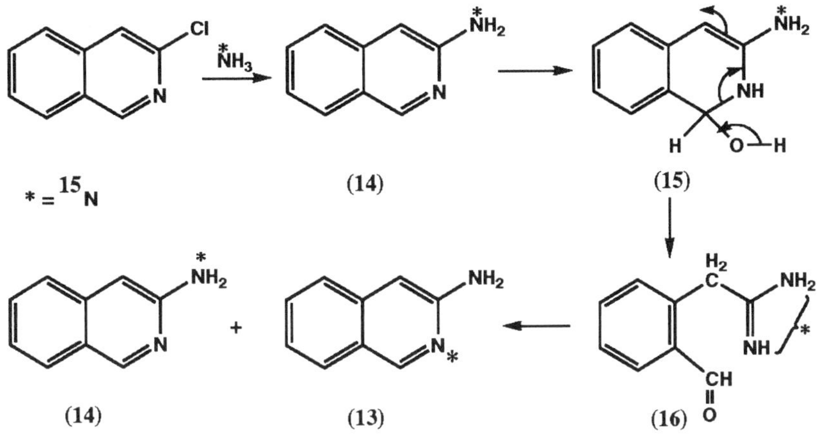

SCHEME II.11

C. S_N(ANRORC) Substitutions in Diazines

1. Pyrimidines

a. *Aminodehalogenation of 6-Halogeno-4-Substituted Pyrimidines*

As we have seen in Section II,A, 6-bromo-4-phenylpyrimidine reacts on treatment with potassium amide in liquid ammonia at −75°C into the corresponding 6-amino compound nearly quantitatively according to the S_N(ANRORC) mechanism (71RTC1239). Extensive investigations have been carried out on the scope and limitations of this mechanism and on the several factors that influence the occurrence of the S_N(ANRORC) mechanism in the aminodehalogenation of 4-substituted-6-halogenopyrimidines.

The nature of the leaving halogeno atom. It has been observed that reaction of *mono*-labeled 6-fluoro-, 6-chloro-, and 6-iodo-4-phenyl[1(3)-^{15}N]pyrimidine with potassium amide/liquid ammonia leads to 6-amino-4-phenylpyrimidine in which the exocyclic amino group is ^{15}N-labeled; the percentages, however, strongly depend on the nature of the halogen atom, i.e., F (73 ± 5%), Cl (93 ± 5%), Br (83 ± 5%), I (13 ± 5%) (Scheme II.12) (72RTC1414).

The general procedure to establish the percentages of participation of the S_N(ANRORC) mechanism was as follows. As indicated before, the amination was carried out with 6-halogeno-4-phenylpyrimidines, being *mono*-labeled for x% on the ring nitrogens (this label is of course scrambled over both nitrogens). These percentages are determined by mass spectrometry, measuring the intensities of the $M + 1/M$ peaks. When the S_N(ANRORC) mechanism is fully operative it should lead to 6-amino-4-phenylpyrimidine (**18**) that contains $\frac{1}{2}(x)\%$ ^{15}N on the ring nitrogen and $\frac{1}{2}(x)\%$ ^{15}N on the 6-amino group (Scheme II.12).

To establish which percentage of the ^{15}N label is present on the ring nitrogen and which on the amino group, the amino compound was converted by acid hydrolysis into 4-phenylpyrimidin-6(1H)-one, which was subsequently converted into 6-bromo-4-phenylpyrimidine (**18a**) by treatment with phosphoryl bromide (Scheme II.12). If the ^{15}N content of the bromo compound **18a** contains $y\%$, the percentage of the 6-halogeno compound that reacts according to the S_N(ANRORC) mechanism can be calculated, i.e., $(y : \frac{1}{2}x) \times 100\%$.

From Table II.1 it is evident that for the 6-fluoro-, 6-chloro-, and 6-bromo compounds, the S_N(ANRORC) process is the main reaction pathway in the aminodehalogenation, which is initiated by an addition of the amide ion at C-2; apparently, addition at C-6 is not favored. 6-Chloro-4-phenylpyrimidine is most inclined to the ring-opening, ring-closure process. It is for that reason that most of the amination studies of the 6-halogenopyrimidines were carried with the chloro atom as leaving group. It has been reported that amino-dechlorination of 6-chloro-4-phenylpyrimidine-3-oxide by potassium amide/liquid ammonia does not involve a ring-opening, ring-closure mechanism (74RTC58, 75RTC277).

The iodo compound reacts for only a very low percentage (13%) according to the S_N(ANRORC) mechanism. It may be due to the low electroneg-

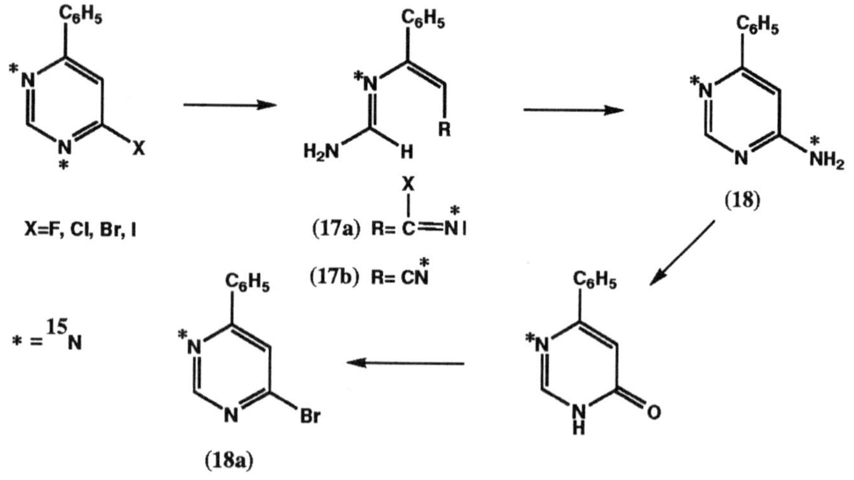

SCHEME II.12

TABLE II.1

PERCENTAGES OF S_N(ANRORC) PARTICIPATION IN THE AMINATION OF 4-SUBSTITUTED-6-HALOGENOPYRIMIDINES IN DEPENDENCY OF THE SUBSTITUENT R AT POSITION 4, THE LEAVING GROUP X AT POSITION 6, AND THE TEMPERATURE

Substituent R	Leaving group X^-	Temperature (°C)	% S_N(ANRORC)
C_6H_5	F	−75°C	79%
C_6H_5	Cl	−75°C	93%
C_6H_5	Br	−75°C	83%
C_6H_5	I	−75°C	13%
t-C_4H_9	Br	−75°C	77%
t-C_4H_9	Br	−33°C	33%
t-C_4H_9	Cl	−33°C	100%
OCH_3	Br	−33°C	31%
OCH_3	Cl	−33°C	100%
NC_5H_{10}	Cl	−33°C	21%
NHC_6H_5	Cl	−33°C	0%
CH_3	Cl	−33°C	0%

ativity of the iodo atom, which activates the double carbon–nitrogen bond much less than the other halogeno atoms for addition of the nucleophilic amide at C-2. The pK_a values for the four different halogenopyridines support this view [2-fluoropyridine (pK_a −0.44), 2-chloropyridine (pK_a +0.72), 2-bromopyridine (pK_a +0.90), 2-iodopyridine (pK_a +1.82)] (55JA3752).

It is also possible that the iodo compound partly reacts according to a nonrearranging radical chain process ($S_{RN}1$-mechanism) (70JA7463, 70JA7465,72ACR139). It generates via an electron transfer process an iodopyrimidinyl radical anion. After loss of the iodide ion, a pyrimidinyl radical is formed, which combines with ammonia and yields a new aminopyrimidinyl radical anion. Transfer of an electron to another substrate in a chain-propagating process yields the 6-amino-4-phenylpyrimidine. That radical anions may be involved in aminations taking place via an S_N(ANRORC) mechanism is evidenced in the Chichibabin amination of 4-phenylpyrimidine, where the presence of radical scavengers more or less stops this process [82MI3], see Section II,C,1,g.

The influence of the temperature. It has been established that the temperature has a dramatic effect on the occurrence of the S_N(ANRORC) process. Whereas participation of the S_N(ANRORC) mechanism in the amino-debromination of 4-t-butyl-6-bromopyrimidine at −75°C was found to occur for 77% according to the S_N(ANRORC) mechanism (79RTC5), it decreased to 33% when the amination was carried out at −33°C. Apparently at −75°C attack of the amide ion on C-2 is clearly favored over attack on C-6 (kinetic vs thermodynamic control) (78TL3841). Notice that 4-t-butyl-6-chloropyrimidine, when aminated at −33°C, reacts for nearly the same percentage [100% (!)] $S_N(ANRORC)$ mechanism as at $-75°C$ (93%). See Table II.1.

The influence of the nature of the substituent at position 4. In order to determine the influence of substituents at position 4 on the $S_N(ANRORC)$ process in the aminodehalogenation, the amide-induced amination was studied with 6-chloropyrimidines containing different substituents at position 4, i.e., the 4-*t*-butyl-, 4-methoxy-, 4-piperidino-, 4-anilino-, and 4-methyl-groups (79RTC5). The results are summarized in Table II.1.

The data obtained with the 4-methoxy compound confirm that the 6-chloro compound is a more appropiate substrate to participate in the $S_N(ANRORC)$ pathway than the 6-bromo compound. In the aminodechlorination of 4-piperidino-6-chloropyrimidine (potassium amide in liquid amonia, at $-33°C$), the $S_N(ANRORC)$ mechanism is only operative for 21%, whereas in the aminodechlorination of 4-anilino- and 4-methyl-6-chloropyrimidine, no ring-opening is involved. It has been assumed that in this strong basic medium, the anilino group as well as the methyl group are deprotonated to form a charge-delocalized anion with an enhanced electron density at positions N-1 and N-3. In this anion the addition at C-2, the initial step in the ANRORC mechanism, is strongly disfavored, making the $S_N(ANRORC)$ process highly unlikely. The nucleophilic replacement in these compounds may follow an addition–elimination process in the anionized species.

This $S_N(AE)$ substitution reaction can be formulated as indicated in Scheme II.13, involving as intermediate the dianion (**19**), being present with the mono anion in an equilibrium that lies far to the left. Deprotonation of substituents containing one or more acidic hydrogens has been observed by

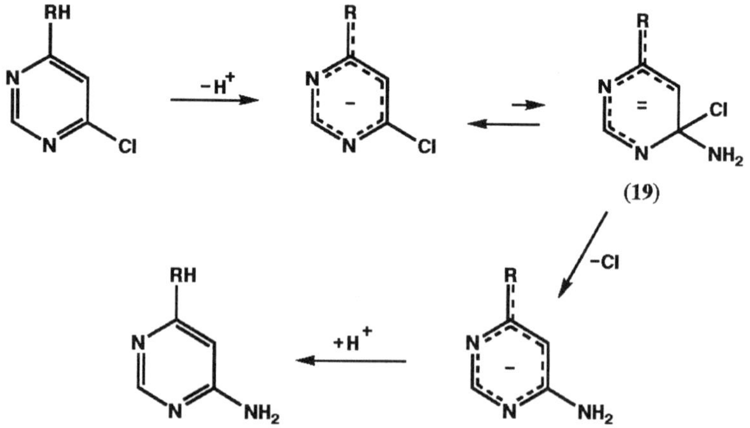

SCHEME II.13

^{1}H NMR spectroscopy in many azines, such as pyrimidines (74RTC231; 75OMR86), pyridines (73JOC658; 81H1041, 81JOC3509; 82JOC963), pyrazines (73JOC658; 81H1041), pyridazines (81H1041), naphthyridines [78JOC1673; 83AHC95], and purines (79JOC3140; 81H1041).

Effect of blocking substituents at position 2 of the pyrimidine ring. Since an important step in the ANRORC process is the addition of the amide ion to the C-2 position of the pyrimidine ring, investigations were carried out with substrates in which position 2 is occupied by a blocking group. It was found that after treatment of 2,4-diphenyl-6-halogeno[1(3)-^{15}N] pyrimidines with potassium amide in liquid ammonia and determining the partition of the ^{15}N label over the ring nitrogen and the amino nitrogen according to the procedure described earlier, the ring opening–ring closure process [S$_N$(ANRORC) mechanism] is operative only in the aminolysis of 6-chloro-2,4-diphenylpyrimidine, and for only 45% (Scheme II.14) (73RTC145, 73RTC442).

More dramatic blocking effects were observed when at position 2 of the pyrimidine ring a *t*-butyl group is present: The replacement of the chloro atom in 2,4-di-*t*-butyl-6-chloropyrimidine by an amino group does not involve ring opening (79RTC5). This is in sharp contrast to the results obtained in the aminodechlorination of 4-*t*-butyl-6-chloropyrimidine at −33°C, which takes place for 100% according to the ANRORC-process (see Table II.1).

These results clearly show that an unsubstituted C-2 position in 6-halogenopyrimidines is an important prerequisite for ring-opening processes that can lead to nucleophilic substitutions involving a degenerate ring transformation.

Structure of the intermediate(s). In the description of the mechanism as

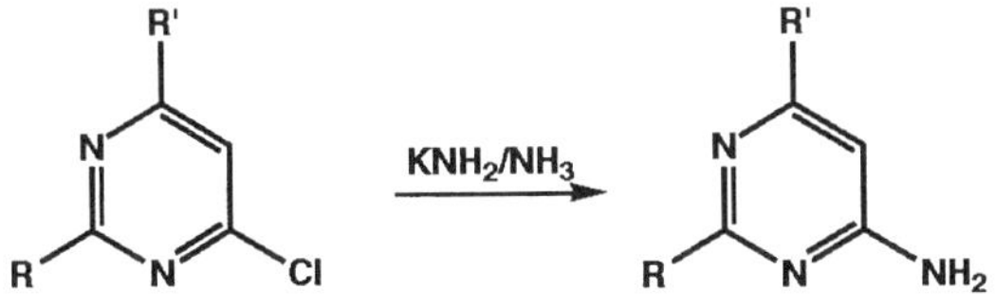

		S$_N$(ANRORC) participation
R = H	R' = Ph	93%
R = Ph	R' = Ph	45%
R = H	R' = t-Butyl	100%
R = t-Butyl	R' = t-Butyl	0%

SCHEME II.14

presented in Scheme II.12, possible open-chain intermediates proposed are the 1-amino-4-cyano-2-azabutadiene (**17b**) or its precursor the iminohalogenide (**17a**). Since it has been shown that a reaction of 6-bromo-4-phenylpyrimidine with the secondary amine system lithium piperidide/piperidine gave as final product the open-chain cyanoazadiene (**2**) (see Scheme II.5), the question can be raised whether in the reaction with potassium amide in liquid ammonia the iminohalogenide (**17a**) or the cyano intermediate (**17b**) is involved in the final ring-closure step. Attempts to isolate one of these intermediates from the reaction of 4-phenyl-6-halogeno- or 2,4-diphenyl-6-halogenopyrimidines with potassium amide were, however, unsuccessful.

In order to clarify the possible existence of these intermediates, 6-chloro-5-cyano-4-phenyl[1(3)-^{15}N]pyrimidine (**20**) (the ^{15}N label is scrambled over both nitrogens) and the radioactive 6-chloro-5-[^{14}C-cyano]-4-phenylpyrimidine (**23**) were synthesized as substrates. Because of the presence of the cyano function at C-5, one can expect that **20** (and **23**) would undergo amination involving an S_N(ANRORC) mechanism. This has indeed been found. When **20** was reacted with potassium amide in liquid ammonia, two products were obtained: as main product, 6-amino-5-cyano-4-phenylpyrimidine (**21**, 75%), and as minor product, α-amino-β,β'-dicyanostyrene (**22**, about 20%) (Scheme II.15).

Both products contained the *same* percentage of ^{15}N enrichment as was present in the starting material (73RTC471). Degradation studies of the 6-amino compound (**21**) showed that 50% of the ^{15}N label is present on the ring-nitrogen N-3 and the remaining 50% on the amino group, proving that the 5-cyanopyrimidine (**20**) has reacted about 100% according to the S_N(ANRORC) mechanism. This is another fine example of a degenerate ring transformation.

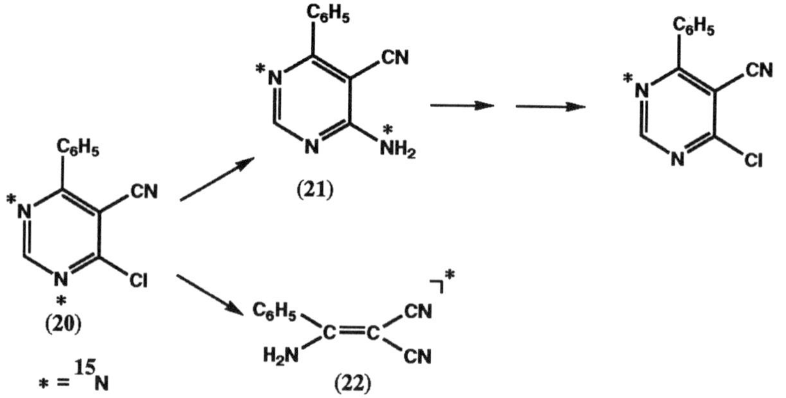

SCHEME II.15

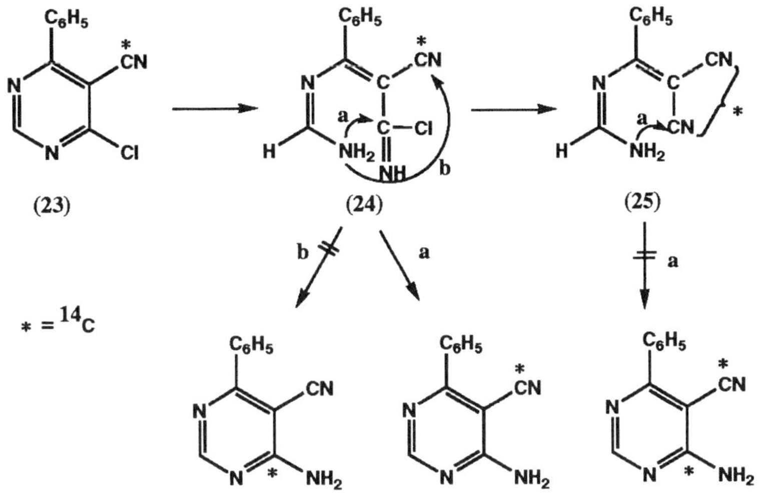

SCHEME II.16

When **23** reacts in this ANRORC process, in fact two intermediates, a cyanoimidoyl chloride (**24**) and a dicyanoazadiene (**25**), can be postulated (Scheme II.16). When the dicyano compound **25** is the transient open-chain intermediate, scrambling of the radioactive carbon over both cyano groups takes place, and consequently after ring closure (route a), incorporation of the ^{14}C-label into the pyrimidine ring should be expected (Scheme II.16).

After isolation of the radioactive 6-amino-5-cyanopyrimidine (**26**) and hydrolysis of **26** with acid into 4-phenylpyrimidin-6-one and carbon dioxide (the last being derived from the cyano function, Scheme II.17), the

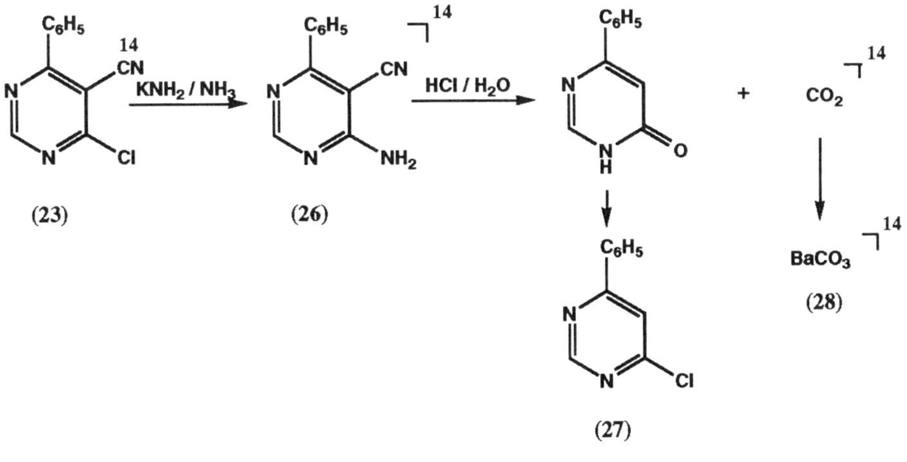

SCHEME II.17

TABLE II.2
RADIOACTIVITY (CI/MMOL) OF THE
4-PHENYLPYRIMIDINES **23, 26, 27** AND
BARIUM CARBONATE **28**.

Compound	Radioactivity
21	0.348
26	0.387
27	0.011
28	0.279

pyrimidin-6(1H)one hardly contains any radioactivity [measured as 6-chloro-4-phenylpyrimidine (**27**)], and the carbon dioxide [measured as barium carbonate (**28**)] has about the same specific ^{14}C radioactivity as measured in the starting material (**23**) (see Table II.2). Since this result shows that the radioactive carbon in **26** is not present in the ring, the conclusion is justified that the formation of the the open-chain dicyanoazadiene **25** as intermediate in the amino-dehalogenation of **23** is highly unlikely.

As already reported in Section II,A, the amination of 6-bromo-5-deuterio-4-phenylpyrimidine with potassium amide in liquid ammonia provides a product in which deuterium is no longer present. Based on the work described previously, it seems reasonable to conclude that this easily occurring ring deuterium–hydrogen exchange takes place in the intermediary imidoyl bromide (**17a**, X = Br) (Scheme II.18) and not in the cyanoazadiene (**17b**). In the strong basic medium a fast equilibrium can be formulated between these open-chain intermediates (**17a**, X = Br, **29**, and **30**) (Scheme II.18).

Further evidence for the intermediacy of the imidoyl halide **24** can be taken from the formation of the aminodicyanostyrene **22**, formed as a minor product in the amination of **20** (Scheme II.19) (73RTC471). As already mentioned, **22** contains the *same* ^{15}N content as the starting material, and based on this experimental result its formation from the imidoyl halide seems now straightforward. In this strong basic solution the tautomeric con-

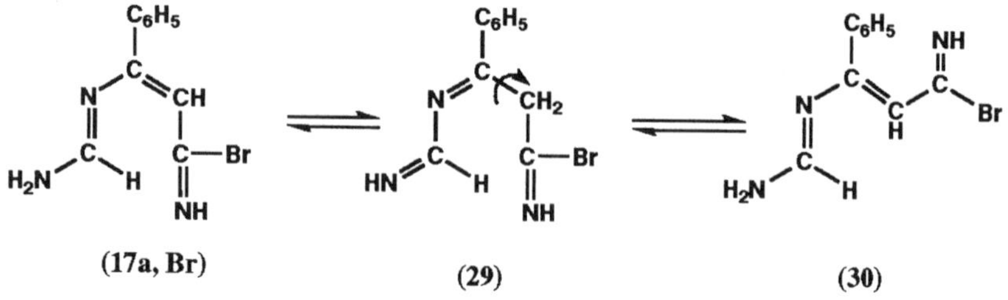

SCHEME II.18

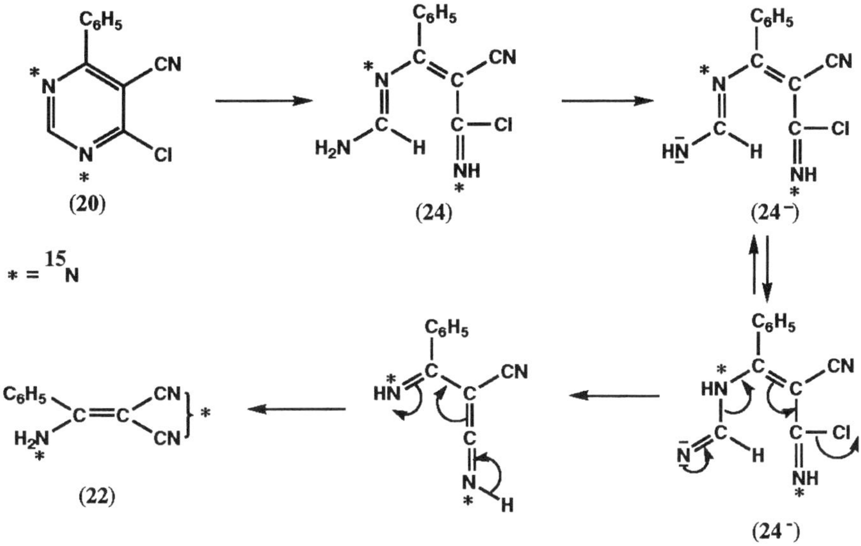

SCHEME II.19

jugate base (**24⁻**) can easily be formed from **24** because of the presence of the cyano group. It may undergo a base-induced fragmentation reaction followed by elimination of hydrogen cyanide, resulting in the formation of the styrene derivative (**22**) (Scheme II.19).

Amination of 6-halogenopyrimidines with primary amides. Reaction of 6-bromo-4-phenylpyrimidine with the primary amine system lithium isopropylamide/isopropylamine at 20°C showed the formation of 6-(isopropylamino)-4-phenylpyrimidine (**34**) (73RTC711) (Scheme II.20): at first sight, a classical aminodebromination reaction. However, when the reaction was carried out at −75°C, then along with the cyanoazadiene (**32**), 6-imino-1-isopropyl-4-phenyl-1,6-dihydropyrimidine (**33**) was obtained. On standing, **32** gradually converted into **33**. Since **33** gives a fast Dimroth rearrangement into **34** (68MI1), it became evident that the "simple" conversion of 6-bromo-4-phenylpyrimidine into **34** in fact involves two ANRORC processes. This complicated sequence of reactions presents an interesting example of a *double* degenerate ring transformation.

Discussion. This section has presented ample evidence that in the aminodehalogenation of 6-X-4-phenylpyrimidines (X = Cl, Br, F), using strong nucleophilic reagents such as the amide or piperidide ion, the initial addition of the nucleophile mainly takes place at position C-2. There is hardly any indication for addition at C-6, i.e., the position where the leaving group is present. So, the S$_N$(ANRORC) mechanism is strongly favored over the

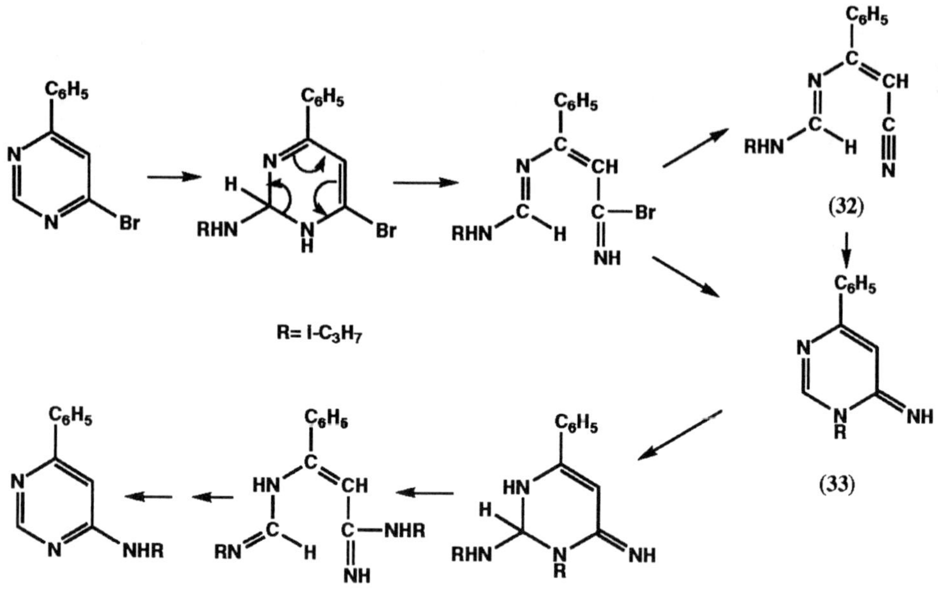

SCHEME II.20

classical $S_N(AE)$ mechanism. This experimental fact is in remarkable contrast to the overwhelming ^1H and ^{13}C spectroscopic evidence that in the *parent* system pyrimidine, the addition of the amide ion at C-4(6) is more strongly favored than addition at C-2 (72JA682). PMO calculations of the stabilization energy (ΔE) for addition at C-6 and at C-2 show that the ΔE for addition at C-6 is considerably higher, confirming the spectroscopic results Moreover, calculation of the heat of formation of both the covalent σ-adducts gave as result that the ΔH for the 2-amino adduct is about 3.3 kcal/mol higher than that for the 6-amino-adduct (95UP1). This means that at $-33°C$ the formation of the 6-amino adduct is about 600–700 times faster than that of the 2-amino adduct.

There have been discussions whether the amide addition at C-6 is charge-controlled or orbital-controlled. Charge density calculations in 4-phenylpyrimidine (MNDO method) predict that the addition of the amide ion would preferably take place at position 2 (95UP1); this, however, does *not* agree with the experimental results. Therefore, the conclusion seems justified that the addition is not charge-controlled. Frontier orbital calculations, using the SCF-PPP method, show that the frontier orbital densities in the LUMO of pyrimidine are zero at C-2 and C-5, making these positions quite unreactive for nucleophilic attack (77UP1). These results strongly indicate that the addition is an orbital-controlled reaction.

One possible reason that in 6-X-4-phenylpyrimidine the addition at C-2 is so favored has to do with steric effects: Addition at C-4 is hampered by the bulky phenyl group, and addition at C-6 by the halogeno atom. Somewhat similar reasoning has been put forward to explain why the primary addition of the methoxide in 2,4,6-trinitrochlorobenzene and related compounds (65AHC180) takes place at C-3 and not at C-1 (kinetic control vs thermodynamic control) (70CRV667; 72T3299; 82CRV223; 91MI1) (Scheme II.20A). It has also been postulated that the addition of the amide ion at C-2 is possibly favored by metal complexation of potassium amide with the azomethine bond between N-3 and C-2 (see transition state **35**), because of more basic character of N-3 (*para* toward X) compared to N-1 (*ortho* toward X). Compare the pK_a values of 2-chloropyridine [pK_a = 0.72 (55JA3752)] with that of 4-chloropyridine [pK_a = 0.4 (64JCS3591)]. A similar transition state has been proposed in the amination of 2-aminopyridine (65AHC145).

SCHEME II.20A

b. *Aminodehalogenation in 4-Halogeno-2-Substituted Pyrimidines*

When reacting the 4-chloro-2-R-pyrimidines (R = CH_3, C_2H_5, C_6H_5, β-$C_{10}H_7$, $N(CH_3)_2$, c-NC_5H_{10}, c-$N(CH_2CH_2)_2O$, $N(CH_3)C_6H_5$) with potassium amide/liquid ammonia, the corresponding 4-amino compounds were obtained in only a small yield. The main product was the ring transformation product 4-methyl-2-R-1,3,5-triazine (66RTC1101; 69RTC426, 69RTC1156). Apparently a carbon–nitrogen skeleton rearrangement has taken place. Although an extensive discussion of this ring transformation is beyond the scope of this book, it is evident that in this pyrimidine-to-1,3,5-triazine conversion the initial addition of the amide ion has to take place at another position than that on which the chloro atom is present. Proof for addition at C-6 is provided by tracer experiments (67RTC15) and ^{1}H and ^{13}C NMR spectroscopy (73RTC1232; 75OMR86). In liquid ammonia containing potassium amide, the aforementioned pyrimidines are present as

2-R-6-amino-4-chloro-1,6-dihydropyrimidinides (**36**); these anionic adducts subsequently undergo carbon–carbon bond fission between C-5 and C-6 (Scheme II.21).

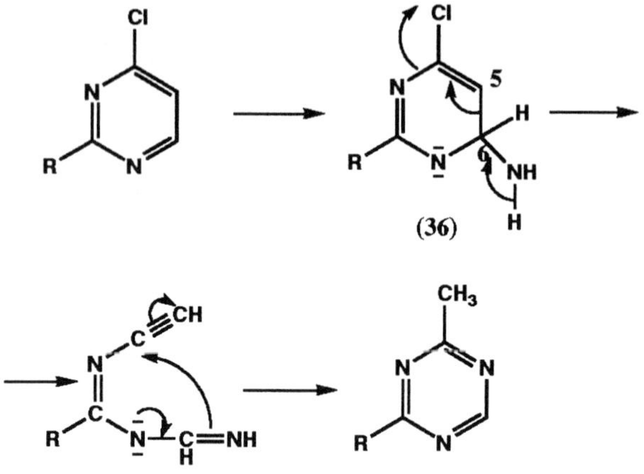

SCHEME II.21

Quite surprisingly, no ring transformation but aminodechlorination was found when 5-R-4-chloro-2-phenylpyrimidines (**37**, R = OCH_3, OC_2H_5, SCH_3, C_6H_5) were subjected to treatment with potassium amide/liquid ammonia. GLC analyses did not show any trace of 2-phenyl-1,3,5-triazines, containing a CH_2R group at position 4(6); only the corresponding 4-aminopyrimidines **39** could be detected (66TL4517; 71RTC105). A careful low-temperature workup of the reaction mixture, obtained from 4-chloro-5-ethoxy-4-phenylpyrimidine (**37**, R = OC_2H_5), however, yields not **39** (R = OC_2H_5) but the open-chain compound N-cyano-N'-(β-ethoxyvinyl)benzamidine (**38a/38b**, R = OC_2H_5, 40%) (Scheme II.22.). Evidently a carbon–carbon fission has occurred. This amidine is the precursor of the 4-amino compound, since **38** (R = OC_2H_5) underwent ring closure on heating into **39** (R = OC_2H_5). The same interesting behavior was also observed with the 2-phenyl-5-methoxy-, 2-phenyl-5-thiomethyl-, and 2,5-diphenyl-4-chloropyrimidine (66TL4517; 71RTC105). By ^{14}C-tracer experiments, using 4-chloro-5-methoxy-2-phenyl[6-^{14}C]pyrimidine (**37***, R = OC_2H_5) as substrate, it could be proven that the cyano function in **38*** (R = OC_2H_5) is radioactive, which leads to the unequivocal conclusion that the carbon–carbon fission has exclusively occurred between C-5 and C-6, involving **40*** (R = OC_2H_5) as intermediate (66TL4517; 71RTC105). Deuterium labeling studies revealed that the hydride transfer from **40*** (R = OC_2H_5) into **38*** (R = OC_2H_5) was proved to be intramolecular (Scheme II.22) (78JOC2682).

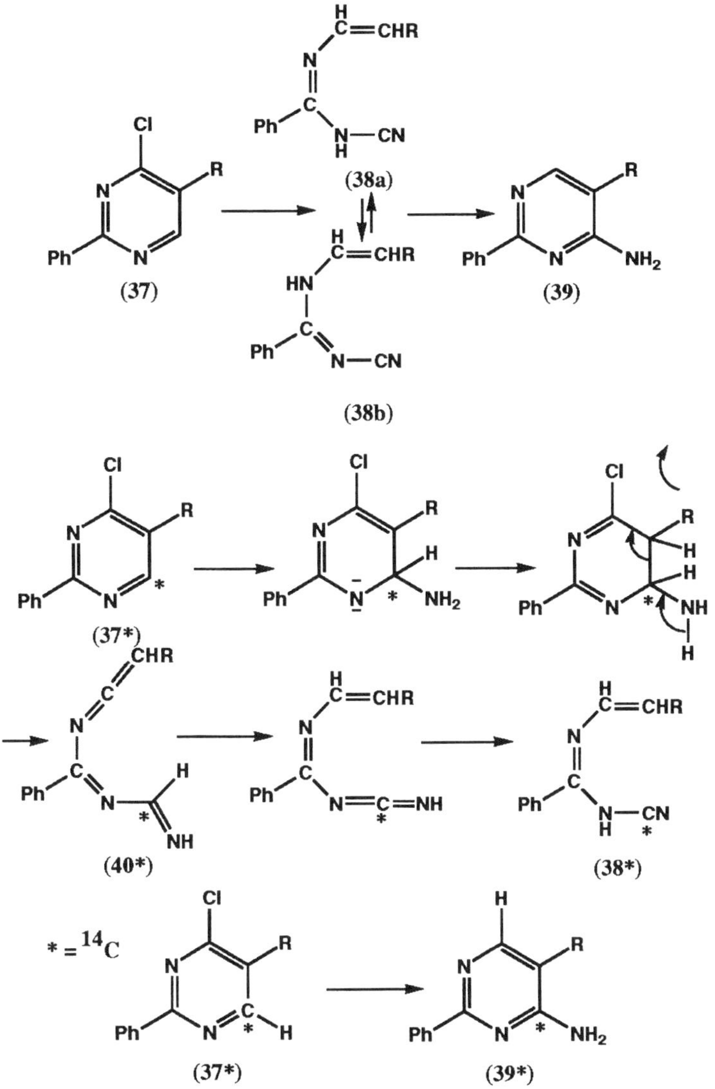

SCHEME II.22

An important consequence of these results is that the overall amination process (**37*** to **39***), which at first sight seems to follow a classical $S_N(AE)$ pathway, is thus in fact a *meta* telesubstitution. This aminodehalogenation is one of the few reactions known that has proved to involve an S_N (ANRORC) mechanism, but that cannot be described as a degenerate ring transformation, since no interchange of atoms with the reagent has taken place. For another example, see Scheme II.7.

c. Aminodehalogenation of 2-Halogeno-4-Substituted Pyrimidines

Extensive studies have been performed on S_N(ANRORC) participation in the amination of pyrimidines that contain a leaving group at position 2. Treatment of 2-X-4-phenylpyrimidines (X = F, Cl, Br, I) with potassium amide in liquid ammonia at $-75°C$ for 30 min gave the corresponding 2-amino-4-phenylpyrimidine in yields between 50 and 80%, depending on the nature of the halogen atom (F 78%, Cl 59%, Br 68%, I 50%) (73RTC1020; 74RTC111, 74RTC325).

In all these reactions a by-product was obtained, i.e., 3-amino-3-phenylacrylonitrile (**40**) being formed in a relative small yield (6–10%) (Scheme II.23). Moreover, it was also observed that the 2-chloro compound, when reacting at $-33°C$ for a short period of time (3–4 min), gave a rather stable second product, i.e., 4-amino-1-cyano-2-phenyl-1-azabutadiene (**41**). This compound could be isolated in about 50% yield when the reaction was carried out at $-75°C$.

To determine whether an S_N(ANRORC) mechanism has played a role in the aminodehalogenation, the reaction was carried out with the double-labeled 2-X-4-phenyl[1,3-^{15}N]pyrimidine. The ^{15}N-enrichment was established by mass spectrometric measurement of the $(M + 2)/M$ peak. The partition of the ^{15}N label over the ring nitrogen and the aminogroup in the 2-amino-4-phenylpyrimidine was established by the usual procedure (see Scheme II.12), i.e., first converting of the 2-amino compound into the corresponding pyrimidin-2(1H)-one, and subsequently converting this compound into the 2-chloro derivative (**42A/42B**) (see Scheme II.24). By mass spectrometric measurement of the $M + 2$, $M + 1$, and M peaks of **42A/42B**, the participation of the S_N(ANRORC) mechanism in the aminodehalogenation of all four 2-halogeno compounds could be calculated.

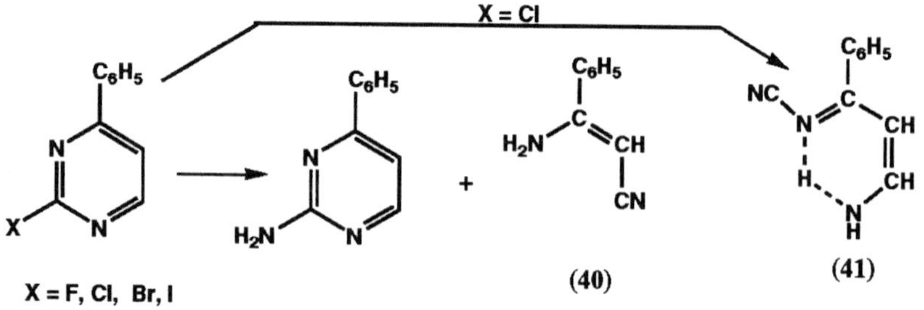

SCHEME II.23

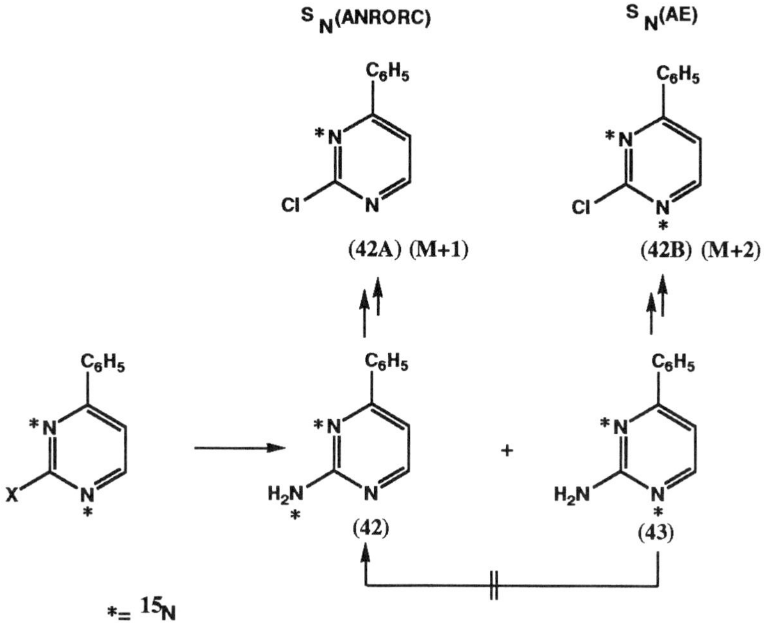

SCHEME II.24

It is evident that in cases where the amination has taken place 100% according to the ANRORC process, no enrichment in the $M + 2$ peak could be measured and that only ^{15}N enrichment of the $M + 1$ peak in the chloro compound could be observed. Based on these mass spectrometric determinations it was established that the 2-halogenopyrimidines react for the greater part according the $S_N(ANRORC)$ mechanism; see Table II.3.

TABLE II.3
PERCENTAGES OF S_N(ANRORC) PARTICIPATION IN THE AMINODEHALOGENATION OF 2-X-4-PHENYLPYRIMIDINE WITH POTASSIUM AMIDE IN LIQUID AMMONIA

Substituent	Temperature	% participation S_N(ANRORC)
$X = F$	$-75°C$	82%
$X = Cl$	$-75°C$	88%
$X = Cl$	$-75°C$	90%
$X = Br$	$-75°C$	88%
$X = I$	$-75°C$	73%

As one can see from the table, the degree to which this ANRORC process occurs is nearly independent of the temperature applied during the amination. For example, the amination of 2-chloro-4-phenylpyrimidine, when carried out at −33°C instead of −75°C, still occurs 90% according to the ANRORC mechanism.

The possibility that the formation of the 2-[^{15}N]amino-4-phenyl[^{15}N]pyrimidine (**42**) occurs by a Dimroth type rearrangement of 2-amino-4-phenyl[1,3-^{15}N]pyrimidine (**43**) can be rejected (see Scheme II.24). Treatment of an independently prepared specimen of 2-amino-4-phenyl [1,3-^{15}N]pyrimidine with potassium amide/liquid ammonia under varying reaction conditions gave no indication of a rearrangement leading to ^{15}N enrichment in the exocyclic amino group.

These results lead to the conclusion that a greater part of the amination starts by addition of the amide ion to position 6, after which ring opening occurs, followed by ring closure (Scheme II.25). In this way the reaction occurs analogously as discussed with the 6-halogeno-4-phenylpyrimidines. Among the four halogens the 2-iodo compound has the lowest ANRORC participation (just as found for the 6-iodo-4-phenylpyrimidine; see Section II,C,1,a)

Further substantive support for the participation of the S$_N$(ANRORC) mechanism can be taken from two important observations: (1) ^{1}H and ^{13}C NMR spectroscopy of solutions of 2-X-pyrimidines in liquid ammonia containing potassium amide firmly proves the presence of the anionic adduct 6-amino-1,6-dihydropyrimidinide (**44**) (74RTC325), the initial adduct in the ANRORC process; and (2) the azadiene intermediate (**41**, which as mentioned above can be isolated) easily converts into 2-amino-4-phenylpyrimidine on treatment with potassium amide in liquid ammonia. Both observa-

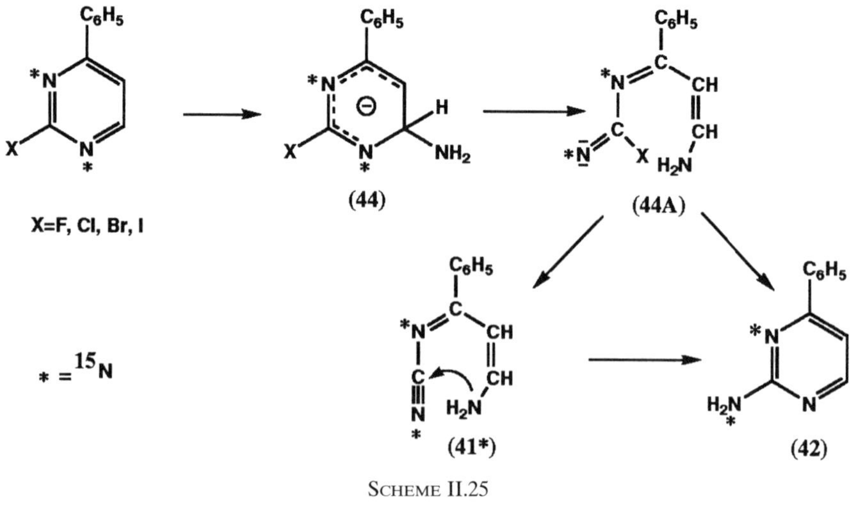

SCHEME II.25

tions provide supporting evidence that **44** and **41** are the precursors of the 2-amino-4-phenylpyrimidine, although it does not exclude the possibility that the iminohalogenide **44A** also acts as a direct precursor of **42**.

All these results provide good evidence that the ring opening is the rate-determining step in the sequence of reaction steps. It is of interest to note that in the reaction of 4-bromo-6-phenylpyrimidine, which reacts for a considerable percentage by the S_N(ANRORC) mechanism (83%, Table II.1), no trace of an open-chain compound could ever be detected. The formation of 3-amino-3-phenylacrylonitrile (**40**), although yield is low, brings forward another interesting feature of the aminodehalogenation reaction. Whereas the formation of the 2-amino compound starts by addition of the amide ion at C-6, it is evident that **40** cannot have been formed by an initial addition at C-6. The presence of the amino group and the phenyl group at the same carbon in **40** seems to suggest that the formation of this compound has occurred by an initial addition of the amide ion on position 4 of the pyrimidine ring, although addition to this position is strongly hampered by the presence of the phenyl group (Scheme II.26). Therefore, this process can

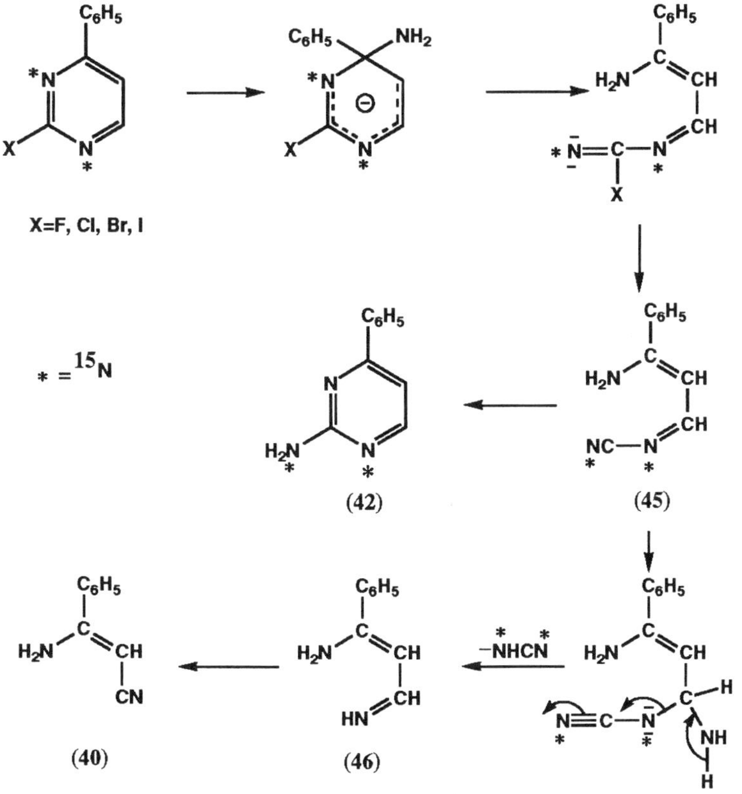

SCHEME II.26

only be of minor importance. Moreover, it was found that compound **40**, being isolated from the reaction mixture obtained from the double-labeled 2-chloro-4-phenyl[1,3-^{15}N]pyrimidine and the amide reagent, does *not* contain ^{15}N-labeling! This means that in the formation of **40** the N_1–C_2–N_3 moiety of the pyrimidine ring is "lost." If, as shown in Scheme II.26, the reaction follows the same pattern of ring opening, 4-amino-1-cyano-4-phenyl-1-aza-1,3-butadiene (**45**) is formed as intermediate, which by loss of double ^{15}N-labeled carbodiimide (aminocyanogen) is transformed into unlabeled 4-amino-4-phenyl-1-aza-1,3-butadiene (**46**). The aminoacrylonitrile **40** is formed from **46** by a Cannizzaro-type oxidation–reduction reaction.

An interesting consequence of the mechanism presented in Scheme II.26 is that, in principle, from the intermediary **45** by ring closure a double ^{15}N-labeled 2-aminopyrimidine **42** can also be obtained! These results suggest that the formation of the 2-amino-4-phenylpyrimidine can occur according to two routes: by an initial addition at C-6 (the main process, Scheme II.25) and by initial addition at C-4 (the minor process, Scheme II.26) (74RTC111). Both routes involve a degenerate ring transformation.

Studies have been carried out on the blocking effect of the phenyl group on position 6 (74RTC227). Amination of 2-X-4,6-diphenyl[1,3-^{15}N]pyrimidine (X = F, Cl, Br) gave the double ^{15}N-labeled 2-amino-4,6-diphenylpyrimidine. By measuring the partition of the ^{15}N label over the ring nitrogen and the amino nitrogen, it was concluded that 2-fluoro-4,6-diphenylpyrimidine has undergone amination *without* ring opening, and that 2-chloro- as well as 2-bromo-4,6-diphenylpyrimidine react for about 70-72% with S_N(ANRORC) participation. Comparing these results with those of the amination of 4-bromo-2,6-diphenylpyrimidine it became evident that the inhibitory effect of the phenyl group at position 2 in 4-bromo-2,6-diphenylpyrimidine is much more pronounced than the blocking effect of a phenyl subsituent on position 6 in 2-bromo-4,6-diphenylpyrimidine (Table II.4).

The results are supported by the results of NMR measurements, showing unequivocally that the addition of the amide ion at C-4/C-6 in the pyrimidine ring is the (thermodynamically) most favored process and more preferred than addition at C-2. Therefore, the hampering effect of the phenyl group at C-2 is more pronounced. All the results mentioned in this section indicate that the aminolysis of highly activated pyrimidines containing a leaving group at position C-2 or C-4(6), a reaction that is usually considered to take place via an addition–elimination process, can also occur via the less conventional S_N(ANRORC) mechanism. As an illustration: It is not unlikely that the aminolysis of 2-chloro-4,6-dicyanopyrimidine into 2-amino-4,6-cyanopyrimidine with ammonia (77KGS821) may at least partly occur by participation of a S_N(ANRORC) mechanism. ^{15}N-Labeling studies are necessary to establish more firmly the reaction course of the aminolysis.

TABLE II.4

PERCENTAGES OF S_N(ANRORC) PARTICIPATION IN THE AMINATION OF 2-BROMO-6-PHENYL- AND 4-BROMO-6-PHENYLPYRIMIDINES AND THE CORRESPONDING 2-BROMO-4,6-DIPHENYL- AND 4-BROMO-2,6-DIPHENYLPYRIMIDINES

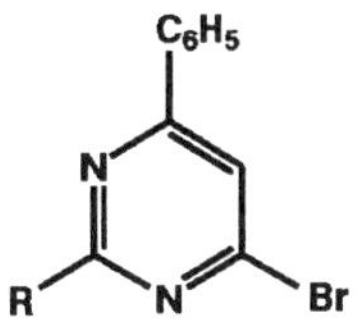

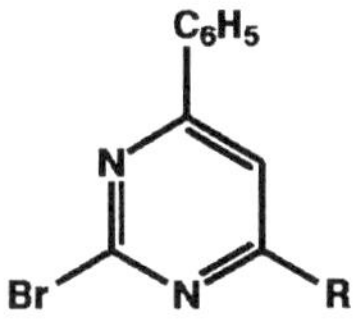

R = H 83%	R = H 88%
R = C$_6$H$_5$ 0%	R = C$_6$H$_5$ 70%

d. Aminolysis of Pyrimidines Containing a Leaving Group at C-2 Different from Halogen

The results of the studies on aminodehalogenation presented in previous sections induced investigations of the influence of C-2 substituents with leaving group mobilities different from halogen on the occurrence of the S_N(ANRORC) process during amination with potassium amide/liquid ammonia. Therefore, a series of 2-X-4-phenyl[1,3-^{15}N]pyrimidines (X = SCH$_3$, SO$_2$CH$_3$, SO$_2$C$_6$H$_5$, SCN, CN, $^+$N(CH$_3$)$_3$) were prepared and subjected to amination with potassium in liquid ammonia (74RTC325). The results are summarized in Table II.5.

As one can see from Table II.5, the compounds with X = SCH$_3$, SO$_2$CH$_3$, and SCN react with high percentages of participation of the S_N(ANRORC) process (73–90%) in the amide-induced aminolysis, while for X = CN,

TABLE II.5

PERCENTAGES OF S_N(ANRORC) PARTICIPATION IN THE AMINATION OF 2-SUBSTITUTED-4-PHENYLPYRIMIDINE BY POTASSIUM AMIDE/LIQUID AMMONIA

Substituent	Temperature(°C)	% S_N(ANRORC)
SCH$_3$	−33	91
SO$_2$CH$_3$	−33	73
SO$_2$C$_6$H$_5$	−33	34
SCN	−33	90
CN	−75	5
N$^+$(CH$_3$)$_3$	−75	10

$^+N(CH_3)_3$, and $X = SO_2C_6H_5$ the amination occurs for a relatively low percentage(<30%) according to the S_N(ANRORC) mechanism.

An interesting difference in reaction behavior was observed between the $SO_2C_6H_5$ group and the SO_2CH_3 substituent. Whereas for $X = SO_2C_6H_5$-group the ANRORC process occurs for only a relative small percentage, for $X = SO_2CH_3$ the ring-opening process is strongly favored. ^1H NMR spectrocopy of solutions of 2-X-4-phenylpyrimidines (X = SCH_3, SO_2CH_3) in liquid ammonia containing potassium amide (74RTC325) showed that both compounds are present in these solutions as their covalently bound 6-amino adducts, i.e., the 6-amino-1,6-dihydropyrimidinides **47** and **48** (Scheme II.27). However, it appeared that in the σ-adduct **48** the SO_2CH_3 group is present in the deprotonated form. This behavior probably explains why 2-phenylsulfonyl-4-phenylpyrimidine reacts differently from 2-methylsulfonyl-4-phenylpyrimidine. The presence of the anionic $SO_2CH_2^-$ group at C-2 strongly disfavors the addition of the amide ion at C-2, while for $SO_2C_6H_5$ the addition at C-2 is promoted because of hydrogen-bridge formation (see **49**) and leads to favoring of the S_N(AE) process (Scheme II.27).

There has been an attempt to correlate that fraction of the 2-substituted 4-phenylpyrimidines that takes part in the S_N(ANRORC) process (X_{ANRORC}) and the inductive and resonance effects of the substituents, present in the substrate on position 2 (see Table II.6). (85T237).

For that purpose the Swain nonresonance field constants F and resonance constants R were used (68JA4328). X_{ANRORC} was calculated from %(chemical yield)/100 × % S_N(ANRORC)/100. The X_{ANRORC} of the thiocyanato and the sulfonylphenyl are not included in Table II.6, since the F and R values are unknown. The 2-sulfonylmethyl substituent is also not included, as the F and R values of the conjugate base (the species present in this strong medium) are unknown.

Based on this set of (limited) data a correlation was established, as given in the following equation:

$$X_{ANRORC} = -0.34F - 1.04R + 0.55.$$

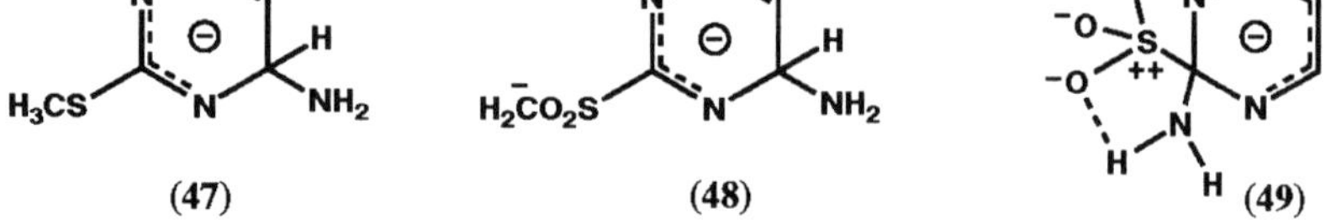

SCHEME II.27

TABLE II.6

PERCENT YIELDS/100 [A] AND PERCENTAGE S_N(ANRORC) MECHANISM/100 [B] OBTAINED IN THE AMINATION OF 2-X-4-PHENYLPYRIMIDINES, X_{ANRORC} [A × B], NONRESONANCE CONSTANTS F, AND RESONANCE FACTORS R OF SUBSTITUENTS X.

X	A	B	A × B	F	R
SCH_3	0.72	0.91	0.65	0.332	−0.186
SO_2CH_3[a]	0.68	0.73	0.50	—	—
CN	0.56	0.05	0.03	0.847	0.184
$N^+(Me)_3$	0.62	0.10	0.06	1.460	0.000
Cl	0.59	0.90	0.53	0.690	−0.161
Br	0.67	0.88	0.59	0.727	−0.176
F	0.78	0.80	0.62	0.708	−0.336
I	0.50	0.73	0.37	0.672	−0.197
H	0.60	0.92	0.55	0.000	0.000

[a] The SO_2CH_3 group is deprotonated and the F and R values are unknown.

The square of the correlation coefficient is a satisfactory 0.90. This equation gives a rather reliable and quantitative description of the influence on the amide-induced replacement of the leaving group in 2-X-4-phenylpyrimidines. The factor +0.55 reflects the fraction of the molecules of 4-phenylpyrimidine (R and F are zero) that undergoes an S_N(ANRORC) aminodehydrogenation. This has indeed been experimentally established (see Section II,C,1,g).

e. *Aminodebromination of 5-Bromopyrimidines*

In Section II,A. it was stated that the slowly occurring amination of 5-bromo-4-*t*-butyl-6-deuteriopyrimidine in liquid ammonia (24 hr, 2 equiv. of potassium amide) gives in about 30% yield the *cine*-substituted product 6-amino-4-*t*-butylpyrimidine, in which no deuterium was present (64TL2093; 65TL555; 68TL9). This result was interpretated as a strong indication for the intermediacy of 4-*t*-butyl-5,6-pyrimidyne (5,6-didehydropyrimidine), which undergoes a one-sided addition at C-6. However, for proton abstraction at C-6, the initial step in the 5,6-pyrimidyne formation, no evidence was found. Moreover, ^1H (74RTC231) and ^{13}C (75OMR86) spectroscopy showed that in solutions of potassium amide/liquid ammonia the 4-substituted 5-bromo compounds are present as their anionic 1:1 amino adducts, i.e., the 6-amino-1,6-dihydropyrimidinides **50**. These results, combined with those obtained in the amination of 2-halogeno- as well as 4-halogenopyrimidines, showing a ring-opening, ring-closure reaction sequence, raised serious doubts about the correct interpretation of the chemistry of the amino-debromination of the 5-bromopyrimidines and induced a reinvestigation.

^{15}N-labeling studies with the *mono*-labeled 5-bromo-4-*t*-butyl[1(3)-^{15}N]pyrimidine (**51**; the label is scrambled over both ring nitrogen atoms) showed that about 49% of the 6-amino compound obtained is ^{15}N-labeled in the exocyclic amino group, i.e., **56** (77RTC101). The retrieved starting material **51** has not undergone any decrease in ^{15}N content. These results indicate that apparently about half of the starting material is converted into 6-amino-4-*t*-butylpyrimidine via an open-chain intermediate, following the S_N(ANRORC) route. The puzzling point, however, is that the formation of the 6-amino compound **56** can never be explained via breaking of the bond between N-1 and C-6 in the σ-adduct 6-amino-1,6-dihydropyrimidinide **50**. If that were the case, after recyclization 2-amino-5-bromo-4-*t*-butylpyrimidine should have been formed, which in fact has not been observed (Scheme II.28).

The ring-opening process, leading to the observed ^{15}N-labeling distribution in **56,** requires the presence of the covalent adduct 2-amino-1,2-dihydropyrimidinide **52**; this σ-adduct, however, could not be detected by NMR spectroscopy. Adduct **52** is quite reactive and undergoes bond-breaking between C-2 and N-1. The resonance-stabilized anionic 1,5-diaza-1,3,5-hexatriene (**53a/53b**) is formed, which loses hydrogen bromide and yields the ynamine **54**. Protonation and ring closure gives the 6-amino compound **56** with the required ^{15}N-labeling partition on ring nitrogen and amino nitrogen.

Supporting evidence that 5-bromopyrimidines can undergo addition at C-2 can be taken from the fact that 5-bromo-4-piperidinopyrimidine gives with potassium amide/liquid ammonia the *tele*substituted product 2-amino-4-piperidinopyrimidine (see Scheme II.29) (78JHC1121).

It has been argued that the loss of the hydrogen bromide in **53,** leading to **54,** has to take place *before* cyclization. This is based on the observation that in the retrieved starting material no decrease of the ^{15}N content was observed. If cyclization would have taken place before loss of the hydrogen bromide, then the 1:1 C-6 amino adduct **55** would have been formed, which is in equilibrium with the starting material, and a decrease of the ^{15}N content in the retrieved starting material should have been found. As indicated previously, this is not observed.

Regarding the formation of the other 51% of the 6-amino compound, in whose formation no ring opening is involved, i.e., **57,** it seems very probable that this occurs via the intermediacy of 4-*t*-butyl-5-bromo-5,6-dihydropyrimidine (Scheme II.28).

Extension of these amination studies to 4-phenyl-, 4-methoxy-, and 4-piperidino-5-bromo[1(3)-^{15}N]pyrimidine revealed that also these com-

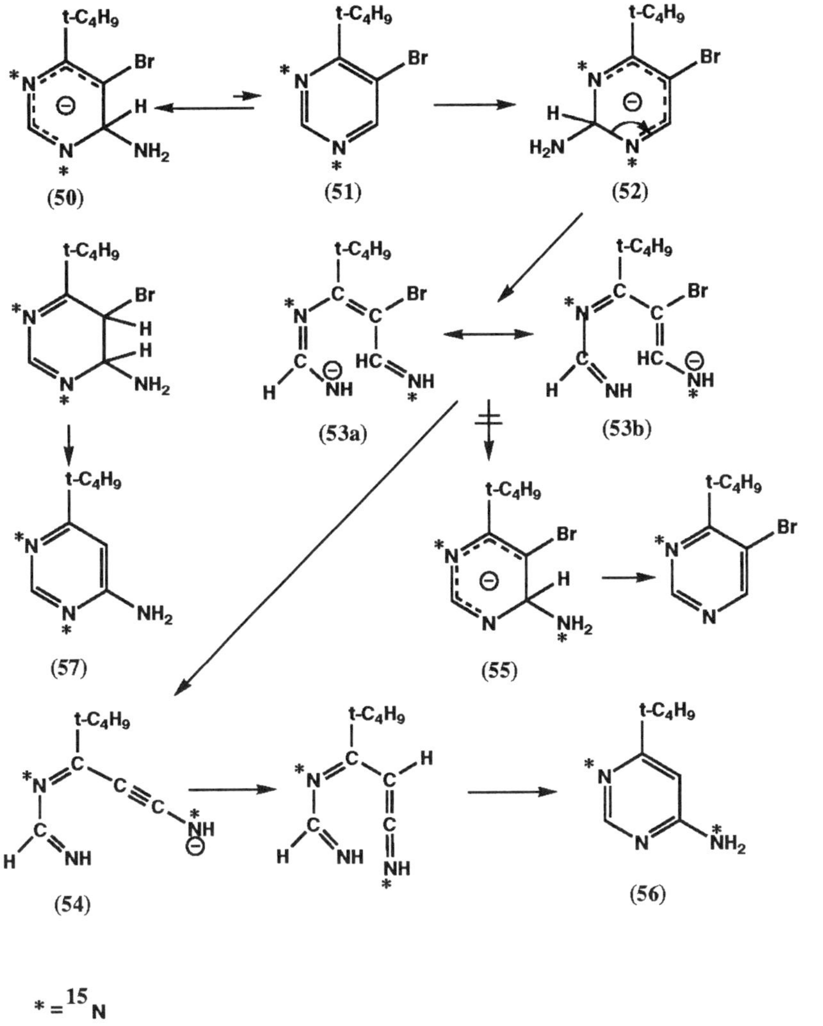

S_N(ANRORC) REACTIONS IN AZINES 43

$* = {}^{15}N$

SCHEME II.28

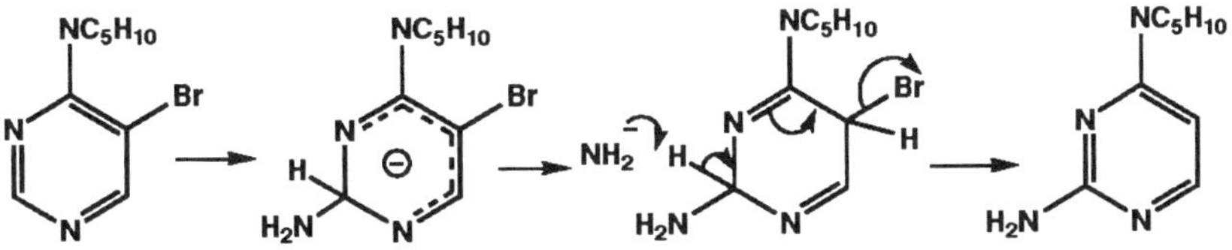

SCHEME II.29

TABLE II.7

PERCENTAGES OF S_N(ANRORC) PARTICIPATION IN THE AMINATION OF 4-R-5-BROMOPYRIMIDINES WITH POTASSIUM AMIDE IN LIQUID AMMONIA

Substituent R	% S_N(ANRORC)
t-C_4H_9	49
C_6H_5	52
OCH_3	26
NC_5H_{10}	28
CH_3	0
$NHCH_3$	0
NHC_6H_5	0

pounds react, although to a different degree, via an S_N(ANRORC) pathway, since it was also observed in these reactions that a portion of the ring nitrogen atoms of the starting material is located in the 6-amino group of the reaction product (Table II.7) (78JHC1121).

As mentioned previously, only in the amination reaction of 4-piperidino-5-bromopyrimidine was an S_NH *tele* amination observed. Besides the 6-amino-4-piperidinopyrimidine, a small amount of 2-amino-4-piperidinopyrimidine was isolated. Its formation is depicted in Scheme II.29. The formation of this compound can be considered as additional proof of the vulnerability of position 2 in the pyrimidine ring for nucleophilic attack, as observed in the formation of **52**.

Amination of 4-methyl-, 4-N-methylamino-, and 4-anilino-5-bromopyrimidine with potassium in liquid ammonia yields the corresponding 6-

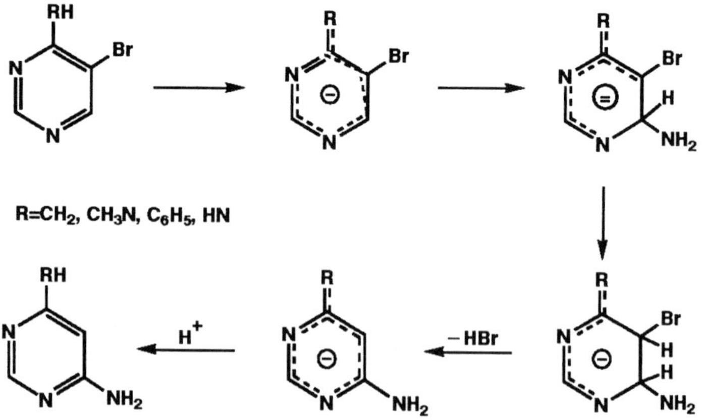

SCHEME II.30

aminopyrimidines; however, ^{15}N-labeling studies revealed that in these conversions no ring-opening was involved (78RTC288). These three substrates have in common that they possess an acidic hydrogen on the atom linked to the ring carbon at position 4. In the strong basic medium, this hydrogen is abstracted and a strongly delocalized anion is formed. Evidence for this anion formation is provided by ^{1}H NMR spectroscopy (74RTC237; 75OMR86). It can be argued that in this anion most of the negative charge is localized in the N-1, C-2, N-3 region, making the nucleophilic attack at position 2 less favorable than addition at C-6, because of electron repulsion between the incoming amide ion and the highly charged region around position 2. The mechanism proposed for the formation of the *cine* product is given in Scheme II.30.

f. *Aminodemethoxylation of Dimethoxypyrimidines*

It has been reported that 4,6-dimethoxypyrimidine and its 5-bromo derivative undergo an aminodemethoxylation to the corresponding 4-methoxy-6-aminopyrimidines on treatment with potassium amide/liquid ammonia (61JCS1298; 65JCS6659). No details were mentioned concerning the mechanism of these aminodemethoxylations. In the light of the results reported in previous sections, investigations were carried out to establish whether this aminodemethoxylation reaction also takes place with a ring-opening, ring-closure sequence of reactions. Treatment of 4,6-dimethoxypyrimidine and 5-bromo-4,6-dimethoxypyrimidine with ^{15}N-labeled potassium amide in liquid ammonia led to the formation of the corresponding 4-methoxy-6-aminopyrimidines **57**, in which *only* the pyrimidine nitrogen atoms of the pyrimidine ring were found to be ^{15}N-labeled; no ^{15}N-label was present on the amino nitrogen.

This partition was determined by mass spectrometry, measuring the ^{15}N-content of the 6-amino compound **57** and that of the 6-chloropyrimidine **58**, formed from **57** on diazotation by action of sodium nitrite in conc. hydrochloric acid. The ^{15}N content in both the 6-amino and the 6-chloro compound is the same, providing clear evidence that in the amino demethoxylation under described conditions the S$_N$(ANRORC) mechanism is the sole operative process. It follows the same pattern as described in previous sections for the aminodehalogenation, i.e., initial addition of the amide ion on position 2 and subsequent ring opening and ring closure (Scheme II.31).

The sensitivity of C-2 in 4,6-dialkoxypyrimidine for nucleophilic attack has also been observed in the hydrazinolysis of 2,4-diethoxypyrimidine, yielding 3-methyl-1,2,4-triazole **59**, Scheme II.31) (70RTC680). One of the intermediates is 3-cyanomethyl-1,2,4-triazole, which in the presence of hydrazine is decyanated.

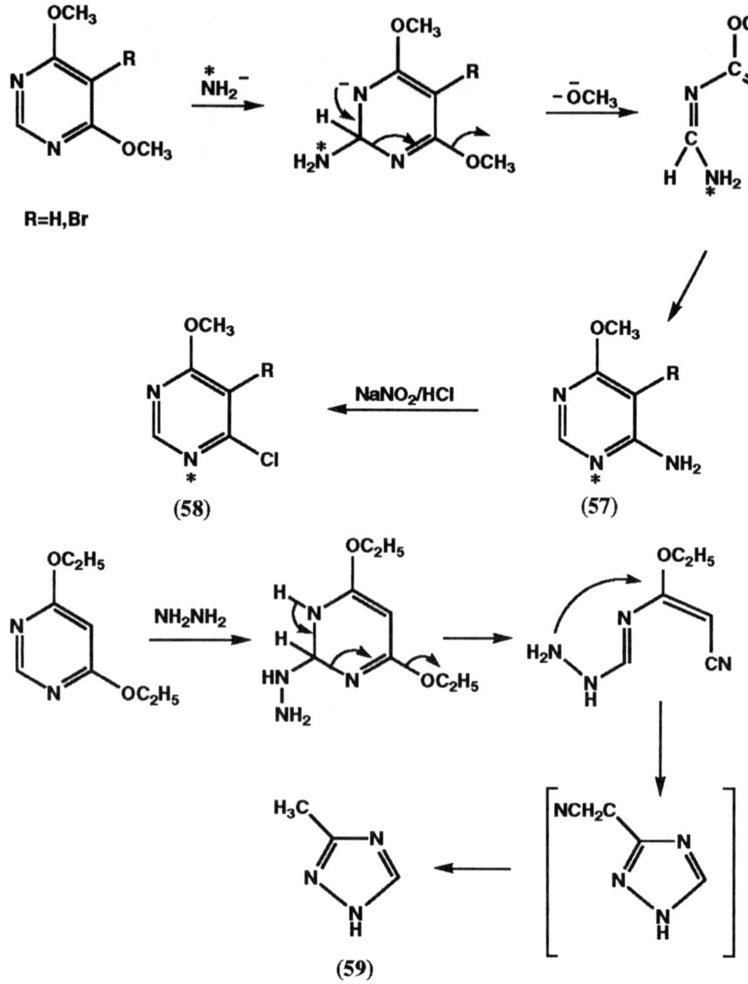

SCHEME II.31

g. *Aminodehydrogenation of Pyrimidines*

The conversion of pyridine into 2-aminopyridine (known as the Chichibabin amination) was considered for a long time to be a typical and outstanding example of an aminodehydrogenation reaction (14MI1; 71MI2; 78RCR1042) and received scant recognition as a method for aminating pyridines and related azines. This nucleophilic replacement of hydrogen by an amino group has been the subject of many investigations and much mechanistic controversy. During the past two to three decades, considerable efforts have been made to modify and to improve the original amination procedure (sodium amide, solvent toluene, 110°C) [78RCR1042; 83AHC305; 85MI1, 85T237; 86CCA89; 87KGS1011; 88AHC2], as well as to develop new methods for substituting hydrogen in azaaromatic compounds by a variety of nucleophiles (88T1). In a recent book titled *Nucleophilic Aromatic Substitution of Hydrogen,* these methodologies are extensively summarized, procedures are given, and mechanisms are discussed (94MI1).

For performing the aminodehydrogenation, two main methods were (are) practiced: heating of the substrate with a metal amide in an inert solvent (toluene, xylene, decaline, *N,N*-dimethylaniline) in the presence of an oxidant (oxygen or oxidizing agent [potassium nitrate (37JOC411)]; or reacting the substrate with liquid ammonia containing amide anions and using as oxidant potassium permanganate [81JOC3805; 82JHC1527; 85S884; 86JHC477; 87JHC1657; 94MI1). This methodology was succesfully applied in the amination of diazines, triazines, and tetrazines (85MI1).

As was extensively shown in previous sections, pyrimidines containing a leaving group can easily undergo aminolysis involving an ANRORC process when reacting with potassium amide in liquid ammonia. This raised the question as to whether pyrimidines that do not contain a leaving group can also undergo an amide-induced amination via a ring-opening process.

4-Phenylpyrimidine. On treatment of 4-phenylpyrimidine with potassium amide in liquid ammonia at $-33°C$ for 70 hr in the presence of potassium nitrate, followed by quenching the reaction mixture by addition of ammonium chloride and workup, two products were isolated: 2-amino-4-phenylpyrimidine (60%) and 6-amino-4-phenylpyrimidine (15%) (79JOC4677). When the reaction was carried out with ^{15}N labeled potassium amide in liquid ammonia and using the combined methodologies of chemical conversions and mass spectrometry as discussed previously (see Section II,C,1,a) it was found that in 6-amino-4-phenylpyrimidine (**62/63**), hardly any ^{15}N label was incorporated in the ring ($\pm 5\%$), but that about

TABLE II.8
YIELDS AND PERCENTAGES OF S_N(ANRORC) PARTICIPATION IN THE CHICHIBABIN AMINATION OF 4-PHENYLPYRIMIDINE

Reaction conditions[a]	Yields **60 + 61**	Yields **62 + 63**
A	60%	15%
B	15%	75%
Reaction conditions	% S_N(ANRORC)	% S_N(ANRORC)
A	92%	5%
B	52%	12%

[a] Reaction condition A refers to aminations *with* ammonium chloride quenching; reaction condition B refers to aminations *without* ammonium chloride quenching.

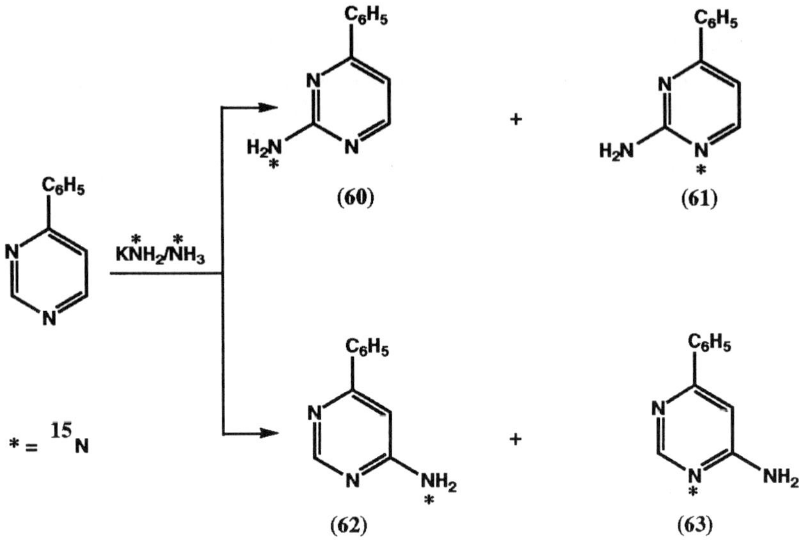

SCHEME II.32

92% ^{15}N-incorporation in the ring of 2-amino-4-phenylpyrimidine (**60/61**) has taken place (see Table II.8 and Scheme II.32).

When 4-phenylpyrimidine was dissolved in liquid ammonia containing potassium amide and the ^{1}H NMR spectrum of this solution was measured after 20 min, it was unequivocally established that two 1 : 1 anionic σ-adducts were formed: 6-amino-4-phenylpyrimidinide (**65**) and 2-amino-4-phenylpyrimidinide (**64**) (ratio 80 : 20) (79JOC4677). ^{13}C NMR spectroscopy confirms the formation of both adducts. On standing of the reaction mixture for several hours, the ^{1}H NMR signals of the 2-amino adduct **64** have disappeared and only the signals of the 6-amino-adduct **65** are then visible in the NMR spectrum. Apparently one deals with the kinetically favored formation of the 2-amino adduct, which converts into the thermodynamically more stable 6-amino adduct. This adduct is quite stable, as standing of this solution for several days does not change the NMR spectrum of the solution. However, when ammonium chloride is added to this solution, hydrogen evolves immediately and the formation of both 6-amino-4-phenyl- and 2-amino-4-phenylpyrimidines is observed (82RTC367).

Kinetically vs thermodynamically favored σ-adduct formation is not an uncommon phenomenon; it has, for example, also been observed in solutions of 2-chloro-3,5-dinitropyridine in liquid ammonia containing potassium amide (85JOC484) and in the σ-adduct formation between quinoline and potassium amide in liquid ammonia (73JOC1947).

A profound effect has been observed of the influence of the ammonium ion on the reaction course. When 4-phenylpyrimidine is treated with potassium amide/liquid ammonia for 20 hr and the reaction is *not* quenched, the same mixture of amino compounds is formed as in the ammonium-quenched reaction; however, the yields are very different, 6-amino-4-phenylpyrimidine (**62/63,** 75%) and 2-amino-4-phenyl pyrimidine (**60/61,** 15%) (Table II.8). The increase of the yield of the 6-amino compound from 15 to 75% and the decrease of the yield of the 2-amino compound from 60 to 15% are especially remarkable.

When amination-without-quenching is carried out with ^{15}N-labeled potassium amide/liquid ammonia and the degree and position of labeling in both amino products are determined, it appears that the incorporation of the label in the pyrimidine ring of 2-amino-4-phenylpyrimidine **61** has decreased from 92 to 52%; see Table II.8. Thus, not only the yield of the 2-amino product is lower, but also the fraction that is formed via a ring opening–ring closure sequence [S$_N$(ANRORC) mechanism].

These results justify the conclusion that quenching of the amination by the ammonium salt is an important factor in determining the degree in which the ANRORC mechanism is operative. Its role can be understood as follows (see Scheme II.33). On quenching, the ammonium cation, being a strong acid in this medium, not only neutralizes the amide ion, but also protonates the ring nitrogen of the 6-aminopyrimidinide (**65**) into the uncharged 6-amino-4-phenyl-1,6-dihydropyrimidine (**66**). This dihydropyrimidine can be converted into 6-amino-4-phenylpyrimidine (**62**), being labeled in the exocyclic amino group, or it may undergo an electrocyclic ring opening to form the intermediary acyclic aminodiazahexa-1,3,5-triene (**67a, b**). On recyclization of **67a,** the ring-labeled 2-amino-4-phenyl-1,2-dihydropyrimidine (**67**) is formed, which is air-oxidized into ring-labeled 2-amino-4-phenylpyrimidine (**61**). It is evident that on ring closure of (**67b**) the amino-labeled **62** is formed. Since one can expect that from **68** by loss of ammonia the starting material 4-phenylpyrimidine, being ring-labeled, may also be produced (see Scheme II.33), experiments were carried out with shorter reaction times, making it possible to retrieve the starting material. It was found that this isolated starting material indeed contains ^{15}N-labeling **in** the ring. These results support the mechanism described in Scheme II.33. In conclusion, the Chichibabin amination of 4-phenylpyrimidine into 2-amino-4-phenylpyrimidine can thus be catagorized as a degenerate ring transformation.

Attempts to detect the uncharged intermediates 6-amino-4-phenyl-1,6-dihydro-4-phenylpyrimidine (**66**), 2-amino-4-phenyl-1,2-dihydropyrimidine (**68**), and the aminodiazahexatriene **67,** by recording the NMR spectra of the solution during the stepwise addition of the ammonium chloride were not successful. The facile ring opening of the pyrimidine ring in **66** is not unexpected. It has also been observed in the reaction of *N*-methylpyrimidinium salts with liquid ammonia, which also involves as neutral covalent

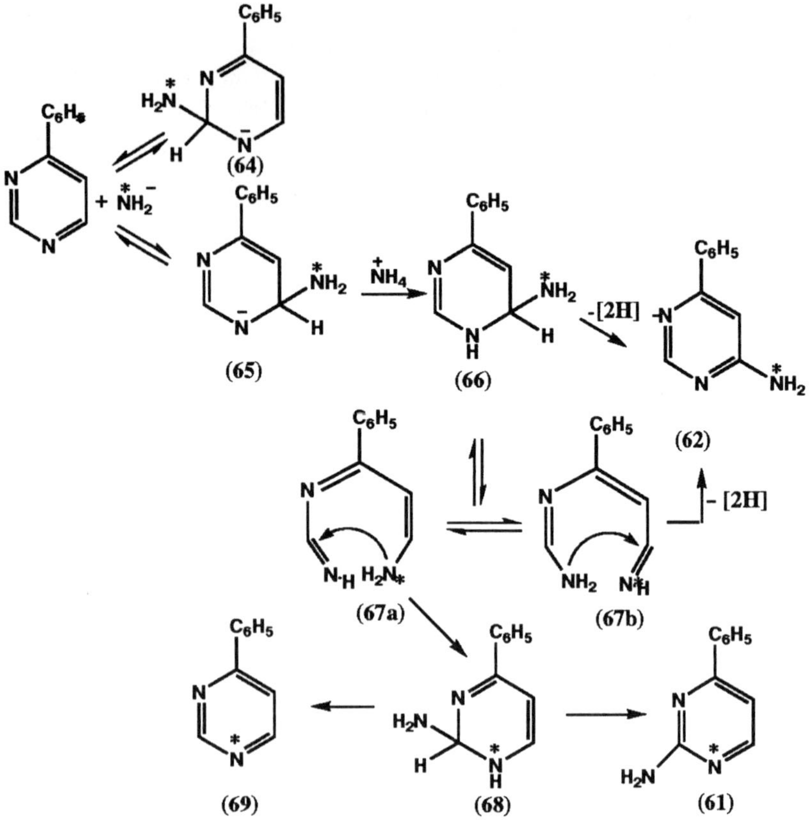

SCHEME II.33

adduct a 6-amino-1,6-dihydropyrimidine derivative (74RTC114) (see Chapter III).

Amination of 4-phenylpyrimidine under more classical conditions (potassium amide, solvent m-xylene, 90°C) also affords a mixture of the 6-amino-4-phenylpyrimidine (35%) and 2-amino-4-phenylpyrimidine (55%). When using ^{15}N-labeled potassium amide, the ^{15}N label was found to be mainly present in the amino group, i.e. formation of **60** and **61**, proving that under these reaction conditions the S_N(ANRORC) mechanism plays only a very minor role in the amination. The strikingly different results obtained in the aminodehydrogenation at low temperature in liquid ammonia and those obtained at the high temperature in boiling xylene can be rationalized as follows. In liquid ammonia the C-2 and C-6 adducts are stabilized by solvation and therefore have a longer lifetime which allows isomerization. In the apolar solvent xylene, stabilization is very weak; therefore, they react immediately into their respective products. It seems that in xylene the rate-determining step is very probably the formation of the σ-adduct, whereas in liquid ammonia the rate-determining step is probably the ring opening of the adduct.

Some evidence has been presented that the amination of azines in the apolar solvent at high temperature may involve a radical anion as precursor of the 1:1 amino σ-adduct (82MI3). This radical anion is formed by an electron transfer from the nucleophile to the heterocycle. From a study of the influence of radical scavengers on the reaction course intriguing effects were found. If azobenzene was added *after* quenching of the reaction mixture with ammonium chloride, the results do not differ from those mentioned before (see reaction conditions A in Table II.8). If, however the azobenzene is added *before* the reaction mixture is quenched with the ammonium chloride, the product mixture has dramatically changed. The total yields of the reaction products are about the same as in the amination of 4-phenylpyrimidine without quenching, but the main product is 2-amino-4-phenylpyrimidine and not 6-amino-4-phenylpyrimidine. Moreover, when the reaction is carried out with ^{15}N labeled potassium amide in the presence of azobenzene, incorporation of the ^{15}N label in the pyrimidine ring is found in none of the products showing the dramatic influence of the scavenger on the S_N(ANRORC) participation in the amino formation. Although these initial results were not fully worked out, the conclusion seems justified that in the S_N(ANRORC) process, a radical and/or radical ion is involved. The occurrence of radical ions in the Chichibabin amination of azines in an apolar solvent is not unprecedented (77CHC210; 78RCR1042).

5-Phenylpyrimidine. Amination of 5-phenylpyrimidine showed about the same results as obtained with 4-phenylpyrimidine (82MI1). Reaction with potassium amide/liquid ammonia for 20 hr at −33°C and quenching of the reaction mixture with ammonium chloride yielded both 2-amino-5-phenylpyrimidine (**70/73,** ±20%) and 6-amino-5-phenylpyrimidine (**72,** ±20%). Investigations by ^{1}H and ^{13}C NMR spectroscopy of solutions of 5-phenylpyrimidine in potassium amide/liquid ammonia clearly showed the presence of the 4(6)-amino adduct; hardly any indication for the presence of the 2-amino adduct was observed (Scheme II.34).

Amination of 5-phenylpyrimidine with ^{15}N-labeled potassium amide revealed that no label was incorporated in the pyrimidine ring of 6-amino-5-phenylpyrimidine; this means that exclusively **72** has been formed.

However, in 2-amino-5-phenylpyrimidine, the pyrimidine ring did contain the ^{15}N label. From the mass spectral data it was calculated that the ratio **70 : 73** is about 20 : 80, proving that about 80% of 5-phenylpyrimidine participates in the S_N(ANRORC) mechanism. Consequently, the ring-labeled 2-amino-5-phenylpyrimidine (**73**) must be formed via ring opening

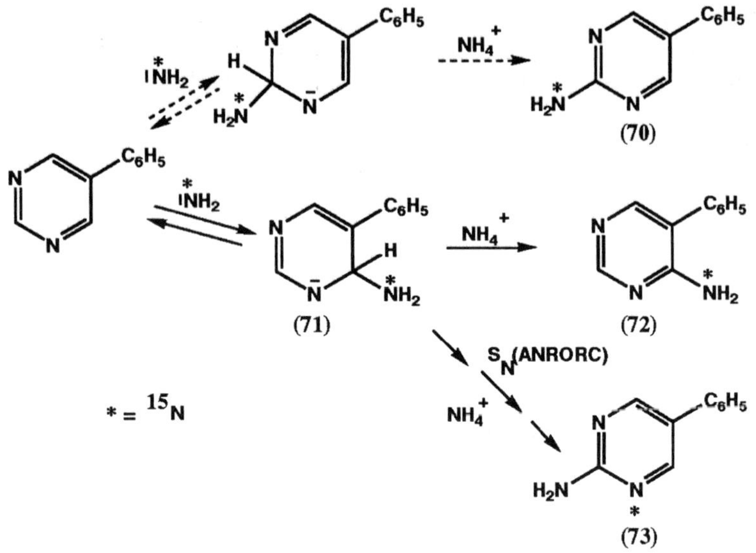

SCHEME II.34

and ring closure, involving the intermediacy of the 6-amino adduct **71** (Scheme II.34).

4-t-Butylpyrimidine. Reaction of 4-*t*-butylpyrimidine with potassium amide/liquid ammonia provided a mixture of both 2-amino-4-*t*-butyl- (25%) and 6-amino-4-*t*-butylpyrimidine (40%) (82RTC367). The NMR spectra clearly indicate the presence of the C-2 adduct as well as the C-6 adduct (ratio 1:9), again showing position 6 as the preferred location to undergo nucleophilic addition. The S_N(ANRORC) participates in the formation of neither the 2-amino product nor the 6-amino product, as shown by the absence of ^{15}N incorporation in the pyrimidine ring of both amino products when the amination is carried out with ^{15}N-labeled potassium amide. This result is in striking contrast with the behavior observed in the aminodehydrogenation reaction of 4-phenyl- and 5-phenylpyrimidine. This result seems to indicate that the σ-adducts are rather stable in the liquid ammonia containing potassium amide and not inclined to ring opening.

It has been reported (39JPJ18) that the Chichibabin amination of 4-methylpyrimidine, using sodium amide in boiling decaline, affords a mixture of 2-amino-4-methylpyrimidine and 2,6-diamino-4-methylpyrimidine. In light of the results mentioned in this section, it certainly cannot be excluded that in this aminodehydrogenation an S_N(ANRORC) mechanism is (partly) involved in the formation of these products.

2. Quinazolines, Purines, and Pteridines

a. *Aminodechlorination of 4-Chloroquinazolines*

A study of the amination of 4-chloroquinazoline with potassium amide/liquid ammonia leads to results similar to those obtained with the phenyl(diphenyl)-4-halogenopyrimidines. Amination of 4-chloro-[3-^{15}N]quinazoline with potassium amide gave 4-aminoquinazoline (**74/75**) in a nearly 100% yield, and it contains the same ^{15}N enrichment as was present in the starting material (73RTC460). It was proven that 53% of the ^{15}N content is located on the 4-amino group in **75** and 47% on the ring nitrogen in **74**, leading to the conclusion that 53% of 4-chloroquinazoline participates in the S_N(ANRORC) mechanism, involving as intermediate the *ortho* (formamidino) iminobenzoylchloride (**76**). This intermediate is formed by an initial addition of the amide ion to C-2 and ring opening between C-2 and N-3 (Scheme II.35). It can be reasonably assumed that the other 47% of 4-aminoquinazoline is formed according to the S_N(AE) process.

The observation that the ratio of the ^{15}N content on the ring nitrogen N-3 and on the exocyclic 4-amino group is about 1 : 1 may suggest the occurrence of a process in which first the chloro atom in the intermediary iminobenzoyl chloride **76** is replaced by the amino group, yielding **77**, followed by ring closure of **77** according to route (a) (Scheme II.35). Since one can expect that in **77** ^{15}N scrambling in the amidine group takes place, it should lead to the experimentally observed 1 : 1 mixture of ring-labeled **74** and amino-labeled compound **75** (73RTC460). However, there is reason to believe that route (a) seems highly unlikely, as one would also expect that **77** is able to undergo a ring closure according to route (b), leading to the formation of 4-aminoquinazoline, which has a *lower* ^{15}N content than the starting material.

As in **74/75** no decrease of the ^{15}N content has been found, the intermediacy of (**77**) is less likely. So, based on this result it is postulated that **74** is obtained by ring closure of **76** and that **75** is formed by the classical addition–elimination process at C-4.

An alternative way to explain the formation of the 1 : 1 mixture of ring-labeled and amino-labeled quinazolines **74/75** might be the occurrence of a scrambling process in 4-aminoquinazoline **74**, formed from the starting material by an S_N(AE) process. This scrambling process should start by addition of the amide ion to C-2 in **74**, followed by ring opening and ring closure. This alternative possibility can however, be excluded, since treatment of an independently prepared specimen of 4-amino[3-^{15}N]quinazoline (**74**) with potassium amide/liquid ammonia, applying the same conditions, does not lead to a decrease of the ^{15}N content on N-3 (73RTC460).

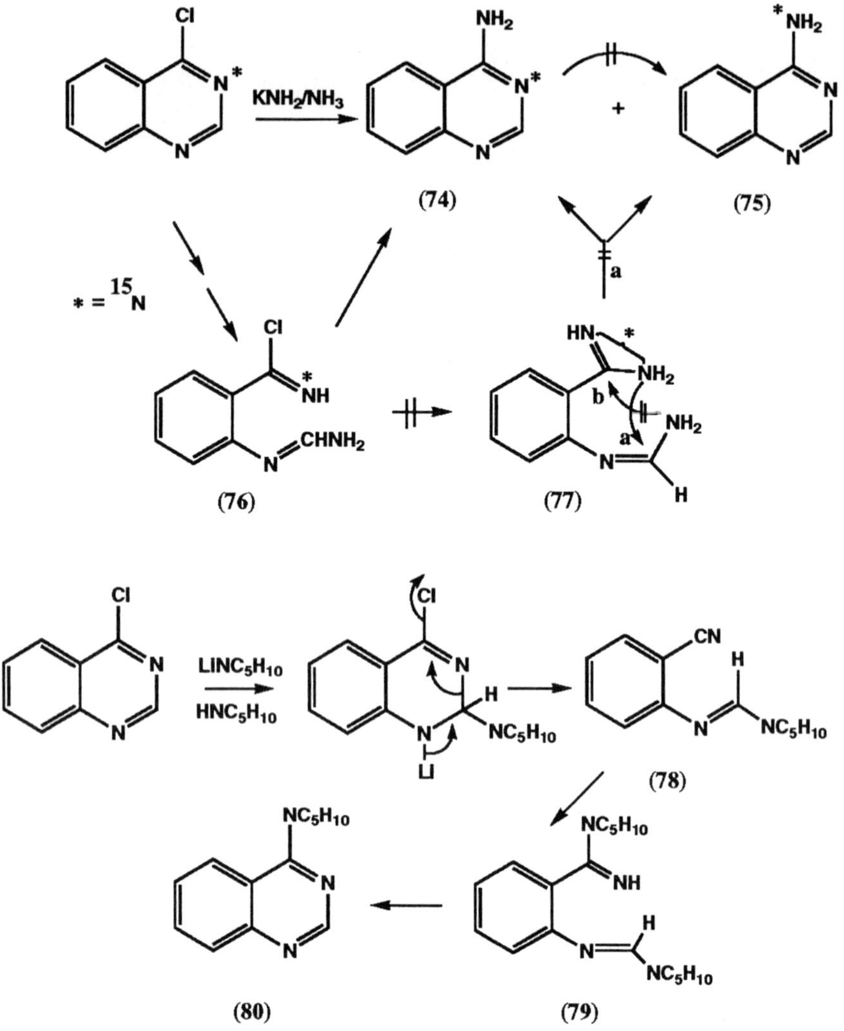

SCHEME II.35

The sensitivity of position 2 in 4-chloroquinazoline for nucleophilic addition was also demonstrated in the reaction with lithium piperidide (73RTC460). Whereas in the amination with potassium amide/liquid ammonia no open-chain intermediate could be isolated, with lithium piperidide/piperidine the open-chain compound *ortho*-(piperidinomethyleneamino)benzonitrile (**78**, 60%) was obtained, in addition to 4-piperidinoquinazoline (**80**, 19%) (Scheme II.35). The formation of **80** from **78** involves addition of lithium piperidine/piperidide to the nitrile group, yielding **79**, which undergoes ring closure into **80** by loss of piperidine.

That the ANRORC process occurs not only with the strong nucleophilic amide agent, but also with the weaker nucleophile ammonia can be demonstrated in the reaction with 4-chloroquinazoline (74RTC227). Heating with ethanolic ^{15}N enriched ammonia at 160°C in a Carius tube gave 4-aminoquinazoline, in which for 19% the ring nitrogen is ^{15}N-labeled. Although the participation of the S_N(ANRORC) mechanism in the aminolysis of 4-chloroquinazoline with ammonia is considerably lower than that observed with the amide ion, it is evident that degenerate ring transformations can (partly) occur with weaker nucleophiles as well.

b. *Aminodehalogenation of 2-Halogenoquinazolines*

It has been established that the conversion of 2-chloro-4-phenylquinazoline into 2-amino-4-phenylquinazoline by treatment with potassium amide/liquid ammonia also occurs with ring opening (74RTC227). This was proved by the experimental result that in the 2-amino compound obtained from 2-chloro-4-phenyl[3-^{15}N]quinazoline, about 70% of the ^{15}N label is present in the amino group, i.e., formation of 2-[^{15}N-amino]-4-phenylquinazoline (**83**); see Scheme II.36.

So, despite the presence of the "blocking" phenyl group at C-4 in the starting material, the initial addition of the amide ion preferentially takes place in the N-3-C-4 azomethine bond at C-4, i.e., formation of **81**. 2-Flu-

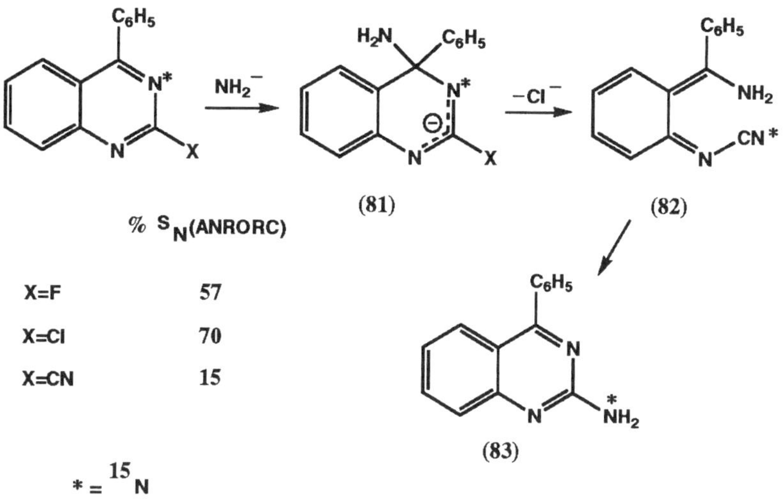

	% S_N(ANRORC)
X=F	57
X=Cl	70
X=CN	15

* = ^{15}N

SCHEME II.36

oro-4-phenylquinazoline undergoes amination in a somewhat lower percentage with ring opening (57%), and 2-cyano-4-phenylquinazoline hardly reacts with ring opening (15%).

The reactivity order Cl > F > CN is in qualitative agreement with that found in 2-X-4-phenylpyrimidine (Cl > F > CN). Attempts to isolate or to detect by spectrometric methods the ring-opened intermediate **82** at $-33°C$ or even at $-75°C$ failed.

Comparison of the percentage $S_N(ANRORC)$ participation of 2-chloro-4-phenylquinazoline (70%) with that of 2-chloro-4,6-diphenylpyrimidine (70%) shows that the reactivity of C-4 in the quinazoline ring is somewhat higher than at C-4,6 in the pyrimidine ring, taking into account that the probability for attack at C-4 in 2-chloro-4-phenylquinazoline is only half of that at C-4,6 in 2-chloro-4,6-diphenylpyrimidine.

Amination with ethanolic ammonia. Amination of 2-chloro-4-phenylquinazoline by heating with an ethanolic solution of ammonia, being enriched with ^{15}N label, at 160°C gave the 2-amino compound, in which the nitrogen in the pyrimidine ring is 34% enriched with the ^{15}N label. It could be proven that this label is present at N-3, in good agreement with the mechanism given in Scheme II.35 for the amination of 4-chloroquinazoline. A similar amination experiment with 2-chloroquinazoline (in this compound C-4 is unsubstituted) gives enrichment of the N-3 in the pyrimidine ring for 31%, again showing that the "inhibitory effect" of the phenyl group at C-4 on the addition of the nucleophile is quite limited. Comparison of the $S_N(ANRORC)$ participation of 2-chloroquinazoline (31%) with that of 4-chloroquinazoline (19%) (Section II,C,2,a) again shows the higher reactivity of C-4 (76MI1).

c. *Aminodeoxogenation of Quinazolin-4-one*

The occurrence of degenerate ring transformations in the amino-dehalogenation of 2- and 4-halogenopyrimidines and quinazolines with strong and weak nucleophiles induced a study of the mechanism of the amino-deoxogenation, which takes place when heating 6-methyluracil and quinazolin-4-one with phenylphosphorodiamidate (PPDA). It has been reported that in these reactions 2,4-diamino-6-methylpyrimidine (69IZV655; 70IZV904) and 4-aminoquinazoline (72JHC1235), respectively, are obtained. It was proposed that the replacement of the oxo group by an amino group occurs via the four-center type transition state **84,** in which a rearrangement takes place leading to a compound that contains, at the carbon atom on position 4 of the pyrimidine ring, both a phosphoro ester and an amino group. By loss of the phosphoro ester group the 4-amino product is yielded (Scheme II.37).

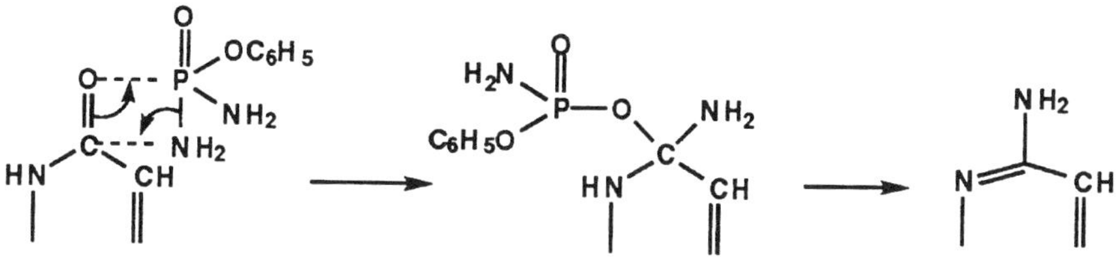

SCHEME II.37

It is open to discussion whether in this reaction the amino group at position 4 is indeed derived from the amino group present in the ester, or whether the nitrogen of the 4-amino group originates from the nitrogen present in the pyrimidine ring. When the reaction was performed with [3-^{15}N]quinazolin-4-one it was found (74TL3201) that a part of the ^{15}N label is present in the 4-amino group. This result provides good evidence that during the aminodeoxygenation a ring-opening process has taken place. This ANRORC process was explained by addition of PPDA to C-2 of the pyrimidine ring, a subsequent ring opening, an intermolecular migration of the phosphor-containing moiety from nitrogen to oxygen, and ring closure (Scheme II.38). There is evidence (74TL3201) (also based on ^{15}N-labeling experiments) that the 4-amino group in 4-amino-

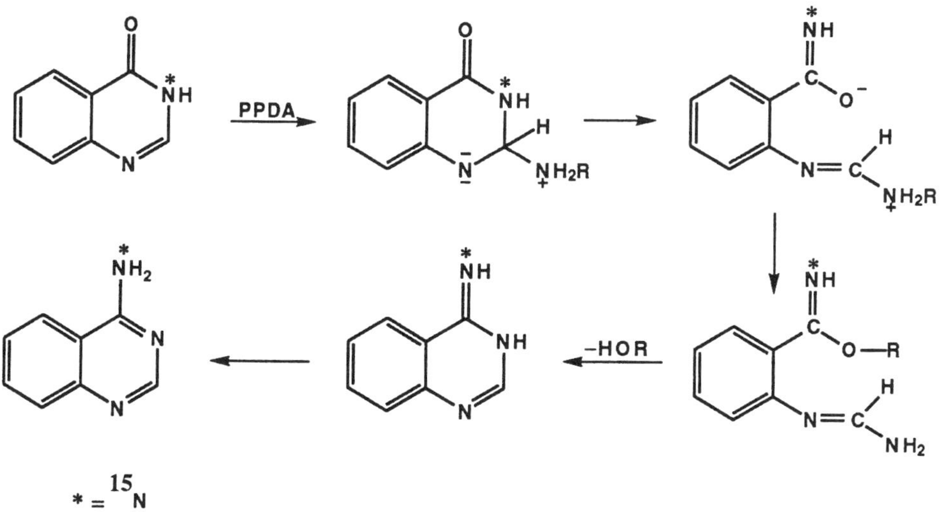

SCHEME II.38

quinazoline can undergo some exchange with the amino group in PPDA. It is suggested that this exchange process also takes place by an ANRORC mechanism.

d. *Aminodehydrogenation of Quinazoline(s)*

The application of the Chichibabin amination to effect a direct amination of quinazoline has been reported. It gives 4-aminoquinazoline (60MI1) as well as 2,4-diaminoquinazoline (59GEP958197). No mechanistic details were discussed, but it can be expected (based on the experience with the amination with 4-phenyl- and 5-phenylpyrimidine) that amination of quinazoline would also involve, at least partly, participation of the $S_N(ANRORC)$ mechanism. Amination with ^{15}N-labeled potassium amide/liquid ammonia will certainly shed some light on the mechanism operative in this Chichibabin amination.

e. *Aminodehalogenation of Halogenopurines*

The first nucleophilic substitution reaction in purines, occurring by a ring opening-ring closure process, was observed by E. Fisher (1898CB542). He found that treatment of 6-amino-2-chloro-7-methylpurine with base does not lead to the formation of the expected 7-methylisoguanine (**88**), but rather leads to the isomeric 7-methylguanine (**86**) (see Scheme II.39). A more detailed study of this reaction revealed that besides **86, 88** is also present in the reaction mixture (62JOC883). The formation of **86** involves the intermediacy of the C-6 adduct and the subsequent formation of the 4-cyanamino-5-carboxamido-1-methylimidazole (**85**). The degenerate conversion of 6-amino-2-chloro-7-methylpurine into **86** can undoubtedly be considered as the first example of a $S_N(ANRORC)$ displacement reaction in heterocyclic chemistry.

A similar degenerate ring transformation was observed when 4-amino-6-chloro-1-methylpyrazolo[3,4-*d*]pyrimidine was treated with dilute alkali. In this reaction as well, the expected product, 4-amino-6,7-dihydro-6-oxo-1-methylpyrazolo[3,4-*d*]pyrimidine, was not obtained, but rather a rearranged isomer, i.e., 6-amino-4,5-dihydro-4-oxo-1-methylpyrazolo-[3,4-*d*] pyrimidine. The formation of this compound takes place according to the same mechanism as that postulated for the formation of **86** (Scheme II.39) (54JOC1570).

Since the formation of the isoguanine derivative **88** has been proved to occur via the intermediacy of 4-ureido-5-cyano-1-methylimidazole (**87**) (having as precursor the C-2 adduct), its formation can thus be described as oc-

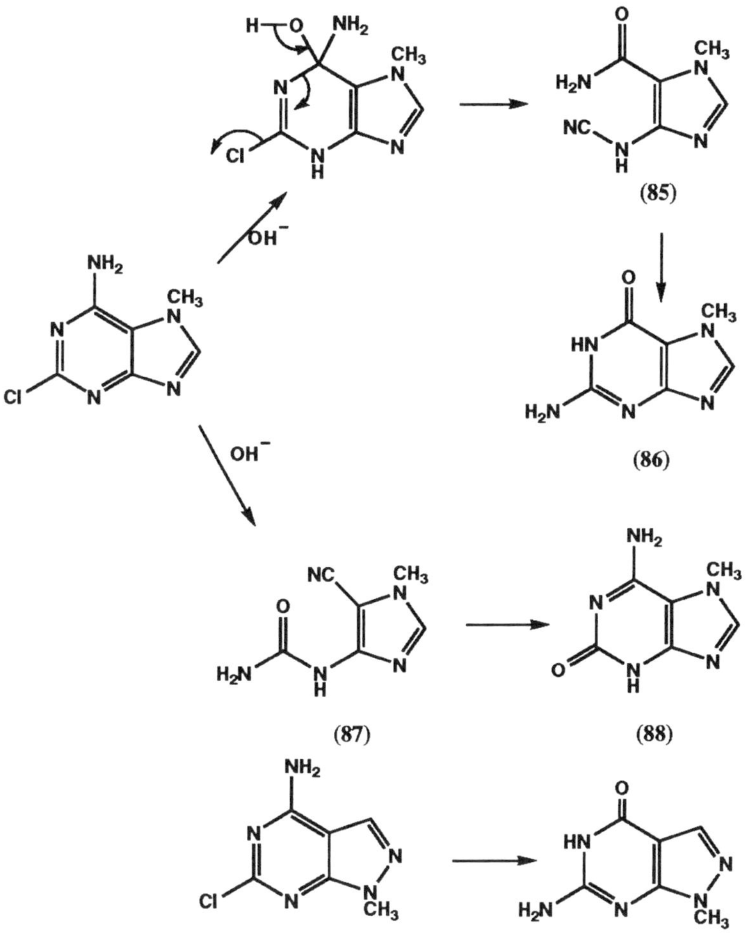

SCHEME II.39

curring according to an S$_N$(ANRORC) mechanism. However, the interesting point is that in contrast to the formation of **86,** the formation of **88** does *not* involve a degenerate ring transformation. Although there is overwhelming evidence that nearly all S$_N$(ANRORC) processes can be classified as degenerate ring transformations, the formation of **88** is an interesting exception to this general rule. A more recent example of an S$_N$(ANRORC) substitution reaction not involving a degenerate ring transformation is the hydroxydechlorination of 2-chloro-5-nitropyridine (see Section II,B,1) and the *tele* amination of 4-chloro-2-phenylpyrimidine into 6-amino-2-phenylpyrimidine (see Section II,C,1,b).

Other examples of nucleophilic substitutions involving a degenerate ring transformation of purines were found when 2-chloro-, 2-fluoro-, and 2-(methylthio)purine treated with potassium amide in liquid ammonia, 2-aminopurine being obtained in good yields (80JOC2942). When the course of the aminolyses was followed by TLC, it became evident that in all three reaction mixtures a compound formed that was insufficiently stable to be isolated and quickly converted into 2-aminopurine. IR spectroscopy of a reaction mixture containing this intermediate and the product 2-aminopurine clearly showed a 2160 cm^{-1} absorption band, characteristic for the presence of an N–CN group. This result proves that an amide-induced opening of the pyrimidine ring occurred, leading to an imidazole derivative that easily underwent cyclization (Scheme II.40).

To substantiate this S_N(ANRORC) mechanism more firmly, amination of 2-R-purine (R = Cl, F, SCH_3) was carried out with ^{15}N-labeled potassium amide/liquid ammonia. It was found (80JOC2942) that in 2-aminopurine the ^{15}N label is exclusively present in the pyrimidine ring nitrogen; no trace of ^{15}N enrichment was found on the amino group on position 2. The incorporation of the ^{15}N label in the heterocyclic ring proves the occurrence of the S_N(ANRORC) mechanism in these amination reactions. It involves addition of the amide ion to position 6 in the purinyl anion (which is unequivocally proved by NMR spectroscopy), yielding the dianionic adduct **89**, and a subsequent ring opening into the negatively charged imidazole derivative **90** by loss of the chloride, fluoride, or thiomethyl ion. This deriv-

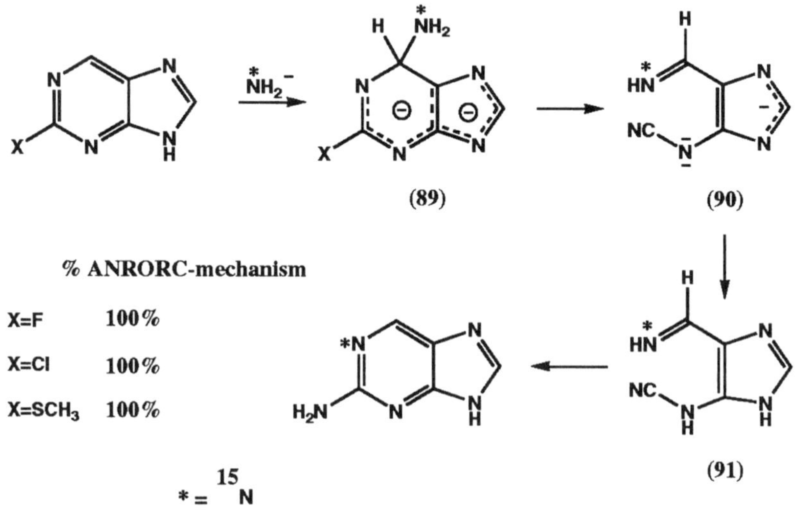

% ANRORC-mechanism

X=F   100%
X=Cl   100%
X=SCH$_3$ 100%

$* = {}^{15}N$

Scheme II.40

ative is rather stable in this basic medium, but working up of the reaction mixture using ammonium chloride (necessary to neutralize the amide ion) yields the neutral species **91**. It is in this neutral species that the cyclization occurs by an interaction of the C-4 iminomethylene group and the cyanamino group at C-5.

Aminodechlorination of 2-chloro-6-phenylpurine with ^{15}N-labeled potassium amide/liquid ammonia gives 2-amino-6-phenylpurine, which contains more than 90% of the ^{15}N label *inside* the pyrimidine ring (80JOC2942). The presence of a phenyl group on position 6 does not prevent the occurrence of the S_N(ANRORC) mechanism. All these results clearly shows that the addition of the amide ion to C-6 in the purinyl anion is the most preferred position, being in agreement with the behavior observed with other nucleophiles [71MI1; 73JCS(PI)2758].

Even when a leaving group is present on position C-2, no addition at C-2 takes place. Another example to illustrate this behavior can be taken from the reaction of 2,6-dichloropurine with potassium amide in liquid ammonia. Besides 2-chloroadenine, 2,6-diaminopurine was obtained (81TH1). It was proved that the 2,6-diaminopurine was not formed from 2-amino-6-chloropurine but from 2-chloroadenine (Scheme II.40A).

However, in view of the results mentioned earlier, direct attack of the amide ion on position 2 seems highly unlikely. An S_N(ANRORC) mechanism, starting with an attack of the amide ion at position 6 containing the amino group, seems to be involved. Adduct formation at a position already occupied by an amino group is not unprecedented. The conversion of 4-amino-2-bromoquinoline into 4-amino-2-methylquinazoline (72RTC841) and of 4-amino-2-bromo-1,5-naphthyridine into 4-amino-2-methyl-1,3,5-

SCHEME II.40A

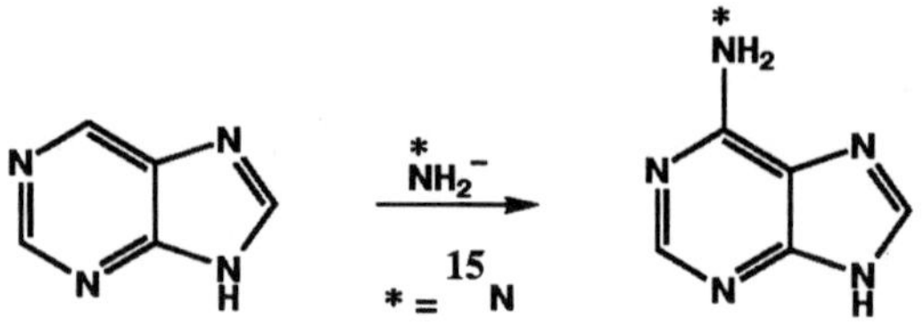

SCHEME II.41

triazanaphthalene (73RTC970) are examples of ring transformations that start by addition of the amide ion to a position occupied by the amino group.

Purine is easily aminated by potassium amide in liquid ammonia, yielding adenine. The initial addition of the amide ion takes place at position 6, in agreement with the general observation that in the anion of purine position 6 is the most favorable one for nucleophilic addition. The formation of adenine does not involve a ring opening–ring closure process, as treatment of purine with ^{15}N-labeled potassium amide does *not* lead to incorporation of the ^{15}N label in the ring (Scheme II.41) (79JOC3140). Thus, the aminodehydrogenation reaction of purine does not involve a degenerate ring transformation reaction. This behavior is in sharp contrast to what has been observed in the amination of phenylpyrimidines.

f. *Aminodehalogenation of 2-Halogenopteridines and Aminodethiomethylation of 2-Methylthiopteridines*

Extensive NMR investigation of the formation of adducts between pteridine and liquid ammonia has shown that two different species are obtained: the covalent 1:1 σ-adduct 4-amino-3,4-dihydropteridine, and the thermodynamically more favored 2:1 σ-adduct 6,7-diamino-5,6,7,8-tetrahydropteridine (Scheme II.42) (75RTC45; 76OMR607). This adduct is also formed in ammonia [71JCS(C)2357].

The easy accessibility of position 4 in pteridines for nucleophilic addition induced a study on the occurrence of the S_N(ANRORC) mechanism in the replacement of a leaving group at C-2 by an amino group.

SCHEME II.42

Aminodehalogenation. 2-Chloro-4,6,7-triphenylpteridine (**92**, X = Cl) was found to give with potassium amide/liquid ammonia a conversion into 2-amino-4,6,7-triphenylpteridine in good yield (**95**, 70%) (78TL2021) (Scheme II.43).

Carrying out the same reaction with ^{15}N-labeled potassium amide/liquid ammonia, the aminodechlorination leads to 100% incorporation of the ^{15}N label in the ring of **95**. This exclusive enrichment of the nitrogen of the ring in **95** means that the replacement of the 2-chloro substituent starts by an exclusive nucleophilic attack at C-4 and formation of the C-4 adduct **93**. Ring opening to 2-iminobenzoyl-3-cyanamino-5,6-diphenylpyrazine (**94**) and subsequent recyclization yield **95**.

Similar observations have also been made with 4,6-diphenyl-2-chloropyrimidine (Section II,C,1,c) and 2-chloro-6-phenylpurine (Section II,C,2,e).

The aminodefluorination occurs to a considerably lesser extent according to the S_N(ANRORC) mechanism (40%); apparently, the competitive addition on C-2, leading to the S_N(AE) substitution, is more favored. This preference for S_N(AE) reactions is characteristic of the fluoro atom; it has been also observed in aminodefluorinations of 2-fluoropyrimidines (Section II,C,1,c).

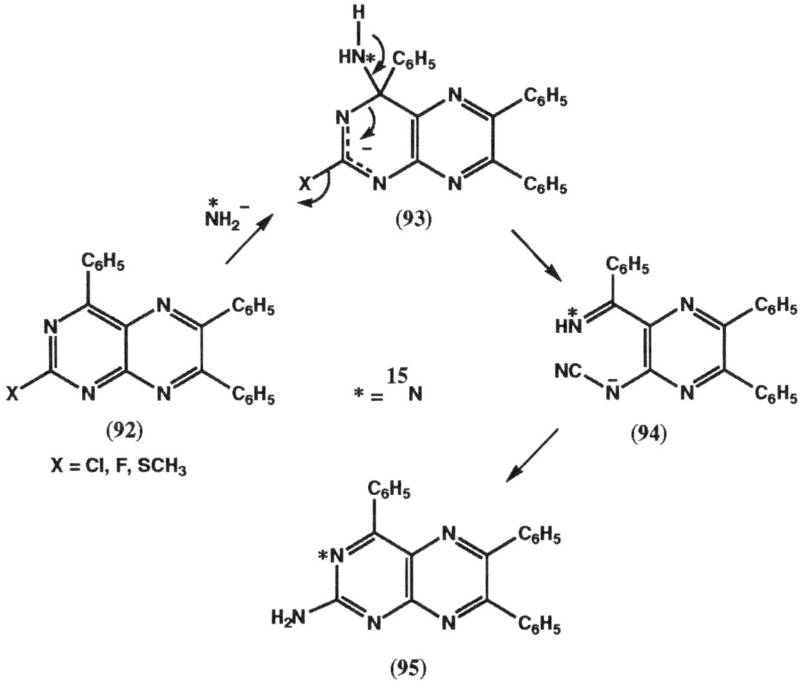

SCHEME II.43

Aminodemethylthiolation. Whereas reaction of 2-methylthiopteridine and 6,7-diphenyl-2-methylthiopteridine with potassium amide/liquid ammonia showed complete decomposition into unidentifiable tarry products, very different behavior was observed when 2-methylthio-4,6,7-triphenylpyrimidine (**96,** $X = SCH_3$) reacts with potassium amide/liquid ammonia. Besides formation of the 2-amino compound **95,** a ring contraction reaction into 6,8-diphenyl-2-methylthiopurine (**99**) took place (Scheme II.44) (75MI1). Study of the mechanism of the formation of **95,** using 4 equiv. of ^{15}N-labeled potassium amide in 25 ml of liquid ^{15}N-labeled ammonia, revealed that 50% of the starting material is converted into the 2-amino compound following the S_N(ANRORC) pathway (Scheme II.43) (75MI1; 77H205). It is assumed that the remaining 50% follows the S_N(AE) mechanism, involving an initial nucleophilic addition at C-2.

It was observed that the participation of the S_N(ANRORC) process in the aminodemethylthiolation is strongly dependent on the concentration of the amide ion (77H205). When 10 equiv. of potassium amide was used instead of 4 equiv. (in the same volume of liquid ammonia, 25 ml), the per-

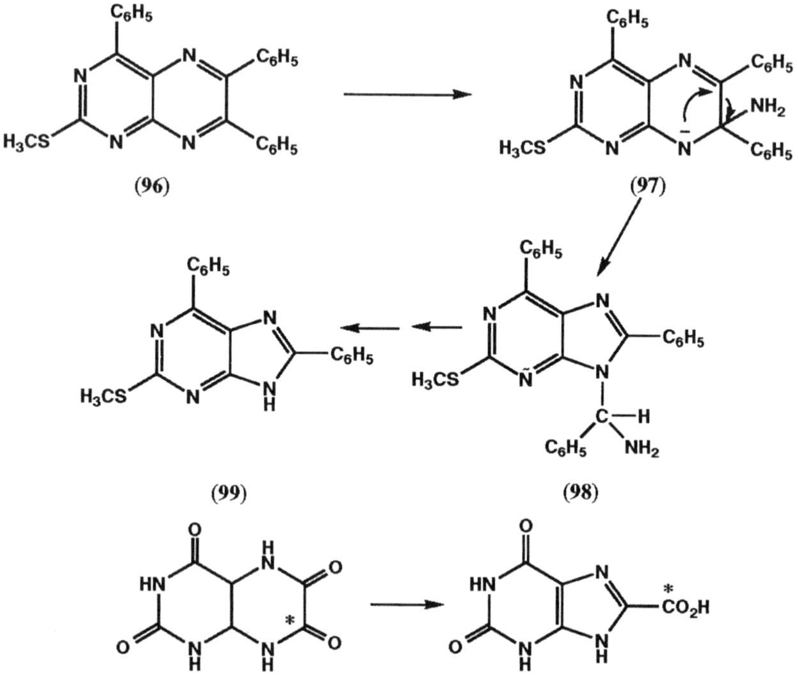

SCHEME II.44

centage of molecules that reacted according to the ring opening–ring closure process [S_N(ANRORC)] increased from 50 to 85%. A similar observation has been made in the amination of 2-substituted quinazolines with ammonia/methanol (Section II,C,2,b).

Although an extensive discussion on the mechanism of the ring contraction of pteridines into purines is beyond the scope of this book, it can only be mentioned that conclusive proof has been obtained (77H205) that the ring contraction starts by addition of the amide ion to C-7 as well as to C-6, although the addition to C-7 is the main pathway (Scheme II.44).

Ring opening of **97** as indicated gives the 9-(α-amino-α-phenylmethyl) purine **98**, which by a base-catalyzed elimination of benzylideneimine is converted into 6,8-diphenyl-2-methylthiopurine **99**. This pteridine–purine transformation has a close resemblance to the enzyme-catalyzed ring contraction of tetrahydropteridine into xanthine-8-carboxylic acid (64MI1), in which reaction it was proved by radioactive labeling that it is exclusively C-7 that is expelled.

Thus, 2-methylthio-4,6,7-triphenylpteridine is a *multi*reactive compound. It shows activity to nucleophiles at four ring carbon atoms: at C-4 [S_N(ANRORC)], at C-2 [S_N(AE)], and at C-6, as well as at C-7 (purine formation). Based on qualitative product studies and ^{15}N-label incorporation studies, the order of reactivity is approximately C-4 > C-2 > C-7 > C-6.

3. PYRAZINES

2-Chloropyrazine with potassium amide in liquid ammonia gives three products: 2-aminopyrazine, and two ring-contraction products, imidazole and 2-cyanoimidazole (Scheme II.45). When using 2-chloro[^{15}N]pyrazine, it was established that the ^{15}N label is exclusively present on the amino group, as proved by the conversion of the ^{15}N labeled 2-aminopyrazine into unlabeled pyrazin-2-one by treatment with nitrosylsulfuric acid (72RTC949; 72RTC449). It is evident that in the aminodechlorination the S_N(ANRORC) mechanism is fully operative. This degenerate ring transformation can be described as starting with addition of the amide ion on position 6, bond breaking between N-1 and C-6, and ring closure of the open-chain imino chloride and/or cyano compound (Scheme II.45).

The reaction is one of the few examples of a degenerate ring transformation in pyrazine chemistry. Extensive investigations on the amination of 2-chloroquinoxaline have shown that in the formation of the 2-amino compound *no* S_N(ANRORC) process is involved (72RTC850).

Although the topic is beyond the scope of this book, it can be mentioned that ^{15}N-labeling studies (73RTC311) provide good evidence that the for-

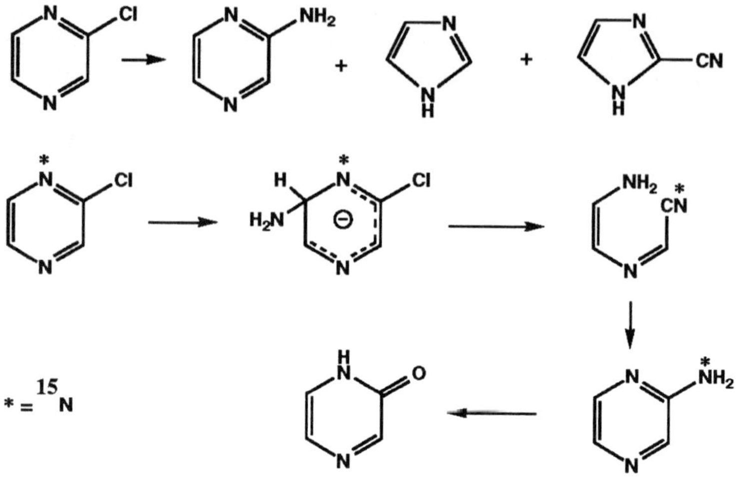

SCHEME II.45

mation of both ring-contracted products occurs by an initial addition of the amide ion on position 3. Bond breaking between C-2 and C-3 gives an open-chain product from which imidazole is formed; bond breaking between C-3 and N-4 leads to an intermediate, from which 2-cyanoimidazole can be formed (Scheme II.46).

Similar amide-induced ring contractions have been reported when heating 2,3-diphenylquinoxaline with potassium amide at 140°C, 2-phenylbenzimidazole being obtained (65JOC2858). A quinazoline–benzimidazole transformation has also been reported when heating 6-bromoquinazoline with ammonia. 2-Bromoquinazoline is a probable intermediate (74RC1233).

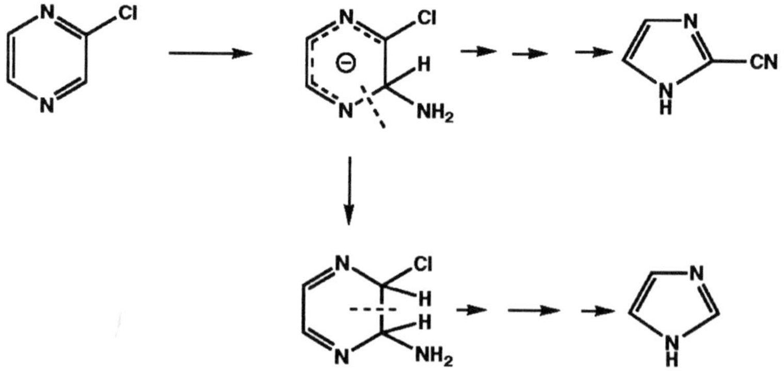

SCHEME II.46

Amination of 2-chloro-3,6-diphenylpyrazine with potassium amide in liquid ammonia yields 2-amino-3,6-diphenylpyrazine **102** (73RTC449). It is of interest to mention that, although addition of the amide ion initially forms the C-5 adduct **101** (as proved by NMR-spectroscopy), this C-5 adduct is stable and does *not* undergo ring opening (73RTC708). It isomerizes slowly into the C-2 amino adduct **100** (via an equilibrium with 2-chloro-5,6-diphenylpyrazine) in which a rapid chloride expulsion takes place (Scheme II.47).

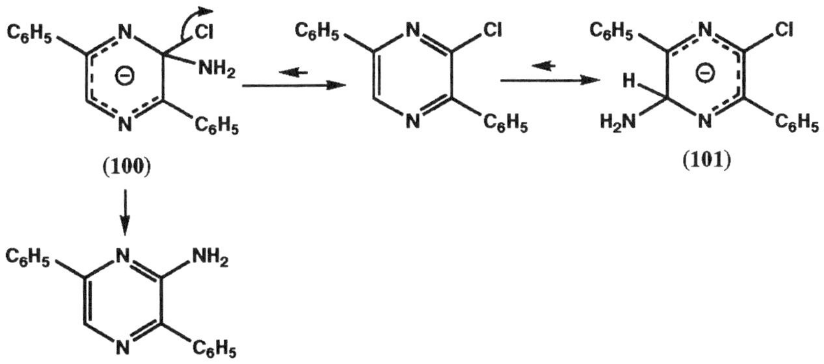

SCHEME II.47

The Chichibabin amination of phenylpyrazine with ^{15}N-labeled potassium amide/liquid ammonia gave two products, 3-amino- and 5-amino-2-phenylpyrazine; in both products the ^{15}N label is only present in the amino group, and no ^{15}N label was found to be incorporated into the pyrazine ring (82MI1). This result proves that in the aminodehydrogenation of phenylpyrazine, no S_N(ANRORC) mechanism is involved. This result is confirmed by the fact that amination of phenylpyrazine in the presence of the radical scavenger azobenzene, a compound that has been found to prevent the S_N(ANRORC) mechanism in the Chichibabin amination of 4-phenylpyrimidine, still yields both aminopyrazines.

4. Pyridazines and Phthalazines

The occurrence of degenerate ring transformation reactions in the hydrazinodehalogenation of halogeno-1,2,4,5-tetrazines (see Section II,D,3.) raised the question as to whether corresponding reactions would also occur with the less reactive halogenopyridazines. Extensive investigations on hydrazination of 3-bromopyridazine and 3-chloro-6-methylpyridazine with ^{15}N double-labeled hydrazine show that in the corresponding 3-hydrazinopyridazines about 25–30% of the ^{15}N label is present *in* the heterocyclic ring (Scheme II.48) (83JHC1259). The incorporation of the ^{15}N-label into the pyridazine ring can be described as pictured in Scheme II.48, involving an initial addition of the hydrazino group at C-6. It appeared that the presence of the methyl group does not influence the degree of S_N(ANRORC) participation. In this solution the methyl group is not deprotonated (as in the case when the amide ion is present). Moreover, its steric influence on the addition is small. For a more extensive discussion on the ring transformations of *ortho* diazines with hydrazine see Section II,D,3.

Extension of this work to the bicyclic system phthalazine showed that hydrazination of [2,3-di-^{15}N]-1-chloro(bromo)-4-*R*-phthalazines (*R* = H, CH$_3$) gave, with hydrazine, 1-hydrazinophthalazines, in which no ^{15}N enrichment is found in the hydrazino group (83JHC1259). This result clearly shows that the hydrazinodehalogenation of 1-halogenophthalazines does not involve an ANRORC process (Scheme II.48). Whether the complete absence of this ANRORC mechanism in the hydrazination is due to the nonoccurrence of addition at C-4 or to the reluctance of a possibly formed covalent σ-adduct to undergo ring opening is unclear. The nonoccurrence of ring opening in covalent adducts has, for example, also been observed in the amination of 2-chloro-5,6-diphenylpyrazine (73RTC449) (see **100** and **101** in Scheme II.47).

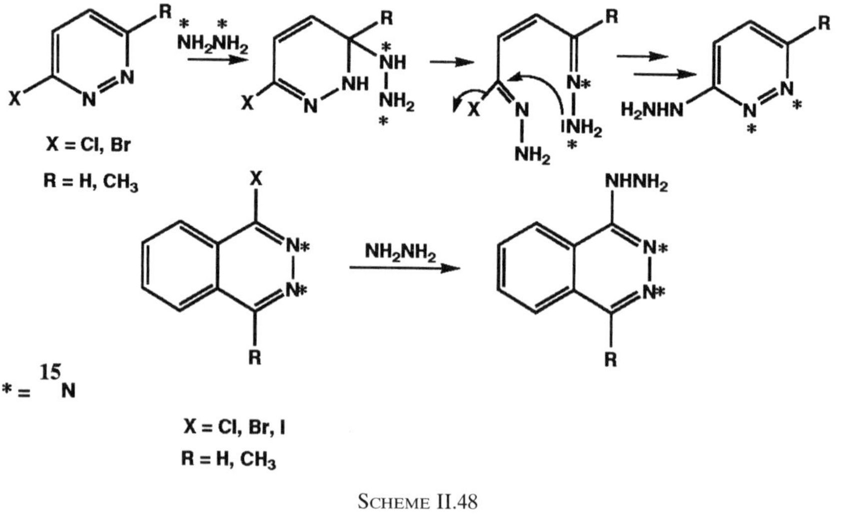

SCHEME II.48

D. S_N(ANRORC) Substitutions in Triazines and Tetrazines

1. 1,2,4-Triazines and Benzo-1,2,4-Triazines

a. *Aminodemethylthiolation of 3-Methylthio-1,2,4-Triazines*

The reaction of 3-(methylthio)-5-R-1,2,4-triazine with potassium amide/liquid ammonia was the first example in 1,2,4-triazine chemistry where the replacement of the methylthio group by an amino group was found to occur with important participation of the S_N(ANRORC) mechanism (75RTC204). This was proved by the experiment that the 3-amino-1,2,4-triazine obtained from 3-methylthio-[4-^{15}N]-1,2,4 triazine (**103**, R = H) was found to contain 93% of the ^{15}N label in the exocyclic amino group, i.e., formation of 3-(^{15}N-amino)-1,2,4-triazine (**106**, R = H). This percentage was calculated from mass spectrometric determinations of the ^{15}N-excess in **103**, in the 3-amino compound **106/107**, and in 2-methyl-3-oxo-2,3-dihydro-1,2,4-triazine **108**, being obtained by treatment of **106/107** with potassium hydroxide and subsequently with methyl iodide. The high percentage of ^{15}N-incorporation in the amino group of **106** proves that a ring-opening process is the main pathway in the aminodemethylthiolation. The amination involves the anionic 1 : 1 adduct 5-amino-3-(methylthio)dihydro-1,2,4-triazinide (**104**) (78RTC273) and the nonisolable open-chain intermediate 4-amino-1-cyano-1,2-diazabuta-1,3-diene (**105**) (Scheme II.49)

There is sound NMR spectroscopic evidence that in the highly π-electron-deficient 1,2,4-triazines the C-5 position is most susceptible (78RTC273) for nucleophilic addition of the amide ion. In fact, a weaker nucleophile such as liquid ammonia (weak in comparison with the amide ion) is also able to give covalent amination at the C-5 position. This behavior is in contrast to that observed with pyrimidines, pyrazines, and pyridazines, which require the strong nucleophilic amide ion to give addition products (88MI1). The availability of the 1,2,4-triazine ring to easily undergo covalent amination is certainly also dependent on the substituents in the 1,2,4-triazine ring; they influence the electron density of the ring. Therefore, it can be expected that there exists a delicate influence of substituents in the 1,2,4-triazine ring and the occurrence of covalent amination; this is illustrated by some data given in Table II.9 (78RTC273).

In studies of the "blocking" effects of substituents on position 5 of the 1,2,4-triazine ring, it was established that the presence of a phenyl or *t*-butyl group in that position has no effect on the course of the reaction (82JHC673). 3-(Methylthio)-5-(phenyl, *t*-butyl)-1,2,4-triazine (**103**, R = C_6H_5, *t*-C_4H_9) undergoes the amination at C-3 almost 100% according to the S_N(ANRORC) mechanism, proving that the initial step of the

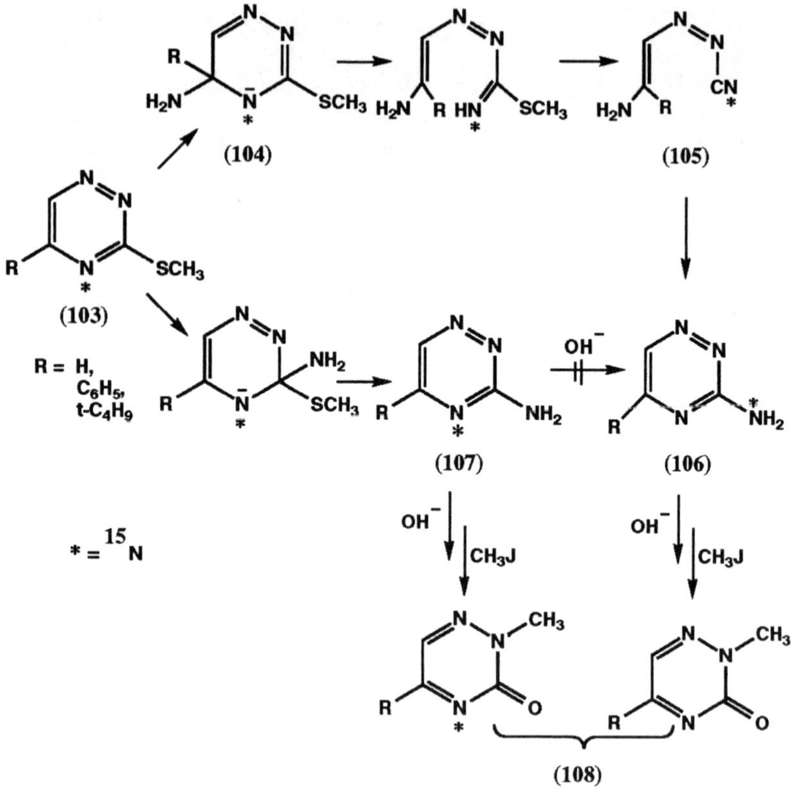

SCHEME II.49

ANRORC process, i.e., adduct formation at C-5, is not prevented by these groups. This is a remarkable effect, certainly when taking into consideration the bulkiness of the *t*-butyl group, which could be expected to diminish the participation of the ANRORC process. It demonstrates that the

TABLE II.9

SUBSTITUENT DEPENDENCY OF THE COVALENT AMINATION IN 3-R-6-X-1,2,4-TRIAZINES

Substituent	Covalent amination
$R = H$, $X = H$	+
$R = SCH_3$, $X = H$	+
$R = OCH_3$, $X = H$	+
$R = NH_2$, $X = H$	−
$R = NH_2$, $X = Br$	+
$R = NH_2$, $X = CH_3$	−

steric effect is fully counterbalanced by the high π-electron deficiency of the 1,2,4-triazine ring.

An alternative explanation for the presence of the ^{15}N label in the *exo-cyclic* amino group of **106** could be the occurrence of a Dimroth-type rearrangement of the *ring*-labeled 3-amino-5-phenyl-[4-^{15}N]-1,2,4-triazine (**107**) into **106**. This rearrangement could occur when **107** is treated with base for conversion into 3-oxo-2,3-dihydro-1,2,4-triazine and subsequently into **108** (see Scheme II.49). This isomerization is, however, very unlikely, since this rearrangement would certainly lead to a 50/50 scrambling of the ^{15}N label over the nitrogen of the ring and the amino group. It should produce a mixture of **107** and **106**, which in fact is not found (Scheme II.50).

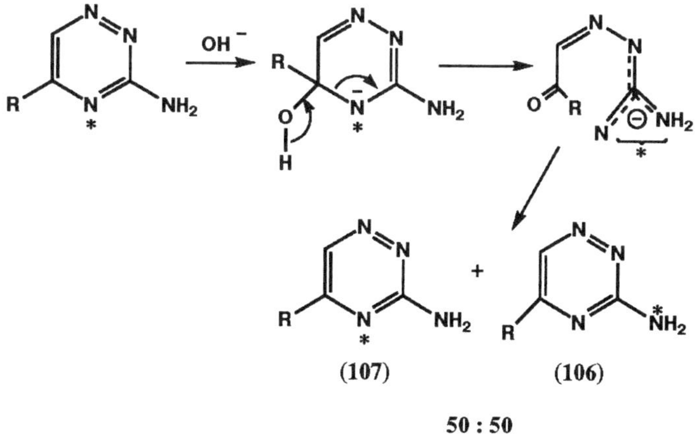

SCHEME II.50

b. *Aminodehalogenation of 3-Halogeno-1,2,4-Triazines*

Reaction of 3-X-5-phenyl-1,2,4-triazine (**109**, X = Cl, Br, I) with potassium amide in liquid ammonia gave rather complex reaction mixtures. As main products were obtained 3-amino-5-phenyl-1,2,4-triazine (**110**, 30–40%) and the ring transformation product 2,4-diphenyl-1,3,5-triazine (**111**, 10–20%), and as minor products benzamide, 6-amino-2,4-diphenyl-1,3,5-triazine (**112**, <1%), the ring-contraction product 3,5-diphenyl-1,2,4-triazole (**113**, X = Cl, Br, I; 1–4%), and, very interestingly, the degenerate ring transformation product 3,5-diphenyl-1,2,4-triazine (**114**, about 2%) (Scheme II.51) (80JOC881; 82JHC673).

When the reaction was studied with the ^{15}N-labeled 3-X-5-phenyl-[4-^{15}N]-1,2,4-triazine (X = Cl, Br) as substrate, it was established by the same methodology as described for the aminodemethylthiolation (see Section

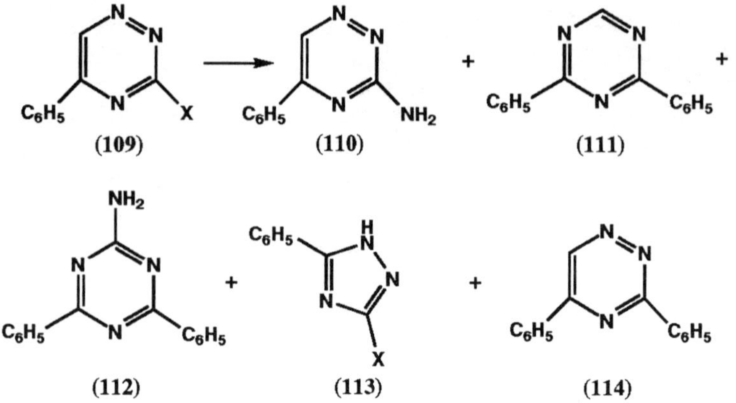

SCHEME II.51

II,D,1,a) that in the 3-amino compound obtained, the ^{15}N label is nearly exclusively present on the nitrogen of the amino group (Cl 96%; Br 93%), proving that the aminodechlorination and aminodebromination have taken place according to the S_N(ANRORC) mechanism. Exploring the aminolysis of 3-iodo-4-phenyl-1,2,4-triazine (**109,** X = I) in a reverse manner, which means reaction of *unlabeled* **109** (X =I) with ^{15}N-labeled potassium amide/liquid ammonia, gave as result that the 3-iodo compound reacts 63% according to the S_N(ANRORC) mechanism. The conclusion from all these experiments is that the formation of the 3-amino compound **110** from **109** (X = Cl, Br, I) can be explained as described in Scheme II.49 for the aminodemethylthiolation.

3-Fluoro-5-phenyl-1,2,4-triazine (**109,** X = F) reacts much faster than the 3-chloro-, 3-bromo-, or 3-iodo-compound. Moreover, the reaction mixture obtained is cleaner than that from the corresponding 3-chloro- or 3-bromo compounds; 3-amino-5-phenyl-1,2,4-triazine (**110**) is formed in good yield. This conversion takes place to only a small extent (18%) via the ANRORC process; the main part of the aminodefluorination seems to involve the S_N(AE) mechanism. This result is consistent with the observation that the aminodefluorination of 4,6-diphenyl-2-fluoropyrimidine follows the S_N(AE) process, whereas 2-fluoro-4-phenylpyrimidine (position 6 is vacant for addition of the nucleophile) reacts for the most part according to the S_N(ANRORC) mechanism (see Section II,C,1,c).

Special attention was paid to the remarkable amide-induced degenerate ring transformation of 3-X-5-phenyl-1,2,4-triazine **109** into 3,5-diphenyl-1,2,4-triazine **114** (80JOC881). It is evident that in this degenerate ring transformation, involving the conversion of the starting material with one phenyl group into a product with two phenyl groups, the presence of the second phenyl group in the 1,2,4-triazine ring of **114** can only be achieved if a fragmentation process occurs in the triazine ring. This fragmentation must lead to a fragment containing a phenyl group. Since benzamide was found as one of the minor products after workup of the reaction mixture, the suggestion was put forward that benzamidine may play an important role as intermediate in this ring transformation.

This compound can either be formed from benzonitrile (**117**), which can give under the applied reaction conditions the potassium salt of benzamidine (**119**) [route (a) in Scheme II.52] (28JA3311; 36BSF1600; 47JCS738) or is formed from the ring-opened compound **118** [route (b) in Scheme II.52]. Both routes involve the fragmentation of the common open-chain compound **115a/115b** involved in the S_N(ANRORC) process of the aminodehalogenation (**109** to **110**). Support for this proposal was provided by ^{15}N-labeling studies. It was established that (1) benzamide, being isolated from the reaction mixture obtained when *unlabeled* 2-chloro-5-phenyl-

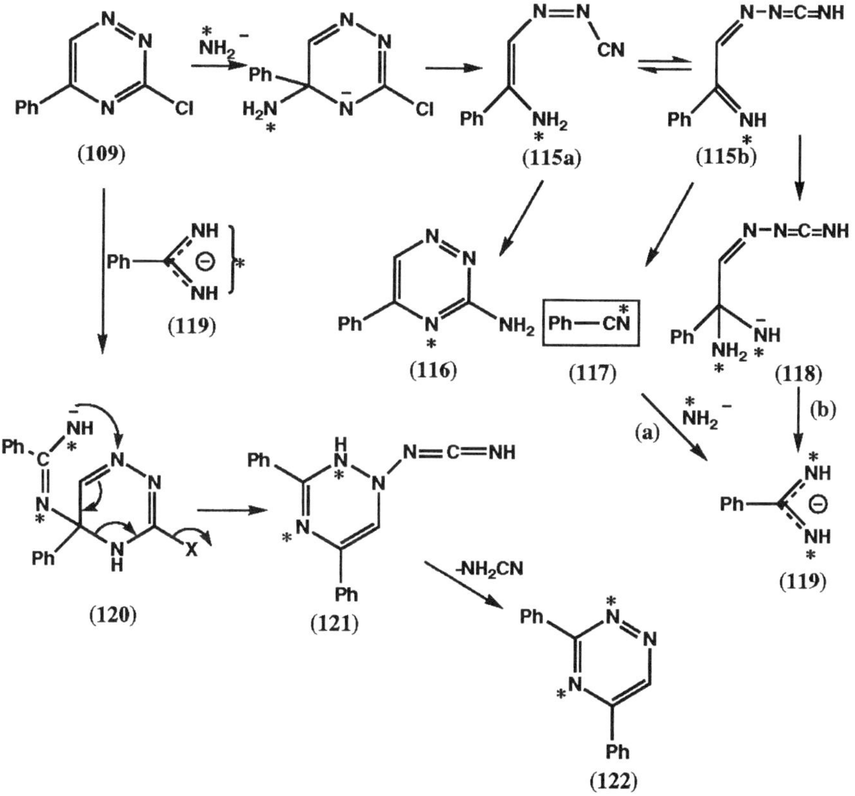

SCHEME II.52

1,2,4-triazine reacts with ^{15}N-labeled potassium amide in liquid ammonia, has the same^{15}N enrichment as the labeled potassium amide (1), and (2) in 3,5-diphenyl-1,2,4-triazine the ^{15}N-enrichment is nearly twice of that in the potassium amide, providing evidence that two nitrogens in 3,5-diphenyl-1,2,4-triazine are derived from the potassium amide. This strongly suggests that benzamidine with its *two* ^{15}N-labeled nitrogen atoms is incorporated into the triazine ring. The pathway suggested for this degenerate ring transformation is pictured in Scheme II.52.

The reaction course can be described as involving addition of the nucleophilic nitrogen in benzamidine to the electron-deficient C-5 position of the triazine ring. In the C-5 adduct **120**, nitrogen–nitrogen bond formation and simultaneously ring opening with elimination of the halide ion occur, yielding intermediate **121**, which on losing aminocyanogen gives the double-labeled product **122**. This mechanism also accounts for the fact that benzamide and 3,5-diphenyl-1,2,4-triazine are not formed in the amination reaction of 3-fluoro-5-phenyl-1,2,4-triazine. As shown previously, this compound reacts for the greater part via a process not involving a ring opening.

c. *Aminolysis of 1,2,4-Triazines Containing at C-3 a Leaving Group Different from Halogen*

The conversion of 3-X-5-phenyl-1,2,4-triazine [X = SO_2CH_3, $N^+(CH_3)_3$] into 3-amino-5-phenyl-1,2,4-triazine was also found to take place partly by an ANRORC process (80JOC881; 82JHC673). This shows the high susceptibility of C-5 in the 1,2,4-triazine ring for nucleophilic attack, despite the presence of groups with a high nucleofugic character at C-3. The percentages are summarized in Table II.10. The reactivity order of different substituents for ANRORC reactivity is SCH_3 > Cl, Br > I > SO_2CH_3 > $N^+(CH_3)_3$ > F. This order will reflect the sensitivity of C-5 for amide addition.

Comparison of this reactivity order with that found in the amination of 2-X-4-phenyl pyrimidines ($SCH_3 \approx Br \approx Cl$ > F > $SO_2CH_3 \approx I$ > $N^+(CH_3)_3$ > CN; see Table II.5 in Section II,C,1,d) shows that these reactivity orders differ considerably. The fluoro substituent, especially, which has in the pyrimidine series about the same reactivity order as the chloro or bromo atom, shows in the 1,2,4-triazine series a low ANRORC activity. Comparison of both series of reactivities is, however, a delicate matter, mainly because the yields obtained for the amino compounds in the 1,2,4-triazine series are much lower than those obtained in the pyrimidine series, because of the occurrence of many side reactions, such as ring contraction, dehalogenation, ring transformations, and degenerate ring transformations

TABLE II.10

Yields Obtained in the Amination of
3-X-5-Phenyl-1,2,4-Triazines and the Percentages
of S_N(ANRORC) Mechanism Involved.

Substituent X	% yields	% S_N(ANRORC)
SCH$_3$	72	100
Cl	40	96
Br	29	93
I	31	63
SO$_2$CH$_3$	65	33
N$^+$(CH$_3$)$_3$	42	34
F	54	18

(Scheme II.51). Some of these reactions are even initiated by the addition of nucleophilic species, being formed by fragmentation of open-chain compounds, which are originally derived from C-5 adducts (see Section II,C,1,b). Thus, the number of molecules that react by addition of the amide ion to C-5 is certainly higher than can be derived from the percentage of the molecules given in Table II.10.

d. Aminodeoxogenation

The replacement of the oxo group in quinazolin-4-one by an amino group, using as reagent phenyl phosphorodiamidate, a conversion that has been proved to occur partly via an SN(ANRORC) process (see Section II,C,2,c), induced a study of the possible occurrence of this process in the aminodeoxogenation of 5-phenyl-[4-^{15}N]1,2,4-triazin-3-one. It was found that this replacement reaction only occurs for a small percentage (about 10%) with participation of the ANRORC process. The remaining 90% follows a nonrearrangement pathway (see Scheme II.37). The corresponding aminodeoxogenation of 5-t-butyl-1,2,4-triazin-3-one does not involve any ring opening reaction.

e. Aminodechlorination of Chlorobenzo-1,2,4-Triazines

As mentioned in Section II,D,1,a, the presence of a t-butyl or phenyl group at position 5 of the 1,2,4-triazine ring does not prevent addition of the amide ion to that position. Therefore, it becomes of interest to investigate whether in the amino-dehalogenation of two *annelated* 3-chloro-1,2,4-triazines, i.e., 3-chloro-1,2,4-benzotriazine (**123**) and 3-chlorophenanthro[9,10-e]1,2,4-triazine (**125**), using potassium amide/liquid ammonia, the S_N(ANRORC) process would be involved. Both compounds

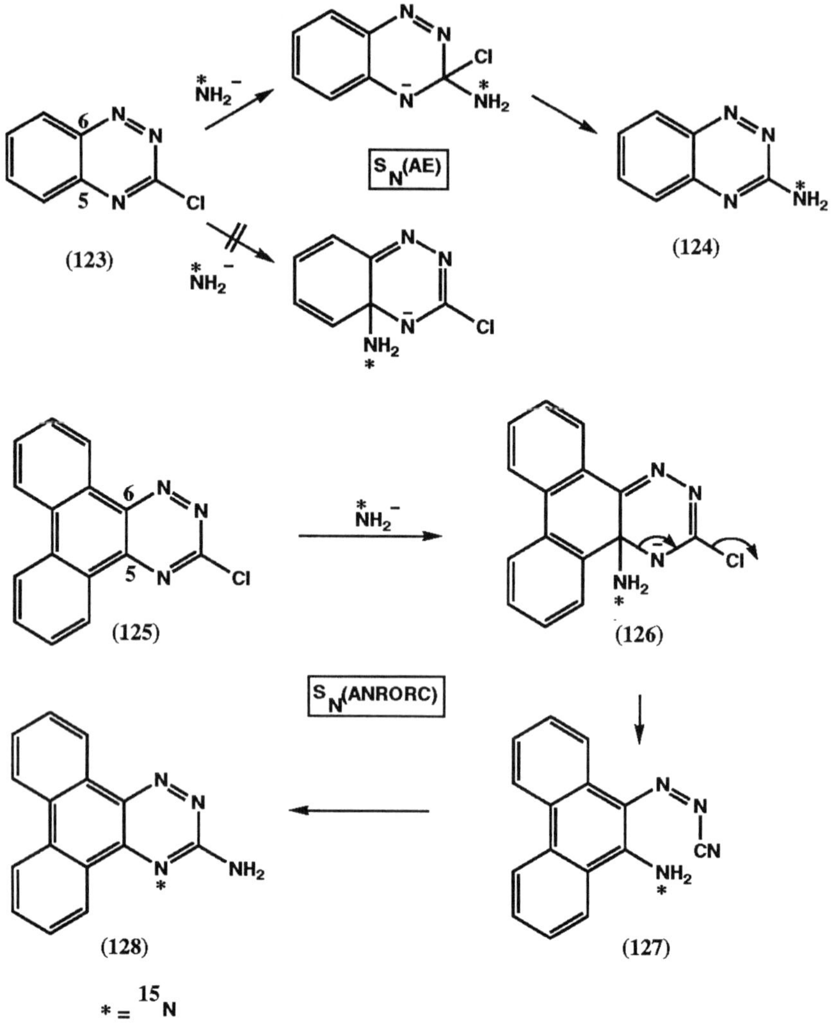

SCHEME II.53

gave in reasonable-to-good yields the corresponding 3-amino compounds (84JHC433). When using ^{15}N-labeled potassium amide/liquid ammonia, it appeared that in the 3-amino-1,2,4-benzotriazine obtained from **123** the ^{15}N label is nearly exclusively present in the amino group, i.e., formation of **124** [S$_N$(AE) process]. As a surprising contrast it was found that in the 3-aminophenanthro[9,10-*e*]1,2,4-triazine (**128**) obtained from **125,** the ^{15}N label is nearly exclusively present in the ring, clearly evidencing the occurrence of the S_N(ANRORC) mechanism in the formation of **128**. These results elegantly demonstrate the easy accessibility of the C-5 position in the 1,2,4-triazine ring of **125** for nucleophilic addition (Scheme II.53).

The different behavior of **123** and **125** toward potassium amide/liquid ammonia can be explained as follows. Addition of the amide ion to C-5 in **123** leads to a considerable loss of resonance energy in the benzene ring. Calculations have shown (66MI1) that position 3 in benzo-1,2,4-triazine has the lowest electron density and therefore the favored position in a charge-controlled nucleophilic addition reaction. In the amide adduct **126**, the 6π aromaticity in the biphenyl part is still maintained. The addition at C-5 in the pattern is in line with the observed high degree of double bond character between C-5 and C-6 in the phenanthrene ring of **125**. The addition at C-5 in the phenanthrotriazine certainly reflects the decreased aromatic character of the central ring. That electronic and not steric effects determine the addition in these systems is convincingly demonstrated by the fact that **123**, which has about an identical steric environment on that particular carbon atom C-5, does not undergo addition (Scheme II.53).

2. 1,3,5-Triazines

a. *Aminodehydrogenation (Chichibabin Amination) of (Di)phenyl-1,3,5-Triazines*

When phenyl-1,3,5-triazine (**129**) reacts with potassium amide in liquid ammonia at $-33°C$ for 40 hr, aminodehydrogenation occurs, leading to the formation of 4-amino-2-phenyl-1,3,5-triazine (76RTC125). This reaction is slow, since after 40 hr starting material can still be recovered. Furthermore, the reaction could not be perceptibly accelerated by adding potassium nitrate. ^{1}H NMR spectroscopy showed that phenyl-1,3,5-triazine, when added to liquid ammonia, containing potassium amide, exists as the anionic σ-adduct 2-phenyl-4-amino-1,3,5-triazinide (76RTC125). When the Chichibabin amination was carried out with ^{15}N-labeled potassium amide/liquid ammonia, 55% of the amino compound is ring labeled, i.e., **131**, and 45% is labeled on the exocyclic nitrogen of the amino group, i.e., **132** (Scheme II.54).

The presence of the ^{15}N label inside the ring of **131** is characteristic of the occurrence of an S_N(ANRORC) mechanism in this aminodehydrogenation reaction. The exocyclic amino labeling can be explained by the S_N(AE) substitution process. So, besides pyrimidines (see Section II,C,1,g), 1,3,5-triazines are also able to react in a Chichibabin-type amination via a ring opening–ring closure sequence. Possible reaction routes are outlined in Scheme II.54.

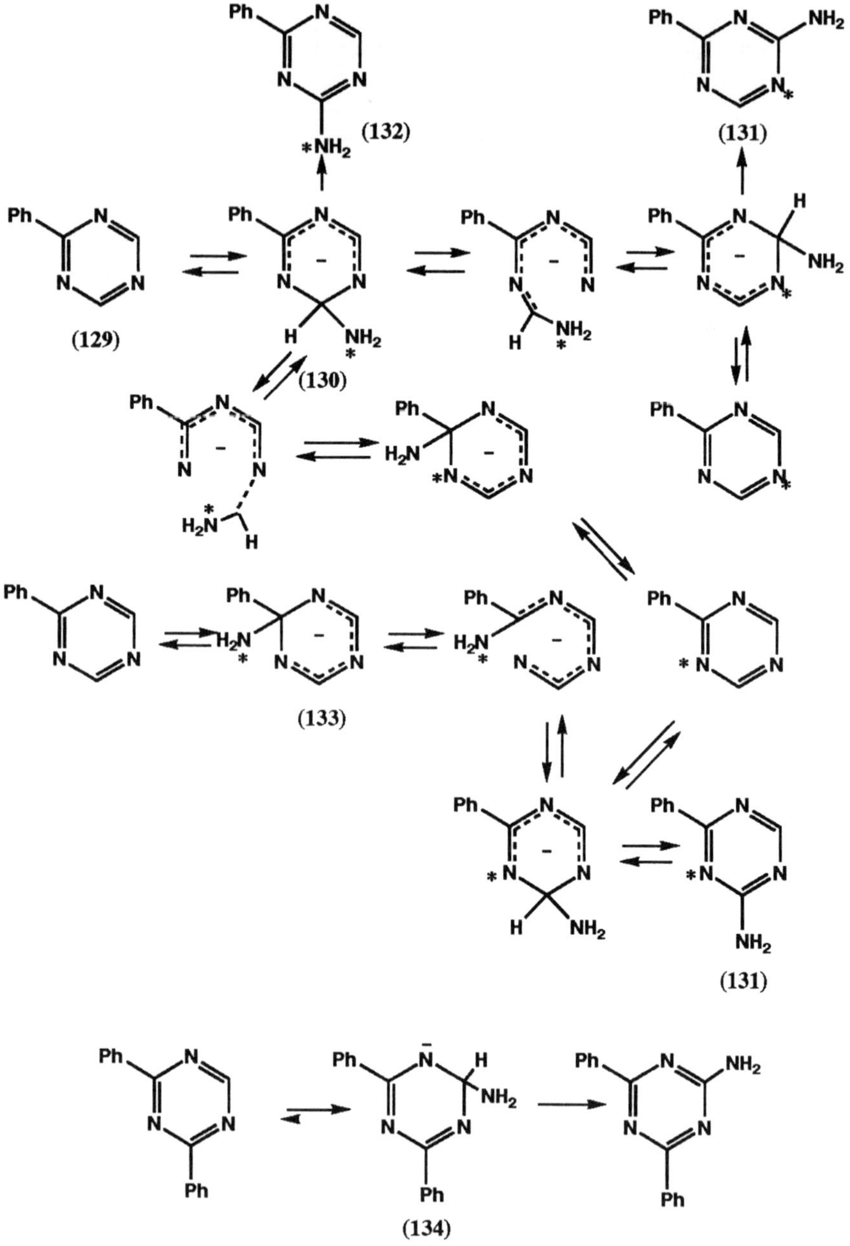

SCHEME II.54

A few characteristic features of the mechanism can be mentioned. First, in principle, two $S_N(ANRORC)$ routes can be visualized. In one, the initial addition takes place on position C-4, i.e., **130**, and can be characterized as an $S_N(A_{(4)}NRORC)$ process. In the other route, the initial addition of the amide ion occurs at the "blocked" position C-2, i.e., **133**, and can be described as an $S_N(A_{(2)}NRORC)$ pathway. Although it is tempting to assume that addition at C-2 is not strongly favored because of the possible steric interference with the phenyl group, there is sufficient evidence that the blocking "power" of the phenyl group on addition of the amide ion is limited. From the data available one cannot establish the relative importance of both ANRORC processes in the amination of the phenyl-1,3,5-triazine.

Furthermore, it was pointed out that, whereas the formation of the amino adduct is fast and the formation of the product slow, it is possible that an equilibrium exists among the starting materials, their 1:1 σ-amino adducts, and their open-chain amidines (Scheme II.54). When this is the case, one may expect that, if the amination of phenyl-1,3,5-triazine is stopped before complete conversion, the retrieved starting material should be ^{15}N-labeled. This has indeed been found. This behavior is in agreement with that observed with the Chichibabin amination of 4- and 5-phenylpyrimidine.

The Chichibabin aminodehydrogenation of 4,6-diphenyl-1,3,5-triazine using potassium amide/liquid ammonia gives in good yield the corresponding 2-amino compound, although the rate of the reaction is quite low (76RTC113). It was proven by NMR spectroscopy that 4,6-diphenyl-1,3,5-triazine easily gives addition at the vacant C-2 position, i.e., **134** (Scheme II.54). When using ^{15}N-labeled 4,6-diphenyl-1,3,5-triazine, it was proven (76RTC113) that the amino group in 2-amino-4,6-diphenyl-1,3,5-triazine being obtained is *un*labeled, clearly showing that the amination reaction does not involve a degenerate ring transformation reaction. Apparently it is not the addition reaction, but rather the ring opening that is prohibited. This result is in remarkable contrast to that obtained in the amination of phenyl-1,3,5-triazine.

b. *Aminolysis of 2-X-4,6-diphenyl-1,3,5-Triazines*

2-*X*-4,6-Diphenyl-1,3,5-triazines (X = Cl, SCH$_3$), when treated with potassium amide/liquid ammonia, give after workup in good yield 2-amino-4,6-diphenyl-1,3,5-triazine. In the 2-amino compound, obtained from the mono-labeled 2-*X*-4,6-diphenyl[(1),(3),(5)-^{15}N]1,3,5-triazine (**135**, X = Cl; the ^{15}N label is scrambled over the three nitrogen atoms of the ring), 80% of the ^{15}N label is present in the amino nitrogen; for X = SCH$_3$ even 100% was found (76RTC113). Evidently, the substitution of group X in both

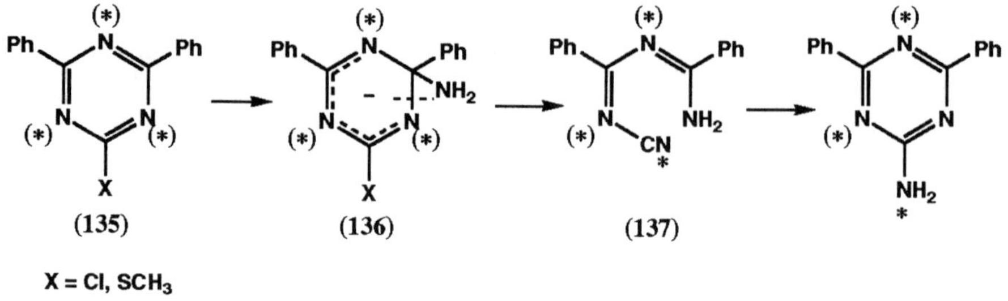

Scheme II.55

(135) X = Cl, SCH$_3$ → (136) → (137) → product with NH$_2$

compounds has taken place with (nearly) exclusive participation of the S$_N$(ANRORC) mechanism.

It is interesting that apparently the addition of the amide ion to the carbon attached to a phenyl group is more favored than addition at the carbon to which the chloro or methylthio group is attached (Scheme II.55). This result indicates that the presence of a good leaving group promotes the ring-opening reaction.

A more detailed investigation of the amination of 2-chloro-4,6-diphenyl-[(1),(3),(5)-^{15}N]1,3,5-triazine (**135**, X = Cl) revealed that the participation of the S$_N$(ANRORC) mechanism is strongly dependent on the potassium amide/substrate ratio (see Table II.11) (76RTC113).

The data in Table II.11 show that the participation of the S$_N$(ANRORC) mechanism decreases with decreasing potassium amide/substrate ratio. When no potassium amide is present, the participation of the ANRORC mechanism is zero and the aminolysis occurs according to the S$_N$(AE) mechanism. Apparently in the amination of the highly π-electron-deficient 1,3,5-triazines, a competition is involved between the strong nucleophilic amide ion, which leads via σ-adduct **136** and the ring-opened compound **137** to product 2-[^{15}N-amino]-4,6-diphenyl-1,3,5-triazine, and the weaker nucleophile liquid ammonia, which replaces by an S$_N$(AE) process the

TABLE II.11

THE PERCENTAGE OF S$_N$(ANRORC) PARTICIPATION IN THE AMINOLYSIS OF 2-CHLORO-4,6-DIPHENYL[^{15}N]1,3,5-TRIAZINE IN DEPENDENCY ON THE POTASSIUM AMIDE/SUBSTRATE RATIO

KNH$_2$/triazine ratio	% S$_N$(ANRORC) mechanism
33	91
6	80
2	6
0	0

chloro atom, yielding ring labeled 2-amino-4,6-diphenyl[^{15}N]1,3,5-triazine (**135**, $X = NH_2$).

3. 1,2,4,5-Tetrazines

In this section special attention will be paid to the occurrence of degenerate ring transformation in reactions between a 1,2-diaza nucleophile (hydrazine) and tetra-aza substrates, i.e., 1,2,4,5-tetrazine. These studies were carried out to investigate whether with the binucleophilic hydrazine, displacement reactions of (substituted) 1,2,4,5-tetrazines would lead to replacement of two vicinal ring nitrogens by two nitrogens of the hydrazine. In the following sections hydrazinodehydrogenation (Chichibabin hydrazination), hydrazinodeamination, and hydrazinodehalogenation are discussed.

a. *Hydrazinodehydrogenation of 1,2,4,5-Tetrazines*

The hydrazination of aza-aromatics, using sodium hydrazide, has been reported, but no detailed mechanism was given (64AG206). On treatment of 3-*R*-1,2,4,5-tetrazines ($R = CH_3$, C_2H_5, t-C_4H_9, C_6H_5) with 3 equiv of hydrazine at room temperature, the corresponding 3-alkyl(phenyl)-6-hydrazino compounds were obtained, although the yields were relatively low (10–15%) (81JOC5102). These hydrazinotetrazines were difficult to isolate, but they could easily be separated from the reaction mixture as their acetone-hydrazone derivatives **138** (Scheme II.56).

To investigate whether in the formation of these hydrazino compounds an S_N(ANRORC) mechanism would be involved, the reaction was studied with ^{15}N-labeled hydrazine(78JHC445). If ring opening occurs, one could

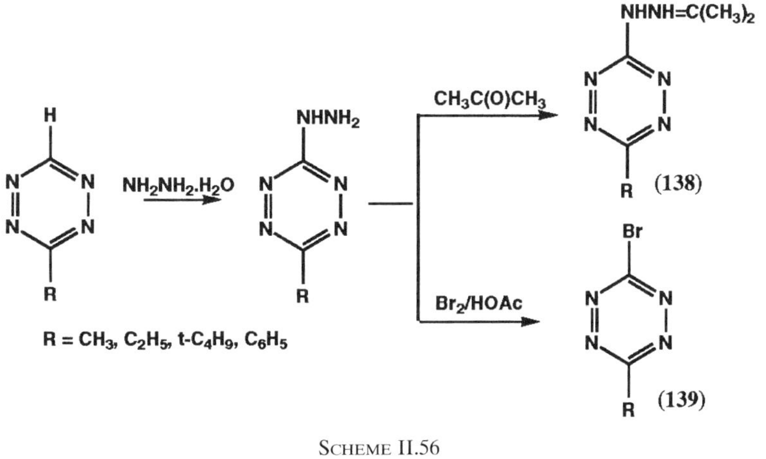

Scheme II.56

expect the incorporation of the ^{15}N label in the tetrazine ring; if that is not the case, the label is only present in the exocyclic hydrazino nitrogen atoms. To establish which percentage of the ^{15}N label is present in the tetrazine ring and/or in the hydrazino group, the 3-R-6-hydrazino-1,2,4,5-tetrazines were converted into their corresponding acetone–hydrazone derivatives **138** and also into the corresponding 6-bromo compounds **139**, using bromine in acetic acid as brominating agent (71KGS571) (Scheme II.56). The enrichment of the ^{15}N label in **138** and **139** was determined by mass spectrometric measurement of their respective $M + 2$ peaks and compared with that of the corresponding unlabeled reference compound.

For R = CH$_3$ and C$_2$H$_5$, at least part of the ^{15}N label of hydrazine is found to be incorporated into the 1,2,4,5-tetrazine ring (26–28%). The remaining part of the ^{15}N label (70–75%) is present in the hydrazino (hydrazone) group. No ^{15}N label was found in the retrieved starting material. For R = C$_6$H$_5$, only 3–4% of the ^{15}N label was present in the tetrazine ring of the 3-phenyl-6-hydrazino-1,2,4,5-tetrazine; for R = t-C$_4$H$_9$, no incorporation of the ^{15}N label in the tetrazine ring of 3-t-butyl-6-hydrazino-1,2,4,5-tetrazine was found. From these results it can be concluded that in the Chichibabin hydrazination of the four 3-alkyl(phenyl)-1,2,4,5-tetrazines investigated, the S$_N$(ANRORC) mechanism is only operative to a very small extent. The mechanism leading to incorporation of the ^{15}N label in the 1,2,4,5-tetrazine ring has the surprising aspect that one must assume that in the hydrazinodehydrogenation addition of hydrazine takes place at the *substituted* carbon C-3 (Scheme II.57). It yields the 2,3-dihydro-3-hydrazino-3R-tetrazine **142**, which undergoes ring opening into **143** and subsequent ring closure into 1,6-dihydro-1-hydrazino-3-R-1,2,4,5-tetrazine **144**. Oxidation gives the required product **145**, being ^{15}N-labeled in the tetrazine ring. The excess hydrazine can function as oxidizing agent, being present in the solution (51MI1). Since in the starting material no incorporation of the ^{15}N label was found, apparently the oxidation step (**144** to **145**) is faster than loss of hydrazine, which should yield starting material that is ^{15}N-labeled (**146**) (Scheme II.57). The reactivity order CH$_3$ = C$_2$H$_5$ > C$_6$H$_5$ > t-C$_4$H$_9$ certainly reflects the increasing steric hindrance for addition at C-3.

In order to obtain additional spectroscopic evidence for this reaction pathway, it was attempted to establish by ^{1}H and ^{13}C NMR spectroscopy whether and which intermediary species could be detected. Comparison of the proton chemical shifts of the 3-R-1,2,4,5-tetrazines (R = alkyl), dissolved in deuteriomethanol, with those measured upon dissolving these tetrazines in a 1:1 mixture of hydrazine hydrate and deuteriomethanol at −40°C shows that H-6 has undergone a large upfield shift of about 8.25 ppm for R = CH$_3$, 8.37 ppm for R = C$_2$H$_5$, and 8.92 ppm for R = t-C$_4$H$_9$, strongly suggesting the formation of 3-R-6-hydrazino-1,6-dihydro-

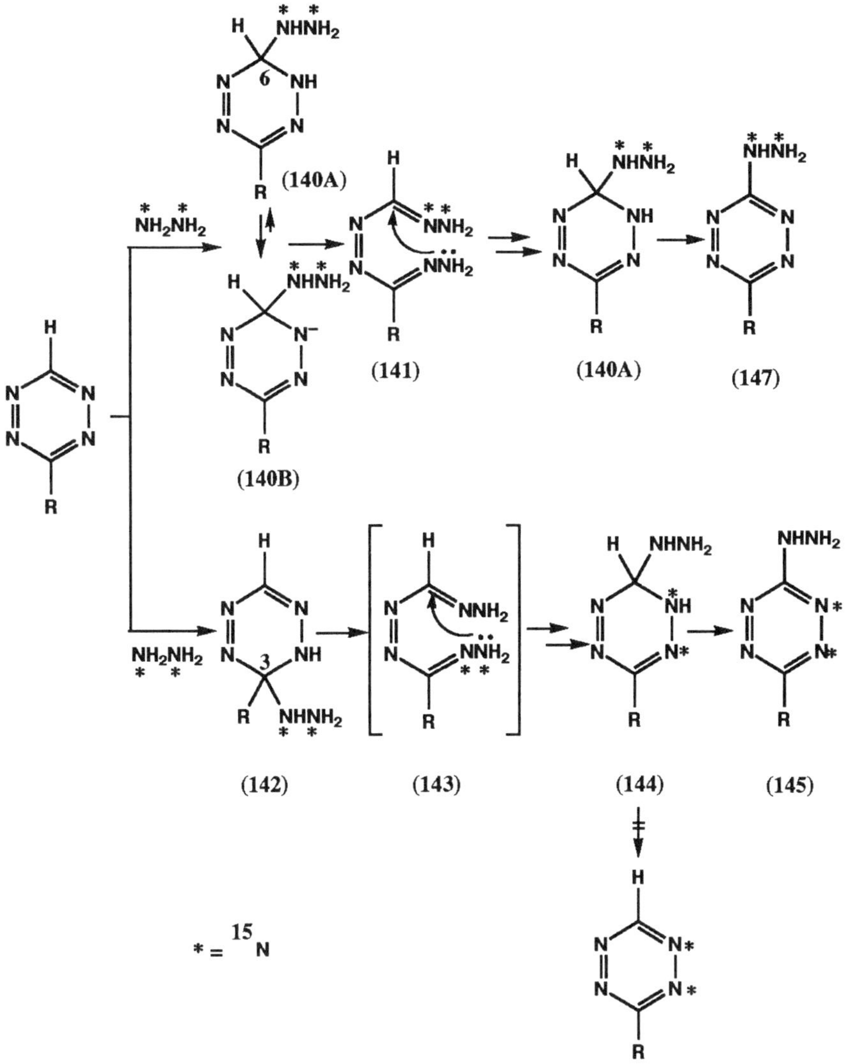

SCHEME II.57

1,2,4,5-tetrazine (**140A**). Evidence was presented that at the pH of the solution, adduct **140A** is present in its anionic form **140B** (81JOC5102). ^{1}H NMR signals of the starting material could not be detected, nor could those of the 2,3-dihydro-3-hydrazino-3-*R*-1,2,4,5-tetrazine **142**. The presence of the 6-hydrazino adduct **140** was further proved by ^{13}C NMR spectroscopic data as shown by an upfield shift of about 60 ppm for C-6 (81JOC5102). These results are surprising, since the observed incorporation of the ^{15}N label in the ring could only be achieved via the intermediacy of the 2,3-dihydro-1,2,4,5-tetrazine derivative **142**! On warming the solutions from -40 to $-20°C$ (for $R = CH_3, C_2H_5$) or from -40 to $0°C$ (for $R = t\text{-}C_4H_9$), the ^1H as well as ^{13}C signals of the adduct disappeared and new signals appeared with quite different chemical shifts. They were attributed to the bishydrazone **141** based on the observation that the chemical shift of H-6 is nearly the same as found for the sp^2 hydrogen in the H–C=N group of N,N-dimethylacetaldehyde hydrazone; also, the chemical shift of the hydrogens of the methyl group in **141** ($R = CH_3$) at 1.88 ppm is in good agreement with that one found for the C-methyl in N,N-dimethylacetaldehyde hy-drazone. The bishydrazone **143**, which is formed from the non-NMR-recognizable 3-hydrazino adduct **142**, has a structure identical to that of **141**, except that the ^{15}N label is present in a different position.

All data obtained from NMR spectroscopy as well as ^{15}N-labeling studies lead to the conclusion that the 6-hydrazino compound, being labeled in the hydrazino group, i.e., **147**, as well as the one being ring-labeled, i.e., **145**, must be formed by a sequence of reactions involving adduct formation, ring opening, and ring closure. This reaction is reported to be the first one in which *both the ring-labeled compound* **145** *and the exocyclic-labeled compound* **147** *follow the $S_N(ANRORC)$ pathway*. It is, however, only the formation of **145** that can be categorized as a degenerate ring transformation. Other examples of $S_N(ANRORC)$ reactions *not* involving a degenerate ring transformation are reported in Sections II,B,1, II,C,1,b, and II,C,2,e.

The remarkably high upfield shift being found on covalent hydrazination of 3-R-1,2,4,5-tetrazines (see above) is of the same magnitude as that observed when 1,2,4,5-tetrazines undergo covalent addition with liquid ammonia (about 8.7 ppm) (81JOC2138, 81JOC3805). These shifts are extraordinarily large, certainly when compared with the upfield shifts usually observed in solutions of diazines and triazines in liquid ammonia (about 4–5 ppm) (78RTC273; 79JOC4677). The unusual upfield shifts have been attributed to the formation of a species with a homotetrazole aromaticity (81JOC2138, 81JOC3805; 82JOC2856, 82JOC2858). It seems not unreasonable to assume that adduct **140** also has such a homoaromatic structure (see Scheme II.58). It is more favored than adduct **142** because **142** has a more crowded structure on the methylene bridge, carrying two large groups (hydrazino and group R). It is easily understood that **142**, if formed, rearranges via **143** into **144**.

Based on this assumption, it can easily be understood why in the case of $R = t\text{-}C_4H_9$ no incorporation of the ^{15}N label in the 6-hydrazino compound was found. The addition at C-3, the required initial step for ^{15}N incorporation in the tetrazine ring, is prevented by the strained conformation in **142** ($R = t\text{-}C_4H_9$).

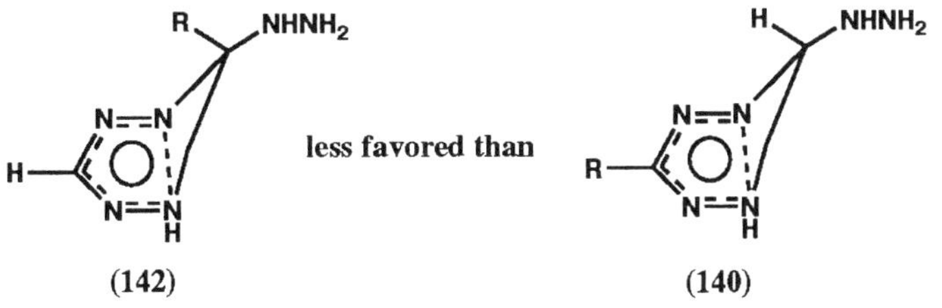

Scheme II.58

(142) less favored than (140)

b. *Hydrazinodeamination and Hydrazinodehalogenation of Amino- and Halogeno-1,2,4,5-Tetrazines*

On refluxing an ethanolic solution of 6-amino-3-R-1,2,4,5-tetrazine (R = H, CH_3, C_2H_5, t-C_4H_9, C_6H_5) containing 2 equiv hydrazine hydrate, the corresponding 6-hydrazino-1,2,4,5-tetrazines are obtained in reasonable yields (81JOC5102). The 6-bromo- and 6-chloro-1,2,4,5-tetrazines react similarly. The conversions, however, are quantitative and they react much faster than the 6-amino compound. When the hydrazinolysis of 6-amino-1,2,4,5-tetrazine was carried out with ^{15}N-labeled hydrazine and the partition of the label over the hydrazino group and the tetrazine ring was established, using the same procedure described in Section II,D,3,a, it was found that for R = H, CH_3, and C_2H_5 the ^{15}N label was only to a small extent incorporated into the tetrazine ring (R = H, 22%; R = CH_3, 25%; R = C_2H_5, 18%). In the case of R = t-C_4H_9, C_6H_5, no incorporation of the ^{15}N label was found (81JOC2138, 81JOC3805). Also, the hydrazino-dehalogenation of 3-methyl-6-X-1,2,4,5-tetrazine (X = Cl, Br) and 3-ethyl-6-bromo-1,2,4,5-tetrazine results in only a small percentage of ^{15}N incorporation into the tetrazine ring (R = CH_3, leaving group Br, 19%; R = CH_3, leaving group Cl, 6.5%) (81JOC5102).

As shown in Scheme II.59, incorporation of the ^{15}N label into the tetrazine ring requires the intermediacy of the dihydrohydrazinotetrazine **150**. A very careful ^1H and ^{13}C NMR spectroscopic analysis of the reaction mixture obtained from 6-amino-3-methyl-1,2,4,5-tetrazine with 2 equiv of hydrazine hydrate, reacting at 50°C for 8 hr, showed different stages of the reaction pathway, i.e., the formation of adduct 6-amino-6-hydrazino-1,6-dihydro-1,2,4,5-tetrazine (**148**, R = CH_3), an open-chain intermediate, and the final product. However, the intermediacy of 2,3-dihydro-3-hydrazino-3-methyl-1,2,4,5-tetrazine (**150**, R = CH_3) could not be detected. Apparently, if **150** (R = CH_3) is formed, it is highly unstable, probably because of the presence of two large groups at the methylene carbon bridge, which destabilizes the homotetrazole aromaticity. The fact that the stability of intermediate **148** (R = CH_3) is somewhat increased may be due to hydrogen bridge formation between the amino and the hydrazino group at C-6 (Figure II.1). There is no reason to believe that intermediate **148** undergoes ring opening

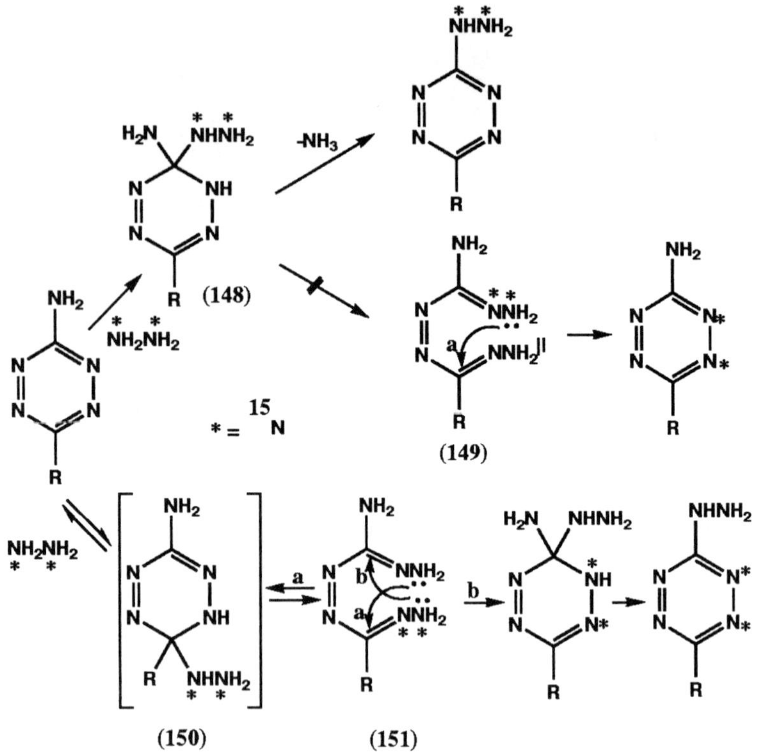

SCHEME II.59

into the open-chain compound **149**. If that should be the case, recyclization in **149** according to route (a) should immediatedly result in the incorporation of the ^{15}N label into the 1,2,4,5-tetrazine ring of the starting material (Scheme II.59). This has not been found experimentally; the starting material isolated after the reaction is unlabeled. The ring-opened compound, observed in the ^1H NMR spectrum, is **151**. All NMR data and the results of the ^{15}N-labeling studies support the mechanism proposed in Scheme II.59.

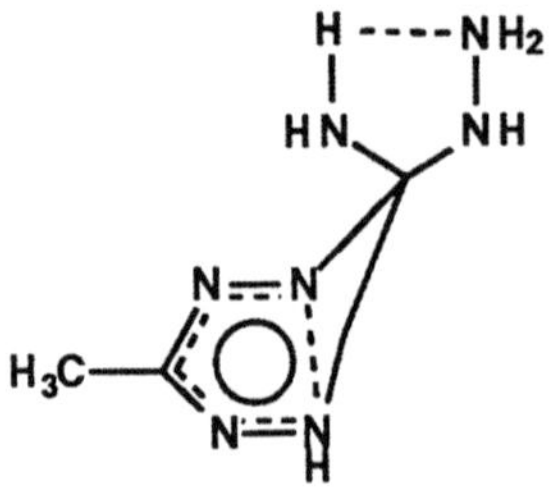

FIG. II.1

Chapter III

S_N(ANRORC) Reactions in Azaheterocycles Containing an "Inside" Leaving Group

The occurrence of degenerate ring transformations in azines, in which an "inside" leaving group is present, has been extensively studied. Especially in pyrimidine chemistry, many interesting examples of degenerate ring transformations are found, and several reviews on this topic are published [73MI1; 78KGS867; 80WCH491; 85T237; 94KGS1649; 95H441]. From the results ot these studies one can distinguish three types of reactions: (i) reactions in which only the ring nitrogen is replaced by a nitrogen atom of the reagent, (ii) conversions in which a carbon–nitrogen moiety of the pyrimidine ring is replaced by a carbon–nitrogen part of a reagent; and (iii) transformations in which a three-atom fragment (nitrogen–carbon–nitrogen or carbon–carbon–carbon) of the azaheterocycle ring is exchanged by the same three-atom moiety of (a) reagent(s).

A. Degenerate Ring Transformations Involving the Replacement of the Nitrogen of the Azaheterocycle by Nitrogen of a Reagent

1. Pyridines

One of the earliest described degenerate ring transformations in pyridine chemistry are the so-called Zincke exchange reactions. They occur when 1-(2,4-dinitrophenyl)pyridinium salts react with alkyl- or arylamines, yielding the corresponding N-alkyl- or N-arylpyridinium salts, respectively, together with 2,4-dinitroaniline (03LA361; 04LA296; 05LA365; 23CB758;41JPC1G). The exchange of N^+-aryl to N^+-alkyl can be considered as an internal S_N(ANRORC) reaction, since it involves as subsequent steps an initial addition of the nucleophile R–NH$_2$ (R = alkyl or aryl) at C2, ring opening, and ring closure (Scheme III.1). The ring closure is facilitated by the presence of the highly electron-deficient C-6 in the intermediary 2,4-dinitrophenylimine group. It yields the 1-R-6-(2,4-dinitroanilino)-1,6-dihydropyridine (**1**), which aromatizes by expulsion of 2,4-dinitroaniline.

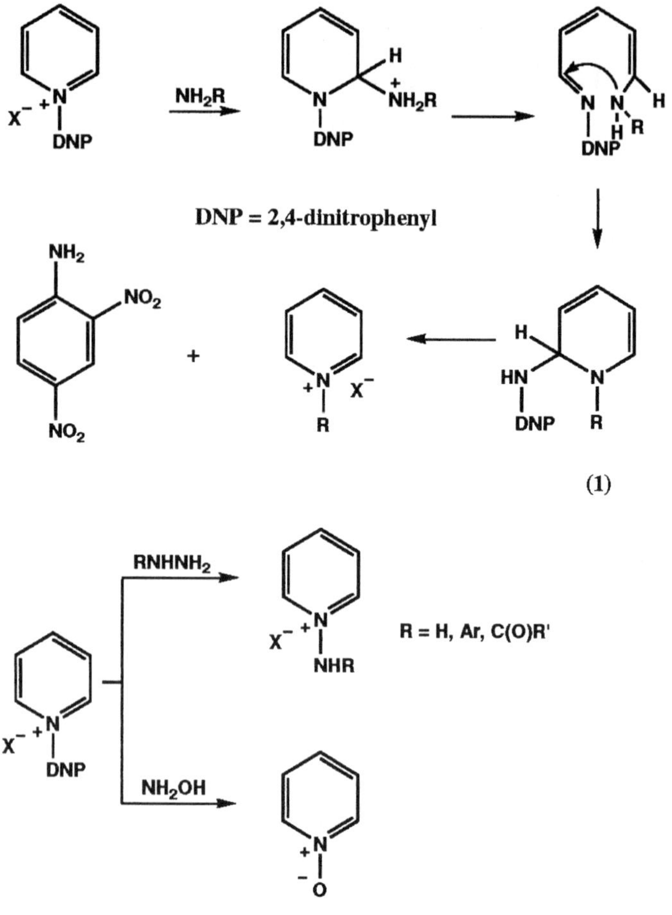

SCHEME III.1

The overall reaction can be described as one in which the "inside" *N*-aryl quaternary ring nitrogen acts as leaving group and is replaced by the R–NH$_2$ group.

This amino exchange methodology has been shown to have considerable synthetic potentialities for the preparation of a variety of N-substituted pyridinium salts. With a variety of arylamines, with heterocyclic amines, and with more complex amines [tryptamine, 7-aminocholesterol (53LA123), etc.], this exchange has been observed. With hydrazine, acyl-, and arylhydrazines, 1-aminopyridinium salts are obtained (66JPR293; 70CI(L)926; 71MI3; 72JHC865), and with hydroxylamine, pyridine *N*-oxides are formed [70CI(L)926; 71MI3] (Scheme III.1).

Other interesting examples of these amino-exchange reactions are the conversion of 3-aminocarbonyl-4-R-1-(2,4-dinitrophenyl)pyridinium salt (**2**) into 3-aminocarbonyl-4-R-1-*t*-butylpyridinium salt by *t*-butylamine (82RTC342) and the preparation of 1-(pentadeuteriophenyl)-3-aminocarbonyl-4-deuteriopyridinium salt from 3-aminocarbonyl-1-(2,4-dinitrophenyl)-4-deuteriopyridinium salt (**2**, R = D) with pentadeuterioaniline (84T433) (Scheme III.2).

These Zincke exchange reactions are not only restricted to pyridinium salts containing the 2,4-dinitrophenyl group as N-substituent, but can also occur with the less activated N^+-phenylpyridinium salts (80S589; 82JOC498). Examples are the degenerate ring transformation of *N*-phenyl-2-(ethoxyxycarbonyl)pyridinium salts (R = *t*-C$_4$H$_9$, C$_6$H$_5$) into the corresponding *N*-methylpyridinium salts on treatment with methylamine in chloroform at room temperature, and the formation of the *N*-ethyl- and *N*-*i*-propyl-2-(ethoxycarbonyl)-4,6-diphenylpyridinium salts, when 2-(ethoxycarbonyl)-1,4,6-triphenylpyridinium salt (**3**) reacts with ethylamine or *i*-propylamine, respectively (Scheme III.3). With *t*-butylamine, no formation of the *N*-*t*-butylpyridinium salt was observed.

It seems very likely that the initial addition of the alkyl amine takes place at the activated N–C(2) azomethine bond and that the reaction sequence occurs in a similar way to that described in Scheme III.1. When the reactions were carried out in boiling ethanol, besides the N^+-phenyl–N^+-alkyl exchange, deethoxycarbonylation occurs. Deethoxycarbonylation is the sole reaction that takes place on treatment of **3** with *t*-butylamine (82JOC498).

An interesting application showing the general usefulness of the Zincke degenerate ring transformation is the immobilization of proteins to

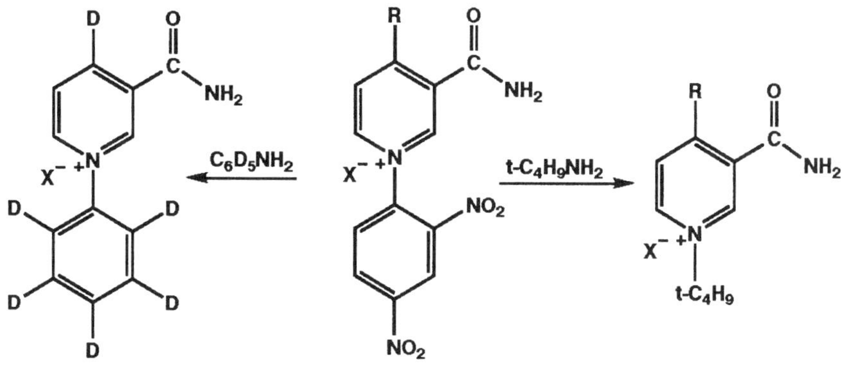

(**2**)

SCHEME III.2

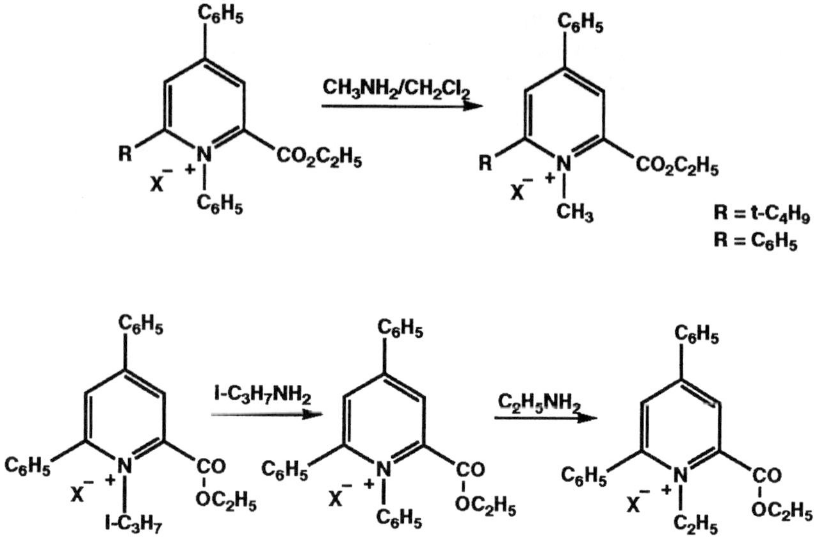

SCHEME III.3

pyridine-containing polymers (80JA2451). When a pyridine polymer reacts with cyanogen bromide, the N^+-cyanopyridinium polymer **4** is highly activated for a nucleophilic ring opening (80S589). When **4** reacts with an aqueous solution of a protein, a N^+–CN to N^+–protein exchange takes place (probably via the intermediacy of a polyaldehyde), leading to a polymeric system, in which the protein is present in an immobilized form (Scheme III.4).

An interesting series of degenerate ring transformation has been found when the N-alkyl-3-R-pyridinium salts (**5**, R = NO_2, SO_2CH_3, $CONH_2$, CF_3, CN) react with liquid ammonia at room temperature (73JOC1949;

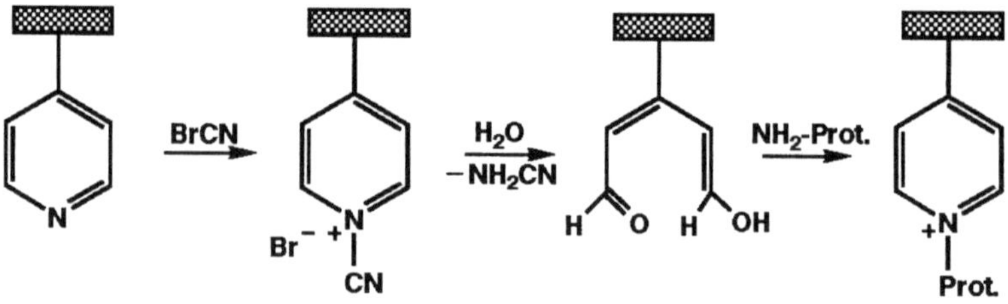

SCHEME III.4

76JOC1303; 84T433). After evaporation of the liquid ammonia the "dealkylated" product 3-R-pyridine was obtained (Scheme III.5). Also, the bipolar compound trigonelline (**7**) was found to be demethylated into nicotinic acid after treatment with liquid ammonia at $-33°C$ (74RTC114). By ^{1}H NMR spectroscopy it has convincingly been shown (84T433) that the pyridinium salt (**5**, R = $CONH_2$, Alkyl = CH_3, C_2H_5, n-C_3H_7), when dissolved in the liquid ammonia, is present as its neutral covalent 1:1 amino σ-adduct **6**. These covalent adducts may follow the regular process of a ring opening and ring closure leading to 2-alkylamino-1,2-dihydropyridine, which aromatizes by loss of alkylamine (Scheme III.5).

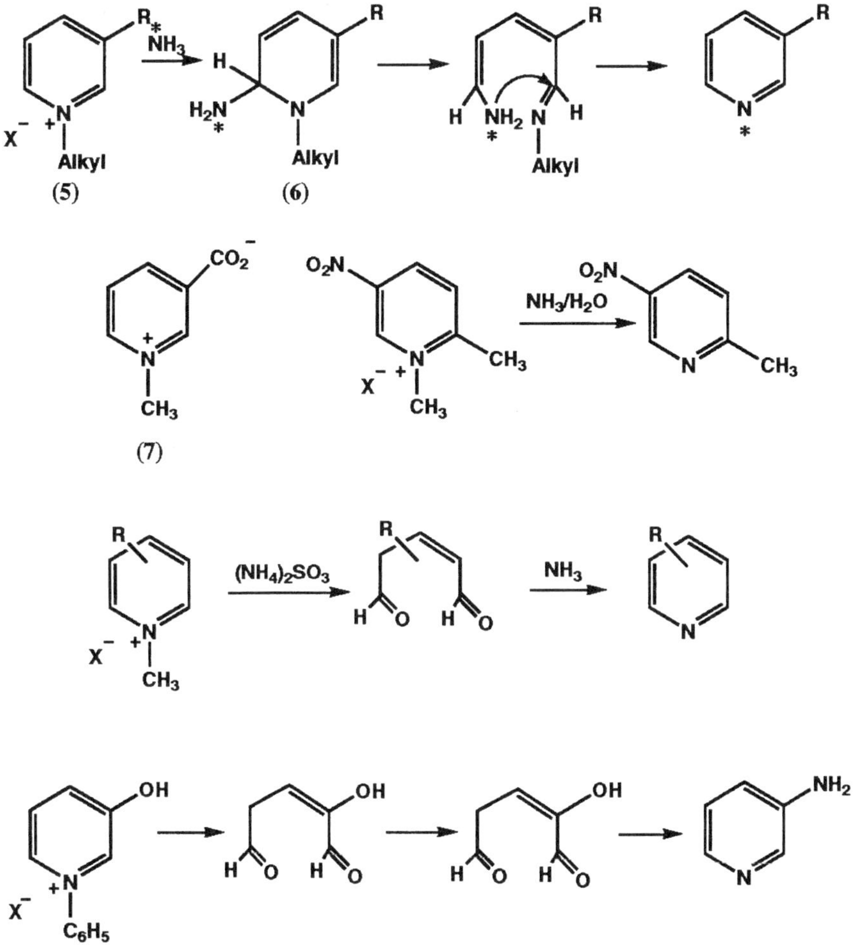

SCHEME III.5

The occurrence of this S_N(ANRORC) process has also been substantiated by ^{15}N-labeling experiments. Reaction of 3-aminocarbonyl-1-methylpyridinium salt (**5**, R = $CONH_2$, Alkyl = CH_3) with ^{15}N-labeled liquid ammonia gives incorporation of the ^{15}N label into the pyridine ring (Scheme III.5) (84T433).

A similar demethylation reaction, also following the ANRORC route, has been found in the reaction of 1,2-dimethyl-5-nitropyridinium salt with *aqueous* ammonia, 2-methyl-5-nitropyridine being obtained (Scheme III.5) (58CLY1131; 80KGS98). The reaction can again be described as taking place by an initial addition of either ammonia or the hydroxide ion to C-6. Worth mentioning is the fact that for **5** (R = $CONH_2$, Alkyl=i-C_3H_7, t-C_4H_9), two adducts are formed, a C-4 adduct and a C-6 adduct, the C-6 adduct being the main adduct. Thus, the site of addition is dependent not only on the nature of the substituent at position 3 (73JOC1949; 76JOC1303), but also on that of the substituent at the ring nitrogen (84T433). For a more extensive discussion on ammonia-induced demethylation reactions, see Section III,A,2,a.

It is of interest to mention that unactivated 1-methylpyridinium salts, when dissolved in liquid ammonia, do not form covalent σ-adducts or ring-opened products in amounts detectable by ^{1}H-NMR spectroscopy (73JOC1947). However, demethylation involving ring opening was found when these unactivated pyridinium salts are heated at 130°C with ammonium bisulfite. Since pyridine reacts with sodium bisulfite and alkali into the ring-opened glutaconic dialdehyde (08CB1346; 10CB2597, 10CB2939), it can be reasonably assumed that in the reaction with the pyridinium salts a glutaconic dialdehyde will also be formed, which under the conditions of the reaction converts further with the ammonium salt (or ammonia) into a pyridine derivative (Scheme III.5). Even the deactivated N-phenyl-3-hydroxypyridinium chloride can be converted into a dephenylated product. The product is, however, not 3-hydroxypyridine (as could be expected), but rather 3-aminopyridine. This amino group is not introduced by an amino–hydroxy Bucherer substitution in 3-hydroxypyridine, but rather by an amino–hydroxy exchange in the intermediate stage of the oxoglutaconic dialdehyde (Scheme III.5).

An interesting question in this connection is whether the pyridine ring opening (see Scheme III.5) is a base-catalyzed process, or whether it can better be described as an electrocyclic ring opening reaction involving four π-electrons and two σ-electrons. Earlier studies on reactions of N-methoxypyridinium salts with various nucleophiles suggest that the opening of the heterocyclic ring only occurs when a hydrogen is attached to the nucleophilic center (65T2205; 69T4291). This is exemplified by the conversion of this N-methoxypyridinium salt into glutaconic dialdehyde *mono-O*-oxime by sodium hydroxide, and this reaction is found to be a second-order with respect to hydroxide (Scheme III.6). It involves as intermediate the 2-hydroxy adduct, which undergoes a base-catalyzed ring opening. However, later experiments show evidence that with secondary amines (no hydrogen is present on the nucleophilic center in the adduct), ring opening can still occur (69T4291; 70RTC129; 74CB3408; 76JOC160). An illustrative example is the formation of the *syn–cis–trans* compound (**8**) when *N*-alkoxypyridinium salts react with the secondary amines pyrrolidine, piperidine, and diethylamine (65JA395; 71JOC175; 76TL4717). (Scheme III.6).

In a communication (97MI1) presenting molecular mechanics calculations of all plausible intermediates in the recyclization reaction of α-picolinium salts with anilines, it has been proved that the diene fragment of isolated products of the pyridine ring opening has the *trans–trans* configuration (67T2775; 75AP594). This geometry contradicts to the concept of an electrocyclic ring opening of covalent amination products. However, the initial configuration (**8**) was detected by ^{1}H NMR spectroscopy in the kinetically controlled ring opening of the *N*-methoxypyridinium salt. The thermal ring opening of the σ-adduct occurs in a disrotatory manner, as is permitted by the Woodward–Hoffmann rules for electrocyclization (65JA395; 81T3423; 95T8599). The communication (97MI1), which presents molecular mechanics calculations of all plausible intermediates in the recyclization reaction of α-picolinium salts with anilines (81T3423; 95T8599), shows that from the two possible variants of pseudobase ring opening—

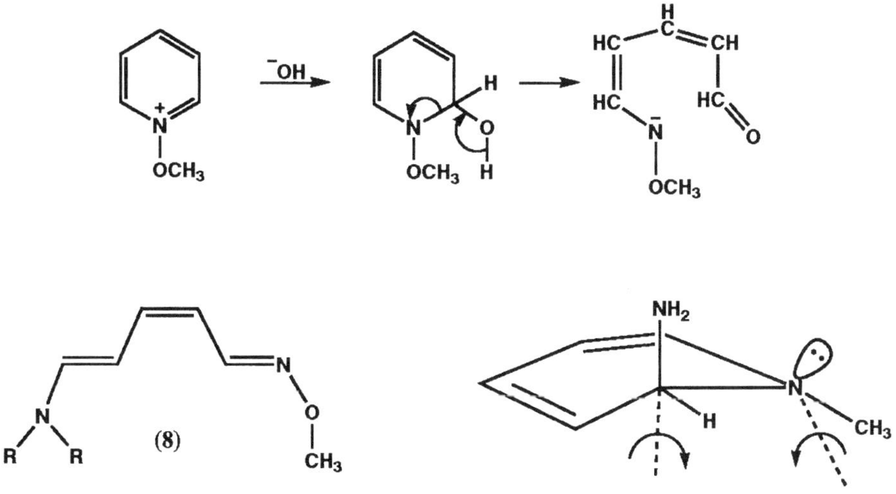

SCHEME III.6

SCHEME III.6A

electrocyclic (a) or ionic (b)—the formation of the enol form is endothermic, but the formation of the ketonic amine is exothermic (Scheme III.6a). Therefore, the ionic variant is strongly preferred. This is also supported by the experience that the formation of the open form was never noted with nucleophiles incapable of ionic ring opening (Cl$^-$, Br$^-$, I$^-$, CN$^-$, etc.) (97MI1). As far as the ring opening of the 6-amino adduct (**6**) in Scheme III.5 is concerned, from the data available so far no definite conclusion about either a base-catalyzed or electrocyclic process can be drawn. However, since ammonia is a weak base ($K_b = 1.8 \times 10^{-5}$), it seems questionable whether it is sufficiently basic to perform the ring opening of **6** by an initial deprotonation of the C-6 amino group in **6**. Therefore, it is quite acceptable that the ring opening occurs by a thermally allowed disrotatory ring opening. The addition of the ammonia is assumed to occur in an axial position. This species is sufficiently long-lived to be involved in conformational equilibrium.

2. PYRIMIDINES

a. *N-Alkylpyrimidinium Salts*

Nucleophile-induced degenerate ring transformatiions have been extensively studied in reactions of pyrimidinium salts. When 1-methylpyrimidinium methosulfate is dissolved in liquid ammonia at $-33°C$ and this solution is kept for 1 hour, after evaporation of the ammonia pyrimidine is formed (Scheme III.6) (74RTC114). The occurrence of a demethylation reaction under low temperature conditions is unexpected, certainly in the light of the very drastic conditions that must be applied in the S_N2-type replacement reactions of the demethylation of *N*-methylpyridinium salts (73SC99; 75SC119; 76JOC2621). When a ^{1}H NMR spectrum of a solution of 1-methylpyrimidinium methosulfate in liquid ammonia was measured, it became evident that not the pyrimidinium salt as such, but an ammonia addition complex was present. This evidence was provided by the observation that H-6 has undergone a considerable upfield shift of about 4.57 ppm (compared with the chemical shift of H-6, when the compound is dissolved in D$_2$O) (74RTC114). This could be attributed to the change of hybridization of C-6 on addition (sp^2 to sp^3). Apparently, covalent amination has taken place between the quaternary salt and ammonia at C-6, leading to the formation of a 1:1 σ-adduct, i.e., 6-amino-1,6-dihydro-1-methylpyrimidine (**9**). This facile addition is certainly due to the strong polarization in the azomethine bond of the quaternary salt. On evaporation of the ammonia, ring opening into the diazahexatriene **10**, and subsequent ring closure, 2-(methylamino)-1,2-dihydropyrimidine is formed, which by loss of methylamine is converted into pyrimidine (Scheme III.7). Further proof for this mechanism was obtained when the demethylation reaction was studied with the double-labeled 1-methyl-[1,3-^{15}N]pyrimidinium methyl sulfate, prepared from double labeled [1,3-^{15}N]pyrimidine (6.8% ^{15}N) and dimethyl sulfate. Mass spectrometric determination of the ^{15}N excess in the pyrimidine obtained after demethylation showed that the $M + 2$ peak is decreased to nearly zero, while the $M + 1$ peak is strongly increased (Scheme III.7) (74RTC114).

The fact that in the liquid ammonia the C-6 amino adduct **9** has a considerable lifetime seems to suggest that the ring-opening reaction is the rate-determining step in the demethylation process. Since this overall process describes the nucleophilic replacement of a quaternary ring nitrogen by an amino group via a ring opening, this "methylammonium–amino"

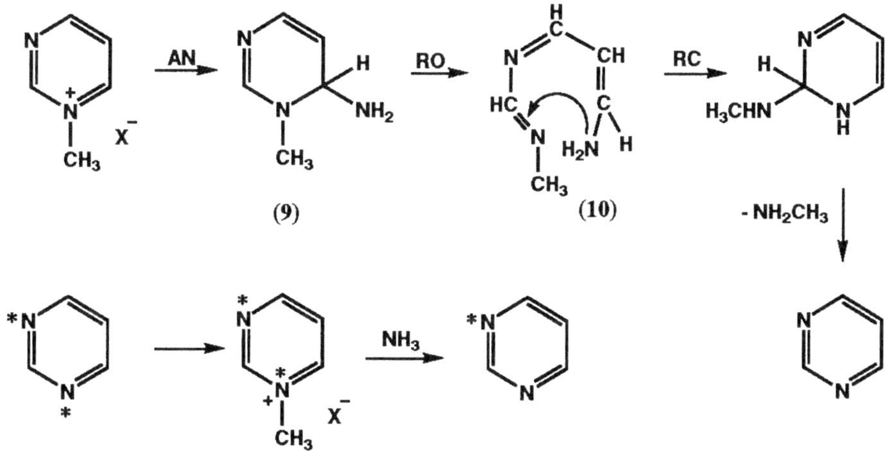

M+2	6.8	0.6
M+1	0	6.3

SCHEME III.7

exchange can be categorized as a degenerate ring transformation, following the S_N(ANRORC) pathway.

In order to get a better understanding of the observed regioselectivity in these demethylation reactions, calculations were carried out, using the PM3 method to establish the charge density on the different atoms in the ring in the N-methylpyrimidinium cation, in particular at C-2 and C-6 [95UP1]. These calculations show that the C-2 position has less negative charge than the C-6 position, predicting a higher reactivity of C-2 than of C-6 for nucleophilic attack. This result is not in agreement with the ^{1}H NMR measurements, leading to the conclusion that the addition is not charge-controlled. Quantum chemical calculations applying the theory of permutation of molecular orbitals have been published on the regioselectivity of nucleophilic addition in nitroquinolines of amide ions or ammonia (87JOC5643; 93LA823), the chloromethyl sulfone anion (96LA641), and the nitromethyl anion (94PJC635). In this approach calculations of the π-electron stabilization energy (ΔE) for all possible sites of attack of the nucleophile (N) on the electron accepting substrate (A) were carried out using the simplified second-order perturbation equation (76MI2)

$$\Delta E \approx 2\left[\frac{C_S^2(\text{LUMO})}{E^N_{\text{HOMO}} - E^A_{\text{LUMO}}} + \frac{C_S^2(\text{LUMO} + 1)}{E^N_{\text{HOMO}} - E^A_{\text{LUMO}+1}}\right]$$

In this equation the C_s represent the respective atomic coefficients for the orbitals of the nucleophile (HOMO) and those of the different atoms in the electron-acceptor (LUMO and LUMO + 1); ($E^N_{\text{HOMO}} - E^A_{\text{LUMO}}$) and ($E^N_{\text{HOMO}} - E^A_{\text{LUMO}+1}$) represent the energy gaps between the HOMO and LUMO (LUMO + 1) levels of nucleophile and acceptor.

On calculating the ΔE values for the positions C-1 to C-6 in the N-methylpyrimidinium ion, using the MNDO, PM3, and AM1 method, the highest ΔE value was found for position 6 (C-6 > C-2 > C-4 > C-5), indicating that position 6 is more favored for nucleophilic attack than position 2. These results are in good agreement with the experimental observations and show that the addition of ammonia to the pyrimidinium ion is probably not a charge-controlled reaction, but an orbital-controlled process. Other observations were similar to those obtained in amination reactions of naphthyridines and dinitrobenzenes (95RTC13). In contrasts, in nitropyridines the addition of the nucleophile seems to be charge controlled and not orbital controlled (91LA875).

Extension of studies on the occurrence of these liquid ammonia-induced demethylation reactions in other pyrimidinium salts has revealed that also 1,2-dimethylpyrimidinium iodide (74RTC114), 1,4,6-trimethylpyrimidinium iodide (74RTC114), 1,2,4,6-tetramethylpyrimidinium iodide (74RTC114), and 1-methyl-4-phenylpyrimidinium iodide (82RTC367) are also easily demethylated under these mild conditions. Measurements of the ^{1}H NMR spectra of a solution of these compounds in liquid ammonia show the occurrence of covalent amination at C-6 (74RTC114). Even in the compounds where position 6 is occupied by a methyl group, covalent amination still occurs. Demethylation has also been reported to occur in aqueous ammonia (75KGS1400).

An interesting demethylation reaction involving a degenerate ring transformation has been observed on treatment of the *N*-methyl quaternary salt of 3-ethoxycarbonyl-6,7,8,9-tetrahydro-4(*H*)-pyrido[1,2-*a*]pyrimidin-4-one (**11**) with ^{15}N-labeled aqueous ammonia. In the demethylated product **13**, the ^{15}N label is incorporated in the heterocyclic ring as well as in the carboxamido group (Scheme III.8) (79TL1337). This result indicates that a ring-opening reaction has taken place. It involves an initial addition of the hydroxide ion (or ammonia) to the activated azomethine bond at the bridgehead carbon, yielding the pseudo base, which is in tautomeric equilibrium with the ring-opened N-substituted piperidone-2 **12**. A Michael addition of the ^{15}N-labeled ammonia to the β position of the ethoxycarbonyl group leads to an amino–methylamine exchange. Cyclization gives the final product; during this process the ethoxycarbonyl group is converted into the ^{15}N-labeled aminocarbonyl group.

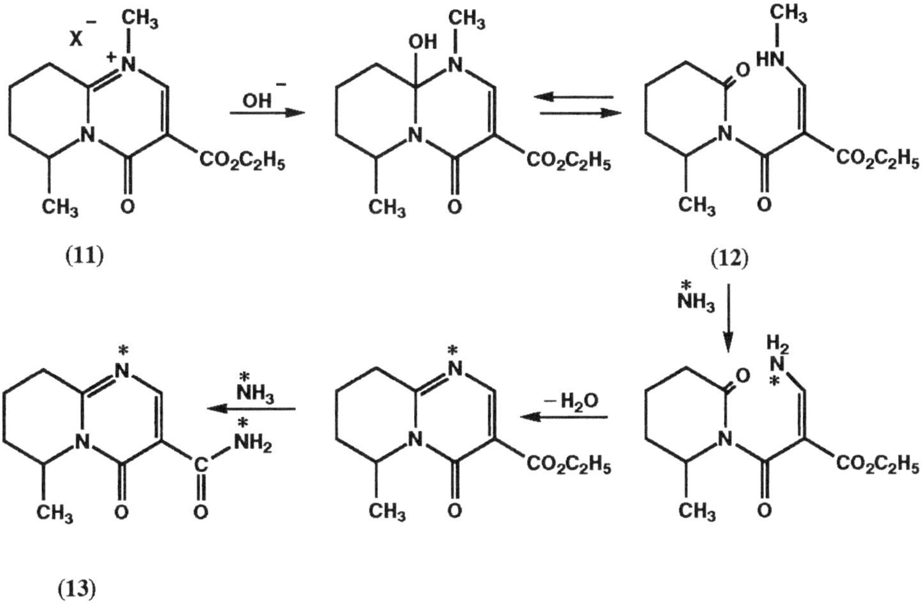

SCHEME III.8

In addition to amino–methylamine exchanges, amino–benzylamine exchanges were found. When N-benzyl salts of pyrimidine, 4,6-dimethylpyrimidine, and 4-t-butylpyrimidine react with liquid ammonia, the respective debenzylated compounds were isolated (Scheme III.9) (82JHC3739).

The N-benzyl-4,6-dimethylpyrimidinium salt, when subjected to treatment with ^{15}N-labeled liquid ammonia, gave ring-^{15}N-labeled 4,6-dimethylpyrimidine. It shows that an S_N(ANRORC) process is also operative in this amino–benzylamine exchange. Furthermore, ^{1}H NMR spectroscopy convincingly shows that these N-benzyl-4,6-dimethylpyrimidinium salts give addition of the liquid ammonia on C-2, i.e., **14** (82JHC3739). This is in remarkable contrast to the observation that the N-benzyl-4-t-butylpyrimidinium salt gives addition at C-6, i.e., **15**! No explanation is offered. An extensive study of the dealkylation reaction in N-methylpyrimidinium salts, containing one or more alkoxy groups at position 4 and/or 6, showed that in these compounds N-dealkylation competes with dealkoxylation. 4-Ethoxy-1-ethylpyrimidinium tetrafluoroborate, when it reacts with liquid ammonia, does not undergo an N-deethylation reaction, but iminodeethoxylation at C-4, yielding 1,4-dihydro-1-ethyl-4-

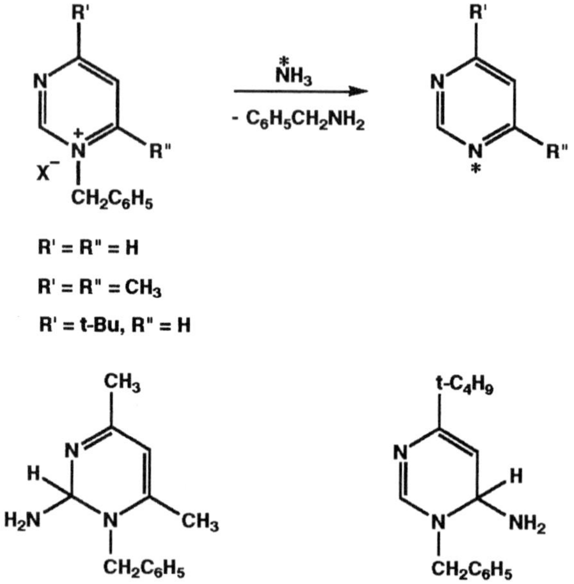

R' = R" = H
R' = R" = CH$_3$
R' = t-Bu, R" = H

(14)   (15)

SCHEME III.9

iminopyrimidine hydrogen tetrafluoroborate (Scheme III.10) (77RTC183). NMR investigations clearly indicated that in the liquid ammonia the 4-ethoxy-1-ethylpyrimidinium salt is present as its C-2 σ-adduct (**16**). The favored formation of the C-2 adduct (**16**) over that of the C-6 adduct is in agreement with the results of charge density MNDO calculations [95UP1]. They predict that in the 4-ethoxy-1-methylpyrimidinium salt, the highest positive charge is at position 4 and that C-2 has a higher positive charge than C-6 (C-4 > C-2 > C-6). Frontal orbital calculations clearly predict a reversed order: the highest reactivity for nucleophilic attack at position 6 (C-6 > C-2 > C-4). Although the charge density calculations have indicated that in 4-ethoxypyrimidinium salts the most favored position for addition of nucleophiles is at C-4, the initial addition occurs at C-2 (formation of **16**). It seems to suggest that the addition at C-2 is a kinetically controlled addition and that addition at C-4 is thermodynamically more favored. Kinetic versus thermodynamic control of a nucleophilic addition reaction in heterocyclic systems is a well-established phenomenon. A kinetic study of the formation of the pseudobase of the 1-methyl-3-nitroquinolinium cation revealed that 1,4-dihydro-4-hydroxy-1-methyl-3-nitroquinoline is the predominant product at equilibrium (72CJC917; 73CJC1965; 74CJC303, 74CJC951, 74CJC962, 74CJC975, 74CJC981; 77RTC68). However, the isomeric pseudobase 1,2-dihydro-2-hydroxy-1-methyl-3-nitroquinoline is the kinetically controlled precursor of the 4-hydroxy isomer (Scheme III.11).

Another example of kinetic vs thermodynamic control is observed in an NMR-spectroscopic study of the ring opening of several 1-methoxy-3-carbamoylpyridinium salts by liquid ammonia. The study shows that

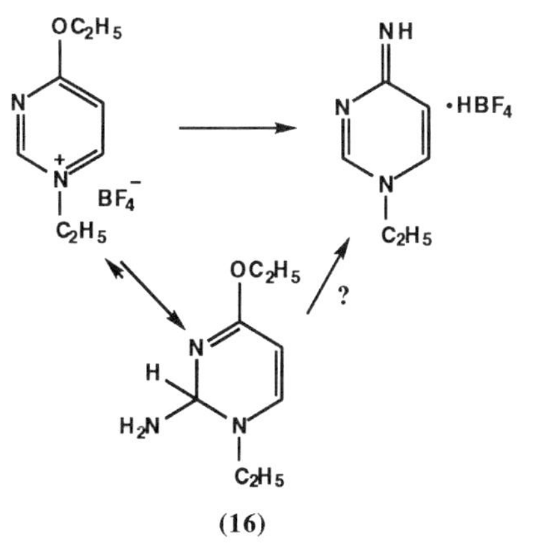

SCHEME III.10

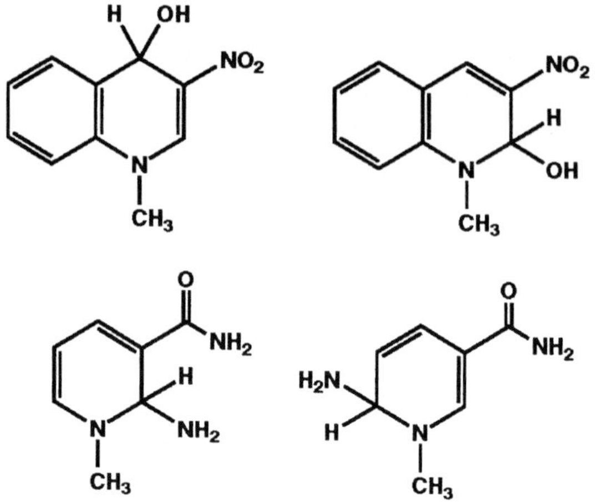

SCHEME III.11

the initial addition takes place at position 2, but that the ring opening occurs by C–N bond breaking in the C-6 amino σ-adduct (Scheme III.11) (76JOC1303).

The 2-phenyl derivative of 4-ethoxy-1-ethylpyrimidinium tetrafluoroborate undergoes with liquid ammonia deethylation at N-1 as well as deethoxylation at C-4, a mixture of 4-ethoxy-2-phenylpyrimidine (**18**) and 1-ethyl-4-imino-2-phenylpyrimidine (**19**), respectively, being obtained (Scheme III.12) (77RTC183).

Investigation of the deethylation with ^{15}N-labeled ammonia showed that in **18** the ^{15}N label is incorporated for about 100% in the pyrimidine ring. Evidently the S_N(ANRORC) mechanism is operative in the replacement of the quaternary ring nitrogen by the amino nitrogen. The initial step in this deethylation reaction will probably be adduct formation on C-6, i.e., **17**. Addition at C-2 is less likely because of the presence of the phenyl group at that position (Scheme III.12).

A more complicated picture arose when the formation of the iminodeethoxylation product, i.e., **19,** was studied with ^{15}N-labeled liquid ammonia. Since it appeared to be difficult to measure by mass spectrometry the ^{15}N content of **19, 19** was converted into the corresponding pyrimidin-4(1*H*)-one (**20**) by refluxing in aqueous potassium hydroxide. Measurement of the ^{15}N-label enrichment in **20** showed that about 40% of the ring nitrogen atoms were ^{15}N-labeled {ratio **20** : **20*** = 60:40}. The occurrence of a degenerate ring transformation in the formation of **20*** seems rather surprising and unexpected, certainly when one takes into account that in the iminodeethoxylation of 4-ethoxypyrimidinium salts, having an *un*-

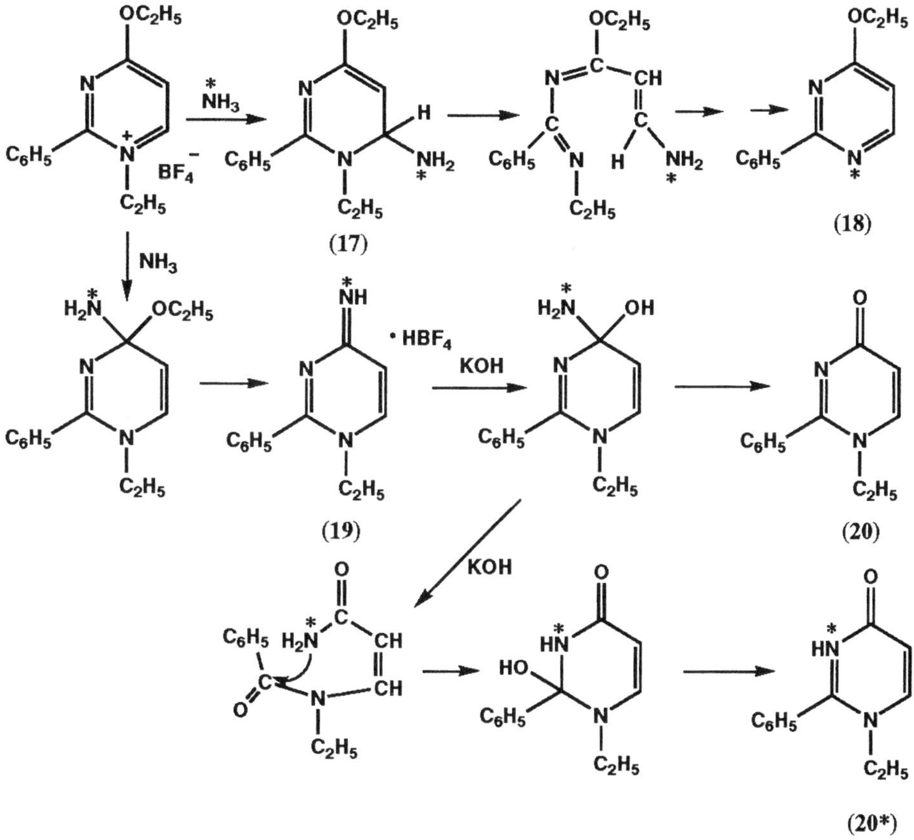

SCHEME III.12

blocked position on C-2, no incorporation of the ^{15}N-label was found. More detailed experimental work has, however, shown that the incorporation of the ^{15}N label takes place, not during the amination reaction, but rather by a Dimroth-type rearrangement (68MI1) of the 4-imino compound **19** during treatment with the aqueous basic solution (Scheme III.12). Thus, the conclusion seems justified that the aminodeethoxylation does not involve an ANRORC process.

Reaction of 4,6-diethoxy-1-ethylpyrimidinium tetrafluoroborate (**21**) showed similar behavior: no N-deethylation, but deethoxylation at position 4 as well as at position 6 was observed, as indicated by the formation of the two isomeric products 1,4-dihydro-6-ethoxy-1-ethyl-4-iminopyrimidine hydrogen tetrafluoroborate (**23**) and 1,6-dihydro-4-ethoxy-1-ethyl-6-

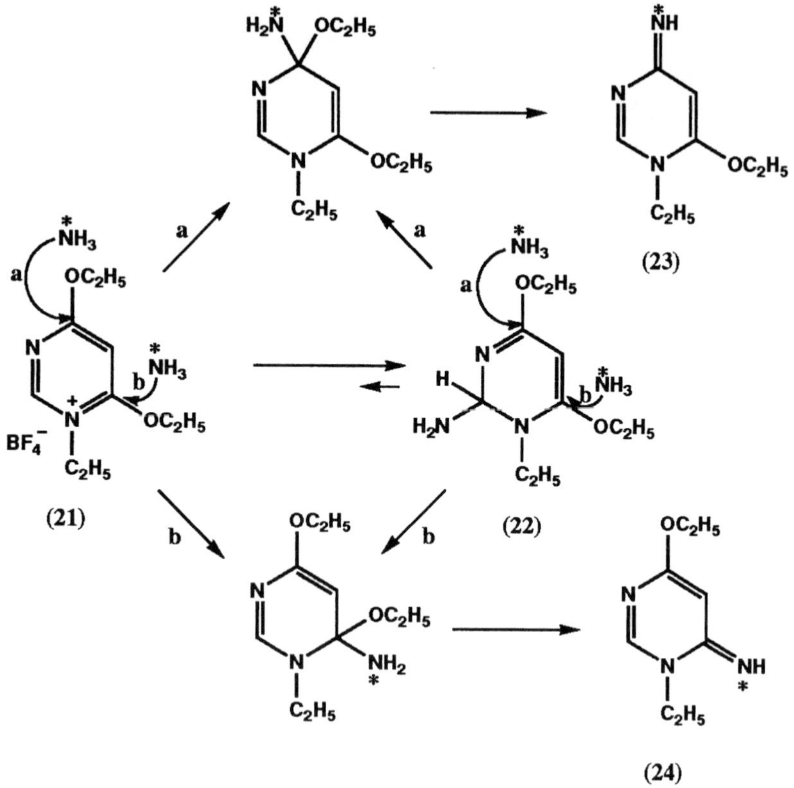

SCHEME III.13

iminopyrimidine hydrogen tetrafluoroborate (**24**) (Scheme III.13). NMR spectroscopy of a solution of salt (**21**) in liquid ammonia convincingly shows the formation of the covalent C-2 adduct (**22**).

Study of the aminodeethoxylation with ^{15}N-labeled liquid ammonia shows that in the 4-imino compound *no* incorporation of the ^{15}N label has taken place, proving that in the replacement of the ethoxy group no ring opening is involved. It is unknown whether the aminodeethoxylation occurs according to routes (a) and (b) in the σ-adduct **22** or in the starting material **21**, which is present in only a small equilibrium concentration with **22** (Scheme III.13). One can expect, however, that despite its low concentration, the aminodeethoxylation reaction takes place in the pyrimidinium salt **21**, being more reactive towards to nucleophiles than the neutral adduct **22**.

The observation was made that the reaction of *N*-ethyl-4-ethoxypyrimidinium salts with liquid ammonia not only gives *N*-deethylation and deethoxylation products, but sometimes also yields 4(6)-(ethylamino) pyrimidines. For example, treatment of 6-ethoxy-1-ethyl-4-phenylpyrimidinium salt with liquid ammonia gives, besides the 6-imino- and the

6-ethoxy-compound, 6-(ethylamino)-4-phenylpyrimidine (77RTC183). Its formation involves as intermediate 6-ethoxy-6-(ethylamino)-1,6-dihydropyrimidine, being obtained by ring opening of the C-2 amino adduct (Scheme III.14).

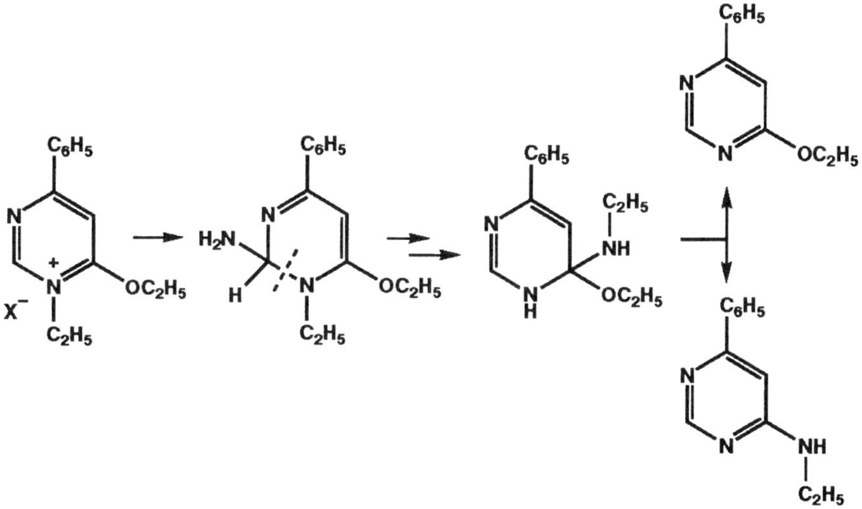

SCHEME III.14

Degenerate ring transformations have also been observed in reactions of 6-ethoxy-4-oxopyrimidinium salts (**25**, R = H, CH$_3$, C$_6$H$_5$) with liquid ammonia (77RTC68). Besides open-chain compounds, 2-R-1-ethyl-4-(ethylamino)pyrimidin-6(1H)-one is formed. The presence of the ethyl-

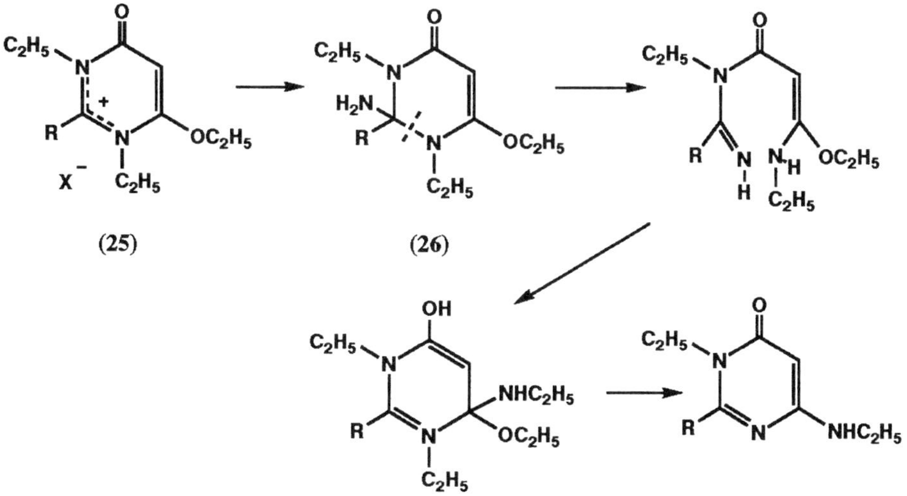

SCHEME III.15

amino group as ring substituent proves the occurrence of the degenerate character of the ring transformation. NMR spectrocopy of solutions of **25** (R = H) in liquid ammonia proves their intermediacy of the C-2 adduct (**26**) (Scheme III.15).

b. *N-Aminopyrimidinium Salts*

ANRORC processes leading to degenerate ring transformations have also been observed in reactions of *N*-aminopyrimidinium salts with liquid ammonia. Treatment of *N*-amino-4,6-diphenylpyrimidinium mesitylenesulfonate (**27**) with liquid ammonia leads to quantitative deamination, yielding 4,6-diphenylpyrimidine. Carrying out the deamination with ^{15}N-labeled liquid ammonia, it was found that 27% of the 4,6-diphenylpyrimidine formed contained the ^{15}N label in the pyrimidine ring (76RTC282). Apparently a part of **27** underwent deamination via the S_N(ANRORC) mechanism, involving the addition, ring opening, and ring closure (Scheme III.16).

This S_N(ANRORC) mechanism is similar to that presented for the

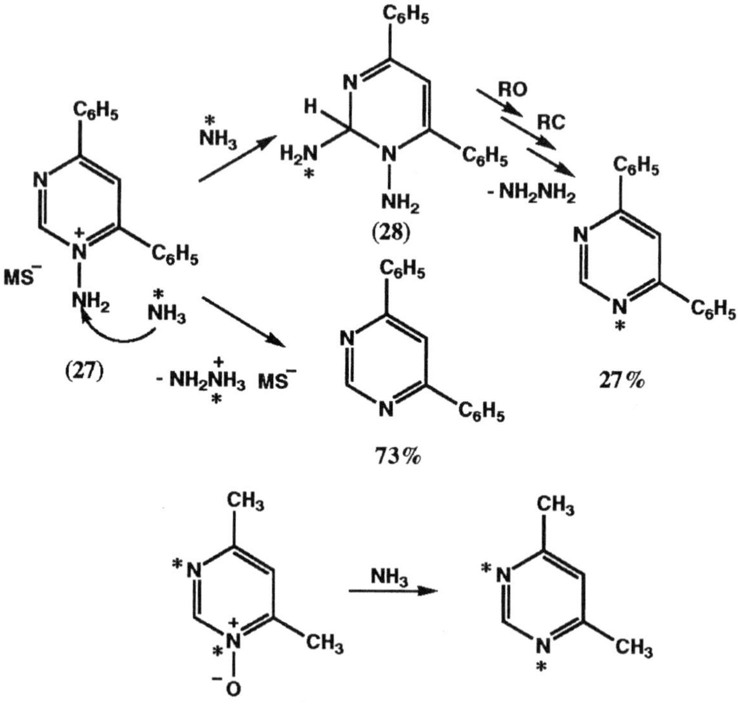

SCHEME III.16

liquid-ammonia-induced quantatitive demethylation of *N*-methylpyrimidinium salts (Section III,A,2,a). It is not evident whether the addition of the ammonia takes place in the *N*-aminopyrimidinium ion or its conjugate base, the *N*-iminoylid.

Attempts to establish the structure of the initial adduct by NMR-spectroscopy failed because of the low solubility of **27**. This makes it impossible to draw a clear conclusion as to whether the ammonia adds to C-6 (as occurs in the case of the *N*-methylpyrimidinium salts) or at C-2. Since ^{1}H NMR spectroscopy of a solution of 4,6-diphenylpyrimidine in potassium amide/liquid ammonia strongly supports the formation of an anionic C-2 adduct (75UP1], it is justified to assume that also in the deamination of **27** by liquid ammonia, a C-2 adduct **28** is involved (Scheme III.16). It is evident that the major part of the deamination (73%) does not involve a ring-opening reaction; the main deamination reaction occurs by an S_N2 attack of ammonia on the *N*-amino group in **27**. A similar mechanism has also been postulated in the deoxygenation of pyrimidine *N*-oxides, when they are heated with liquid ammonia (Scheme III.16) [77UP2].

Deamination into 2,4,6-trimethylpyrimidine, together with a simultaneous formation of 3,5-dimethyl-1,2,4-triazole, occurs when *N*-amino-2,4,6-trimethylpyrimidinium mesitylenesulfonate reacts with liquid ammonia

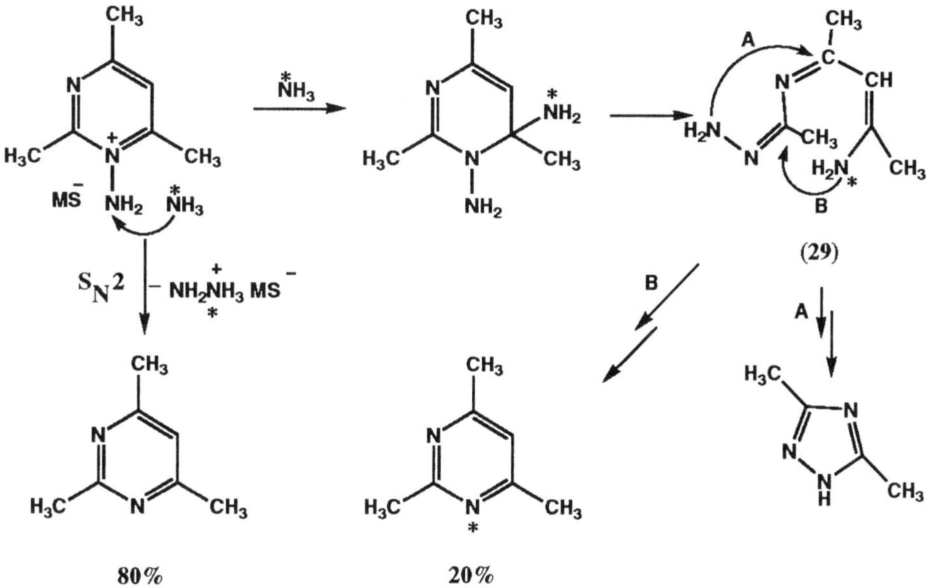

Scheme III.17

(76RTC282). Applying the ^{15}N-labeling technique with ^{15}N ammonia, it was established that just as with compound **27,** only a small percentage (20%) of the ^{15}N label is incorporated in the pyrimidine ring [S_N(ANRORC) mechanism]; the remaining 80% reacts by an S_N2 attack of the ammonia on the N-amino group (Scheme III.17). The simultaneous formation of the 1,2,4-triazole proves the intermediacy of the open-chain compound **29,** which can recyclize by route A into 3,5-dimethyl-1,2,4-triazole or by route B into ^{15}N-labeled 2,4,6-trimethylpyrimidine. It is assumed that just as observed with 1,2,4,6-tetramethylpyrimidinium salts, the addition of the ammonia occurs at C-6.

A very interesting case of a degenerate ring transformation has been observed when the mesitylenesulfonate (MS$^-$)(**30**) reacts with ^{15}N-*double*-labeled anhydrous hydrazine for a short period of time. When after the reaction starting material is recovered, the compound shows a considerable enrichment of the ^{15}N label, i.e., the formation of **30**** (83JHC415)!

This degenerate transformation of **30** into **30**** can be explained when the incorporation of the ^{15}N label into the pyrimidine ring occurs by ring closure of the open-chain species, which is formed by ring opening of the covalent hydrazino adduct, probably at position 6 (Scheme III.18).

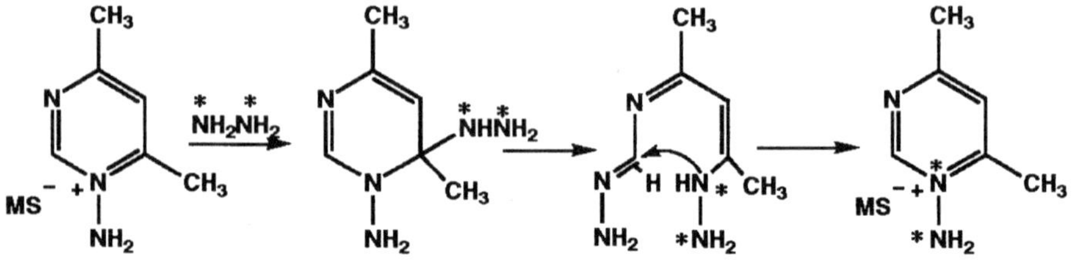

SCHEME III.18

It has been postulated that in the hydrazine solution (pH 8) the N-amino-4,6-dimethylpyrimidinium ion is partly deprotonated into its conjugate base, the resonance-stabilized ylide **31,** so that in fact in this solution two equilibrated active species, **30** and **31,** are present (62TL387; 63AG604). The formation of an ylide intermediate was experimentally supported by the isolation of the dimer 2,4,7,9-tetramethyldipyrimido[1,2-*b*;1′,2′-*c*]hexahydrotetrazine (**32**), which is formed when the pyrimidinium salt **30** is treated with liquid ammonia (Scheme III.19).

One can expect that the *N*-amino 4,6-dimethylpyrimidinium ion has about the same reactivity as the *N*-methyl-4,6-dimethylpyrimidinium ion.

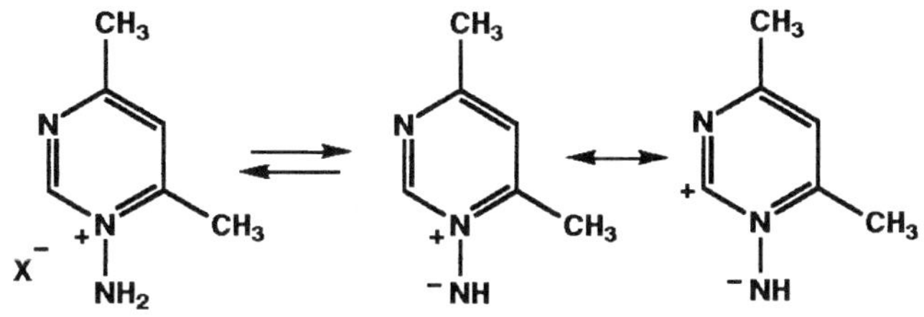

SCHEME III.19

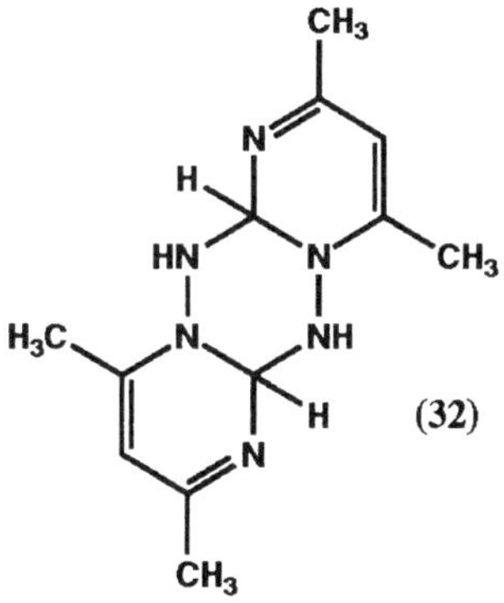

Whereas in *N*-methyl-4,6-dimethylpyrimidinium ion the covalent addition takes place at C-6, it is assumed that **30** also undergoes covalent hydrazination at C-6. However, the formation of dimer **32** shows the high sensitivity of C-2 in **30** for addition of nucleophiles, and it leads to the daring suggestion that it is the resonance-stabilized ylide **31** that probably is the active species undergoing addition at C-2 (Scheme III.19). It was calculated (80UP1) that the reactivity at C-2 in the N-ylide **31** is greater than that at C-2 in the *N*-aminopyrimidinium salt **30**.

Degenerate ring transformations, involving the replacement of the *N*-amino group in *N*-aminopyrimidinium salts by an *N*-oxide group, have been observed in the reaction of *N*-aminopyrimidinium mesitylenesulfonates [**33**, R = H, R' = CH_3; R = H, R' = C_6H_5; R = R' = CH_3; Z = $OSO_2C_6H_2(CH_3)_3$] with hydroxylamine, the corresponding pyrimidine *N*-oxides **35** being obtained (Scheme III.20) (76TL3337).

This nonoxidative method, a useful extension of the more classical oxidative methods of synthesizing pyrimidine *N*-oxides, is of interest especially for the preparation of pyrimidine *N*-oxides that contain substituents being sensitive to oxidation. The reaction involves the intermediacy of

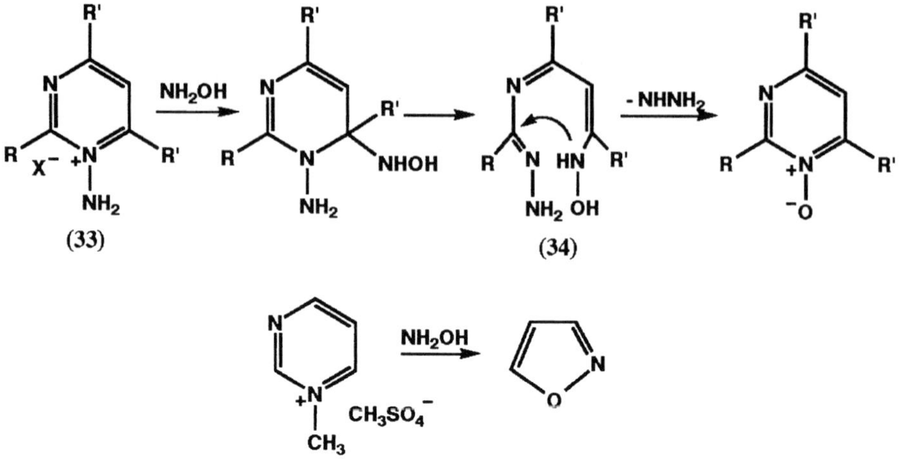

SCHEME III.20

1-amino-6-hydroxylamino-1,3-diazahexa-1,3,5-triene (**34**). In this intermediate the nitrogen lone pair of the hydroxylamino group is able to perform the cyclization.

Note that *N*-methylpyrimidinium methosulfate reacts very differently with hydroxylamine. Instead of pyrimidine *N*-oxide formation, an overall replacement of the N_1–C_2–N_3 fragment of the pyrimidine ring by the N–O fragment of the reagent takes place, leading to ring contraction into isoxazoles (Scheme II.20). For a more detailed discussion on this ring contraction, the reader is referred to the original literature (74RTC225).

c. *N-Arylthiopyrimidones, N-Arylpyrimidinium Salts, and Quinazoline (Di)ones*

Pyrimidine *N*-oxide formation has also been reported to take place when 4,6-dimethyl-1-aryl 2(1*H*)-pyrimidinethiones (**35**) react with hydroxylamine hydrochloride in the presence of sodium hydroxide, the 2-arylamino-4,6-dimethylpyrimidine *N*-oxides **37** being obtained (81CPB2516). The rearrangement can be explained by an initial addition of the hydroxylamine to C-6, subsequent ring opening into **36**, ring closure by attack of the nitrogen of the hydroxylamino group on the thiocarbonyl of the thioamide group, and aromatization by loss of hydrogen sulfide. This transformation is illustrated in Scheme III.21.

This degenerate ring transformation is generally applicable, since with different aryl substituents at positions 1, 4, and 6, aryl-substituted pyrimidine *N*-oxides are indeed formed. In a very similar reaction, 4,6-

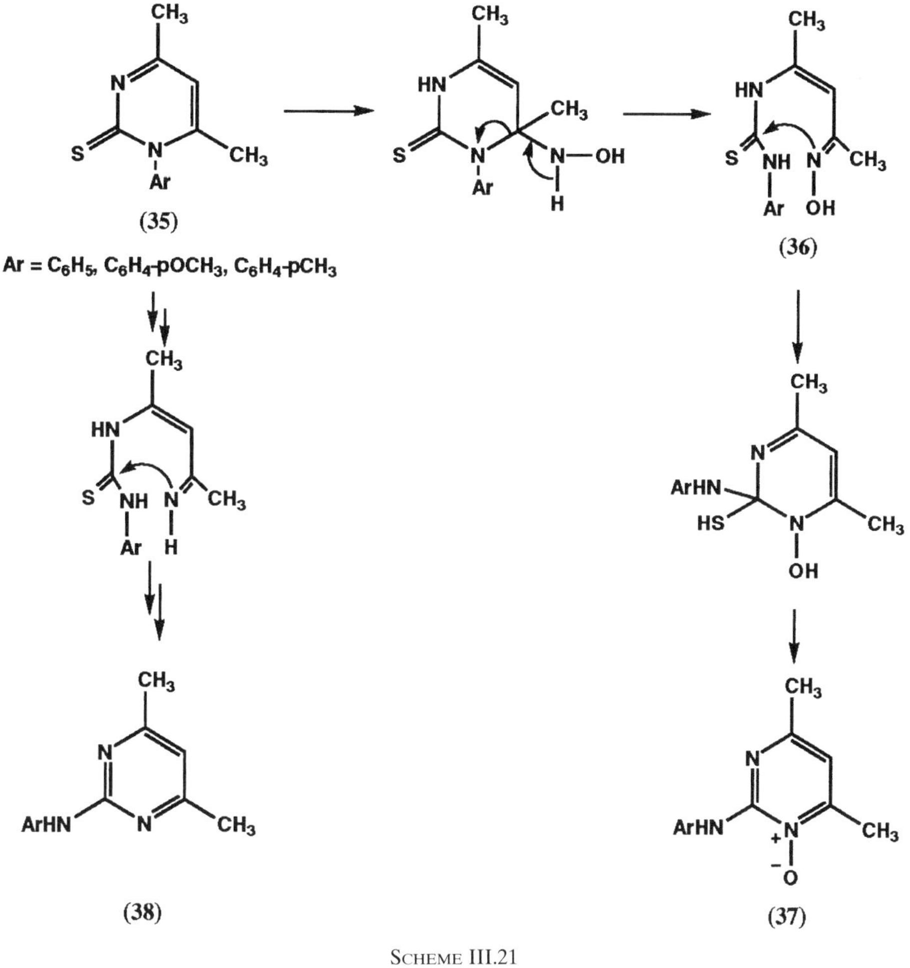

SCHEME III.21

dimethyl-1-aryl-2(1*H*)pyrimidinethione (**35**), when heated with ammonia at 85°C for 20 hours in a sealed tube, provided 2-arylamino-4,6-dimethylpyrimidine (**38**) (82CPB1942). Also, this reaction requires the intermediacy of a ring-opened product. When treated with methylamine, the ring-opened product could indeed be isolated.

The reaction of 1,4-diaryl-2-methylthiopyrimidinium iodide (**39**) with hydrazine did not lead to the corresponding 2-hydrazino compound, as expected, but to the rearranged 1-amino-2-arylaminopyrimidinium salt (**40**) (90JHC1441). This transformation also occurs via an S_N(ANRORC) process, involving the open-chain intermediate **41** (Scheme III.22). It cannot

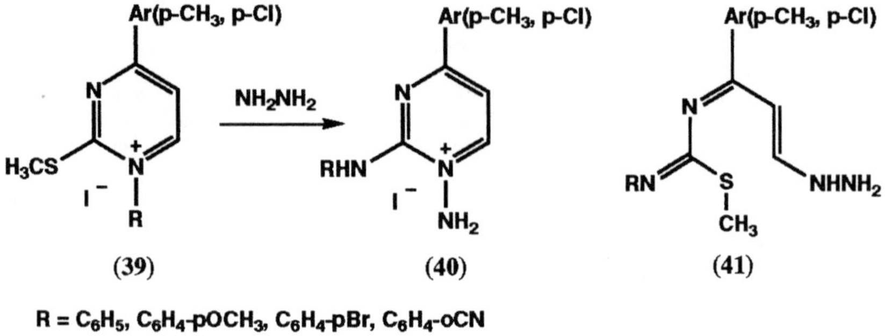

110 REPLACEMENT OF THE NITROGEN OF THE AZAHETEROCYCLE

Scheme III.22

R = C$_6$H$_5$, C$_6$H$_4$-pOCH$_3$, C$_6$H$_4$-pBr, C$_6$H$_4$-oCN

be excluded that first a hydrazinodemethylthiolation takes place at position 2 and that the 2-hydrazinopyrimidine formed undergoes a Dimroth-type rearrangement into the 1,2-diaminopyrimidinium salt.

Degenerate rearrangements have also been found during hydrazinolysis of 2-thioalkyl-3-phenylquinazolin-4(3*H*)-one (**42**). The product obtained is 2-anilino-3-aminoquinazolin-4(3*H*)-one (**43**) (85JHC1535) [and not, as previously reported, the isomeric 2-hydrazino-3-phenylquinazolin-4(3*H*)-one (70IJC1055)]. The reaction course of this transformation is illustrated in Scheme III.23. This 2-hydrazino compound can, however, be isolated (together with **43**) when hydrazine reacts with 2-thioxo-3-phenylquinazolin-4(1*H*,3*H*)-one (64ZOK1745; 85JHC1535).

Quite similarly, the reaction product obtained from 2-benzylthio-3-phenylquinazolin-4(3H)-thione and hydrazine should be assigned the structure 2-anilino-3-aminoquinazoline-4(3*H*)-hydrazone instead of the reported 2-hydrazino-3-phenyl quinazoline-4(3*H*)hydrazone (Scheme III.23) (70IJC1055).

Hydrazine-induced rearrangement has also been observed with 3-phenyl-2,4(1*H*,3*H*)-quinazolinedione, 3-amino-2,4(1*H*,3*H*)-quinazolinedione being obtained (Scheme III.24) (84JHC1403). It is suggested that the reaction proceeds by attack of hydrazine on the amino carbonyl group at C-4 (84JHC1403). If the phenyl group is absent only under very drastic conditions, the amino group could be introduced at C-3 in 2,4-(1*H*,3*H*)quinazolinedione. The activating role of the phenyl group has been ascribed to its electron-withdrawing character enhancing the electrophilicity of the adjacent carbonyl group at C-4, facilitating its reaction with nucleophiles.

The rearrangement of 3-dialkylaminomethyl-6*H*-imidazo[1,2-*c*]quinazolin-5-one into its 2-dialkylaminomethyl isomer on refluxing in acidic methanol is an interesting example of a degenerate ring transformation (96H2607) (Scheme III.25). In the isomerization, the nitrogen at position 4 is "interchanged" with the N-1 of the imidazole ring. The reaction can be

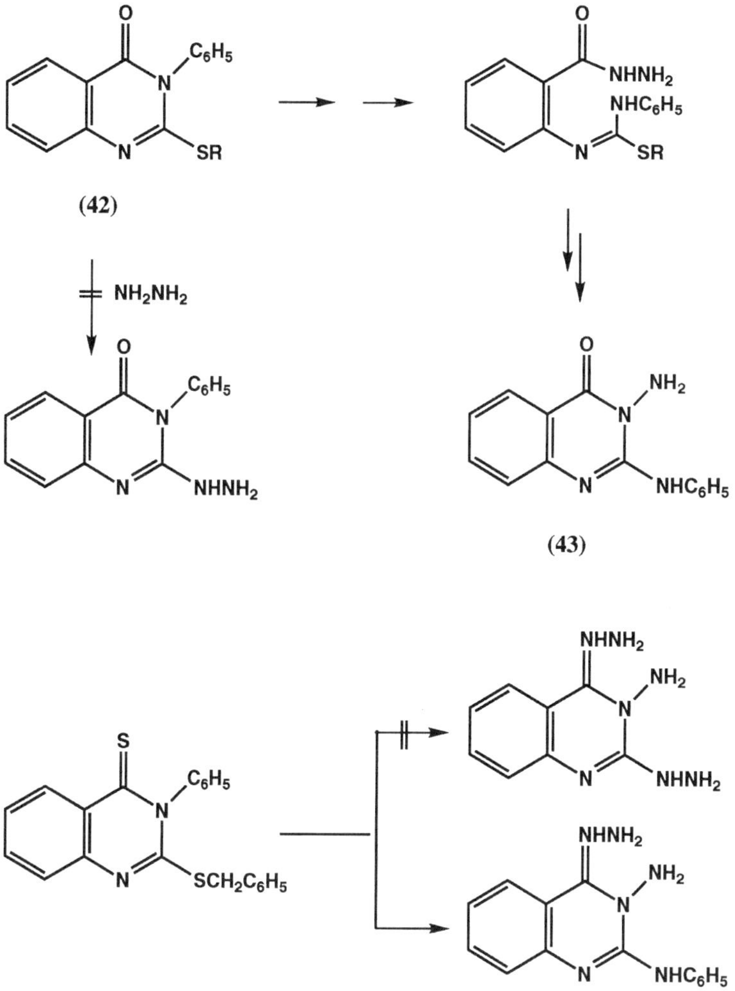

SCHEME III.23

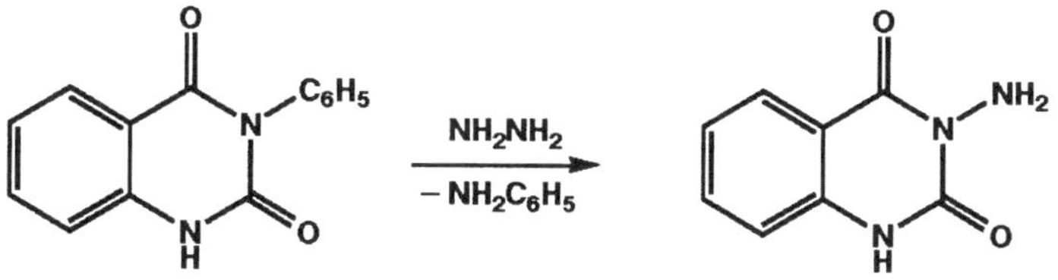

SCHEME III.24

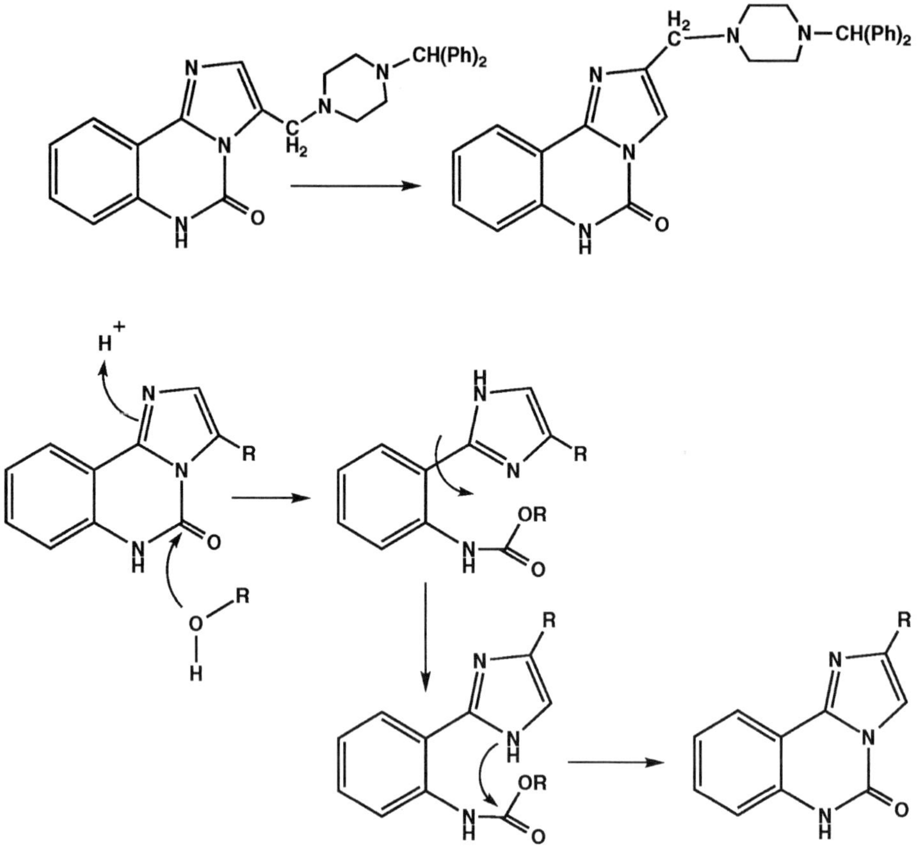

SCHEME III.25

described as a solvolytic attack at the carbonyl function and a subsequent ring opening in which the imidazole ring acts as an "internal" leaving group. Rotation around the phenyl–imidazole single bond and ring closure by N-1 leads to the formation of the 2-substituted product. The isomerization is irreversible, leading to the conclusion that the 2-substituted product is thermodynamically more stable. This was ascribed to a 1,3-allylic strain in the 3-substituted product, due to the eclipsing position of the carbonyl group and the substituted methyl group at C-3 (89CRV1841). In the 2-substituted product this steric strain is considerably lower.

d. *N-Nitropyrimidones*

Interesting examples of degenerate ring transformations have been described in the preparation of specific [3-^{15}N]pyrimidine nucleosides and [1-^{15}N]purine nucleosides when reacting the corresponding *N*-nitropyrimidines (or *N*-nitropurines) with ^{15}N-labeled ammonia, alkylamines, and hydrazine (95JA3665).

A solution of 5'-*O*-acetyl-2'3'-*O*-isopropylidene-3-nitrouridine [**44**, prepared from 5'-*O*-acetyl-2'3'-*O*-isopropylideneuridine and nitronium trifluoroacetate (91JOC7038)] in acetonitrile gave, when treated with an acetonitrile–water solution of ^{15}N-ammonium choride, potassium hydroxide, and triethylamine for 5 days, 5'-*O*-acetyl-2'3'-*O*-isopropylideneuridine (**45**) in 73% yield. It contained the ^{15}N label at position 3. Deprotection of the *O*-acetyl and *O*-isopropylidene groups by methanol/hydrochloric acid gave [3-^{15}N]uridine (**46**) (Scheme III.26).

It is evident that because of the presence of the strong electron-withdrawing nitro group at N-3, the vicinal positions C-2 and/or C-4 are activated for nucleophilic addition. The incorporation of the ^{15}N label can in principle be considered to occur via an ANRORC process, which is initiated by addition at C-4 or C-2. NMR spectroscopic evidence convincingly showed that the initial nucleophilic addition takes place at C-4 and *not* at C-2 (95JA3665). When compound **44** reacts with benzylamine, a ring-opened intermediate could be isolated that is stable at room temperature but undergoes ring closure on heating with potassium hydroxide in acetonitrile at about 60–70°C. ^{13}C spectroscopy of the open-chain intermediate obtained from **44** and [^{15}N]benzylamine as well as from 5'-*O*-acetyl-2'3'-*O*-isopropylidene-3-nitro[3-^{15}N]uridine (**44***) with unlabeled benzylamine shows that the intermediate is a salt, to which structure **47** was assigned (Scheme III.27). It unequivocally proves that the initial addition of benzylamine to 3-nitrouridine has taken place at position 4, and that the ring is opened by fission of the N(3)–C(4) bond. The ring closure of **47** into the 3-benzyl[3-^{15}N]uridine derivative **48** easily takes place, since nitroamine (or

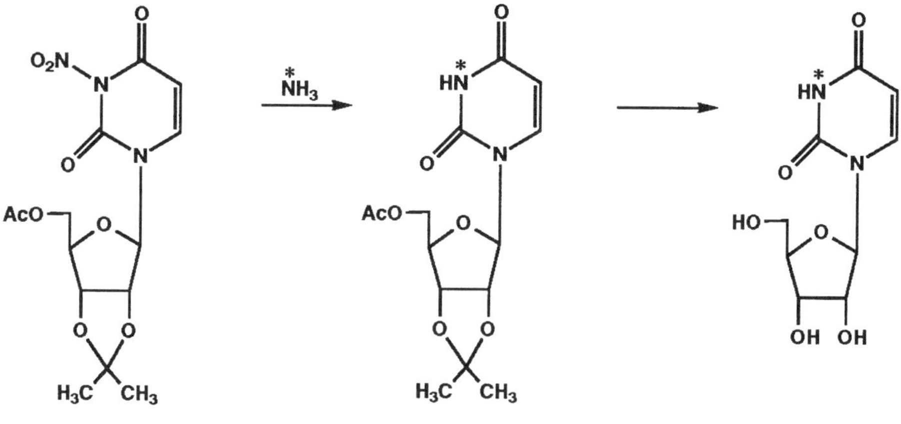

Scheme III.26

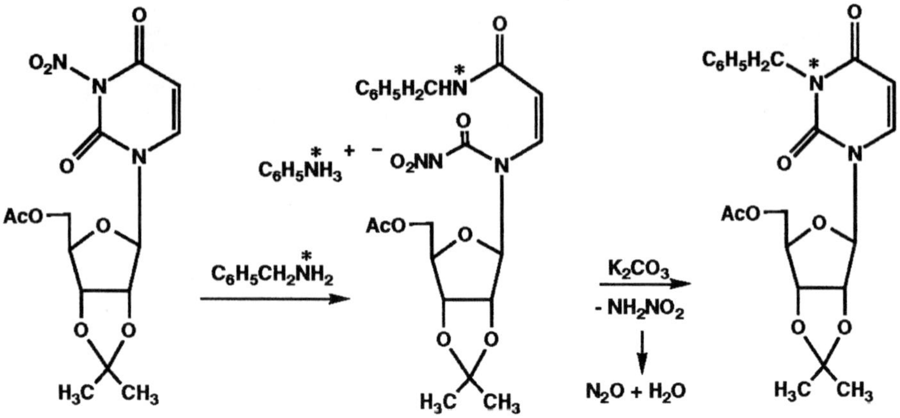

SCHEME III.27

its anion) is an appropiate leaving group because of its easy conversion in nitrous oxide and water [83JCS(D)261].

Reaction of **44** with hydrazine also leads to a facile replacement of the 3-nitro group by an amino group, yielding in almost quantitative yield 5'-O-acetyl-3-amino-2'3'-O-isopropylideneuridine (**50**) (Scheme III.28). Although several methods for the preparation of N-aminouridines from uridines are known (75CPB844; 92JHC1133), the indirect aminodenitration by a degenerate ring transformation reaction provides us with a very elegant method to introduce an amino group at N-3 and at the same time a ^{15}N label in the pyrimidine ring. Reaction, for example, with [^{15}N$_2$]hydrazinium hydrogen sulfate and potassium hydroxide affords the double-labeled compound 5'-O-acetyl-2',3'-O-isopropylidene-3-(^{15}N-amino) (3-^{15}N) aridine (**50***) (Scheme III.28).

That compound (**50**) is indeed an N-amino derivative (featuring a N–NH$_2$ arrangement), and not a seven-membered heterocycle triazepine (characterized by an NH–NH arrangement), was clearly proven by ^{1}H and ^{15}N spectroscopy. Based on this result it is evident that it is the α-nitrogen of the moiety (C(O)N$_\alpha$H–N$_\beta$H$_2$) in the open-chain carbohydrazide **49** that is involved in the ring closure. It is suggested that the β-nitrogen in **49** is largely protonated by the rather acidic hydrogen of the H–NNO$_2$ group, forming an intramolecular hydrogen bridge (Scheme III.28).

Similar reactions were also observed with 1-nitropurine derivatives. Treatment of 2',3',5'-tri-O-acetyl-1-nitroinosine (**51**) with ^{15}N-labeled ammonia gave [1-^{15}N]-2',3',5'-tri-O-acetylinosine (**52**), which after deprotection yielded [1-^{15}N]inosine (**53**) (Scheme III.29) (95JA3665).

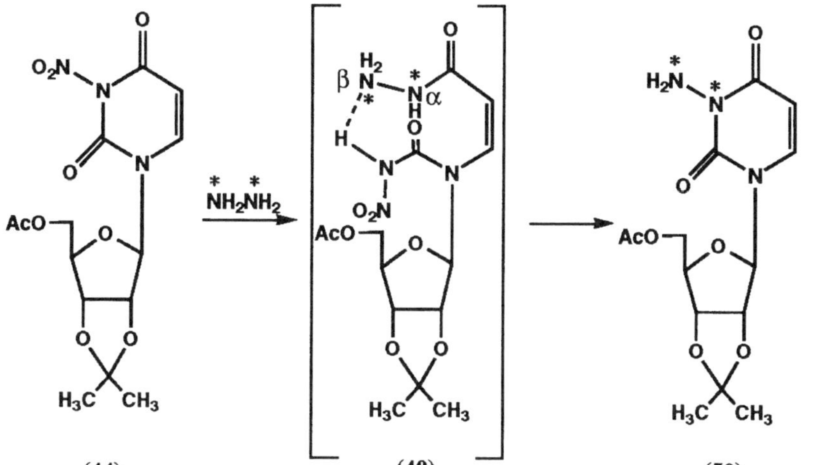

S_N(ANRORC) REACTIONS IN AZAHETEROCYCLES 115

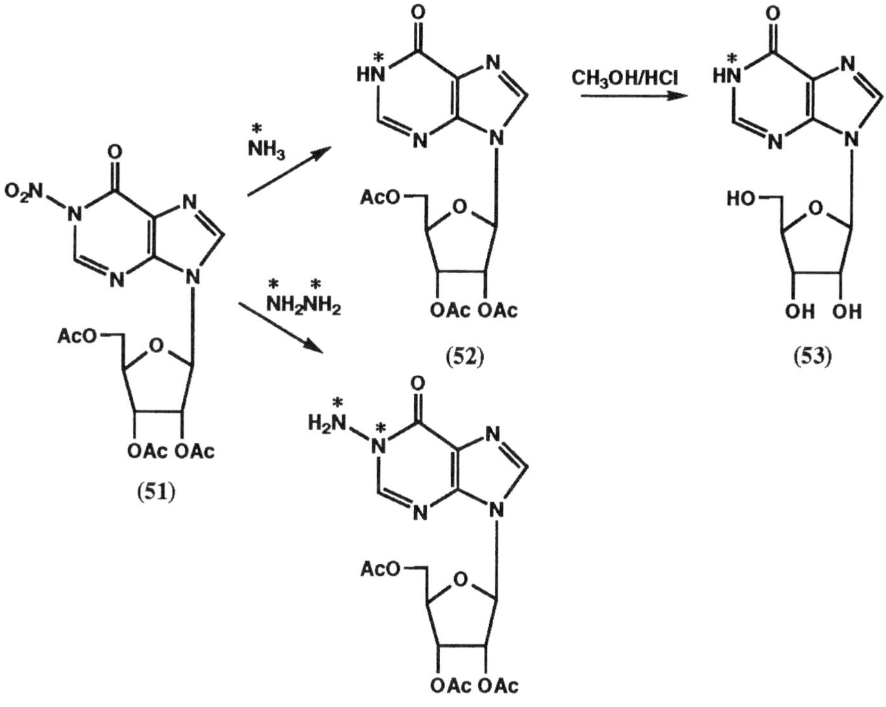

SCHEME III.28

SCHEME III.29

116 REPLACEMENT OF THE NITROGEN OF THE AZAHETEROCYCLE

With [$^{15}N_2$]hydrazinium hydrogen sulfate and potassium hydroxide, the 2',3',5'-tri-*O*-acetyl-1-(^{15}N-amino) (3-^{15}N) inosine **54** is obtained (Scheme III.29). The reaction follows the same reaction pathway as described in Scheme III.28; addition of the nucleophile at C-6, ring opening between C(6) and N(1), and ring closure with elimination of nitrous oxide and water. This S_N(ANRORC) reaction provides us with an good entry to ^{15}N-ring-labeled purines.

The replacement of the N–NO$_2$ group by an R–NH$_2$ has been found to occur more smoothly with the 1-nitroinosines (**51**) than with the 3-nitrouridines (**44**). Reaction of **51** with methylamine gave ring opening al-

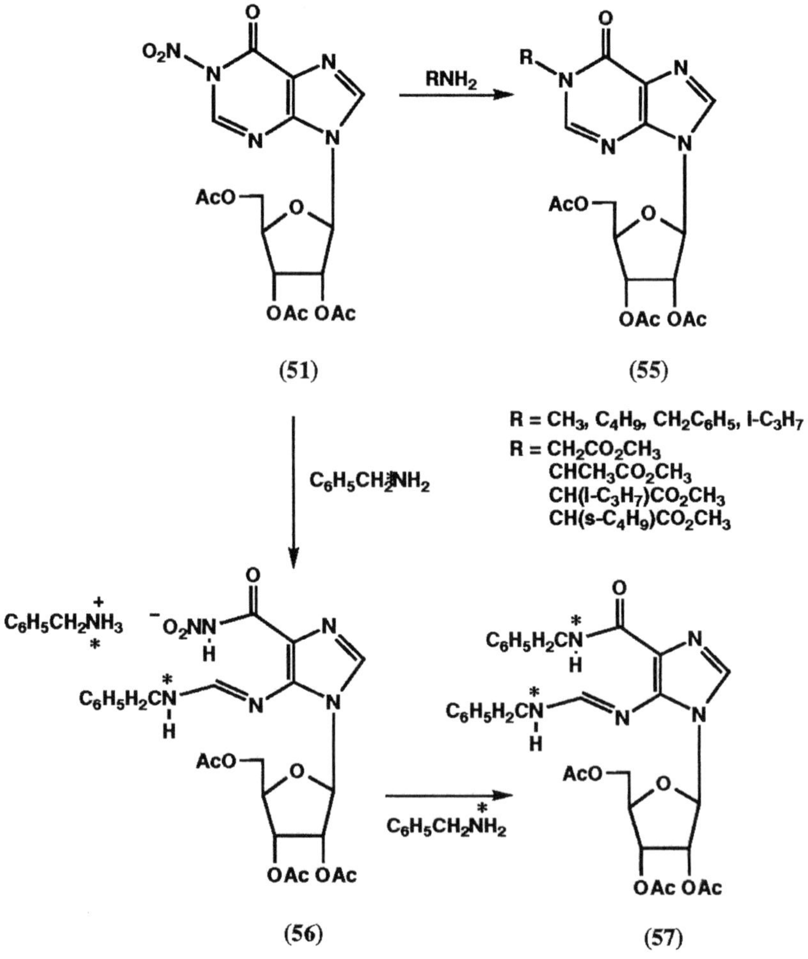

R = CH$_3$, C$_4$H$_9$, CH$_2$C$_6$H$_5$, I-C$_3$H$_7$
R = CH$_2$CO$_2$CH$_3$
CHCH$_3$CO$_2$CH$_3$
CH(I-C$_3$H$_7$)CO$_2$CH$_3$
CH(s-C$_4$H$_9$)CO$_2$CH$_3$

SCHEME III.30

ready at $-78°C$, and at room temperature the ring-opened product underwent a smooth ring closure into the 1-methylinosine derivative (**55**).

Reaction with all kinds of alkylamines and arylamines and with amino acid esters easily leads to replacement of the N–NO$_2$ group. As can be expected, when chiral amino acid esters are used as reagents, the reaction occurs with retention of the configuration (Scheme III.30). See also the discussion on similar reactions with *N*-nitroimidazoles in Section III,A,5.

In a more detailed study of the reaction of **51** with butylamine and benzylamine, it was observed that trifluoroacetic acid was beneficial for the ring closure of the open-chain compounds, yielding the 1-t-butylinosine and 1-benzylinosine derivatives. Based on NMR spectroscopic studies, it was established that in the intermediate, obtained from 2',3',5'-tri-*O*-acetyl-1-nitroinosine with [^{15}N]benzylamine, the N–NO$_2$ group is attached to the C(O) group, i.e., **56**, justifying the conclusion that the attack of the benzylamine (and presumably that of the other amines) takes place at C-2 in the 1-nitroinosines. Further reaction of (**56**) gives the dibenzylamino compound **57**.

e. *Photostimulated Degenerate Ring Transformations of Thymines*

Interesting examples of degenerate ring transformations have been observed when an aqueous solution of thymine (**58**, R = H) or thymidine (**58**, R = desoxyribose) is irradiated (>254 nm) in the presence of *n*-butyl- or *t*-butylamine at ambient temperature (81JA1598; 81TL2365). The 1-R-thymines (R' = *n*-C$_4$H$_9$, *t*-C$_4$H$_9$) are obtained in good yields. Urea derivatives (**59**) are the primary photoproducts; they have been isolated when the

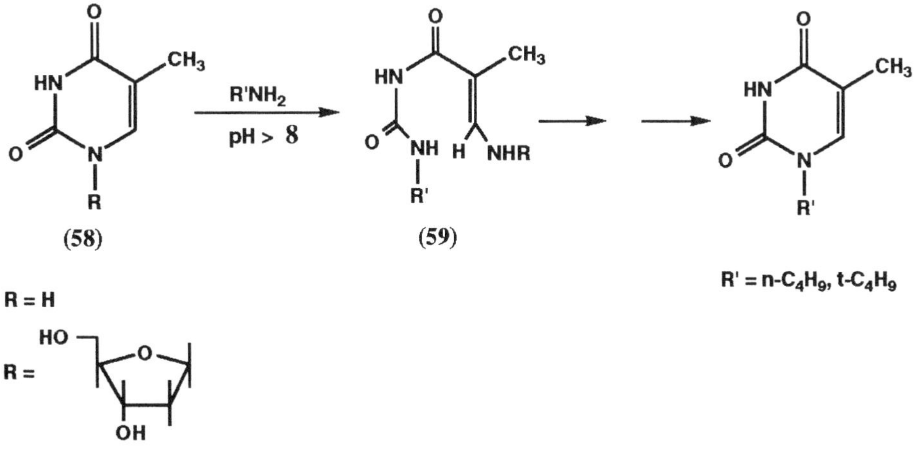

SCHEME III.31

photoreaction is carried out at low temperature (0–5°C). Brief heating of the urea derivatives in water at about 70°C or acid treatment gave the final product (Scheme III.31).

The photoreaction was found to occur only when the reaction was carried out in an aqueous solution at pH > 8. This seems to indicate that an ionized form of the photoexcited species is involved in the photoreaction. Furthermore, it is suggested that this photoinduced ring transformation reaction occurs via the singlet state (**58***) and not via the triplet state. This suggestion came from the experimental result that no ring transformation was observed, but there was an intermolecular cycloaddition into a dimer, when the photoreaction was carried out in acetone (81TL2365). In this anionic singlet state the addition takes place at C-2 (and not, as was originally suggested, at C-6), followed by ring fission of the C(2)–N(1) bond into the urea derivative **59**. This ring-opened compound (**59**) undergoes cyclization into the final product via the intermediacy of 6-alkylamino-5,6-dihydrothymine (**60**) (Scheme III.32). Nucleophilic addition to C-2 in ground-state molecules of pyrimidines is less documented than addition at C-(4,6), but it is unknown which position (C-2 or C-6) in the *excited* state of pyrimidine and its derivatives is favored for nucleophilic addition.

The photostimulated nitrogen exchange has been successfully employed for synthesizing a variety of N-1 substituted thymines. For example, reaction of thymidine with 1,3-diaminopropane leads to the formation of 1-amino-4-(1-thymyl)propane (Scheme III.33), and similarly with lysine the thymyl group is introduced at the ε-position of lysine. Of interest is also the

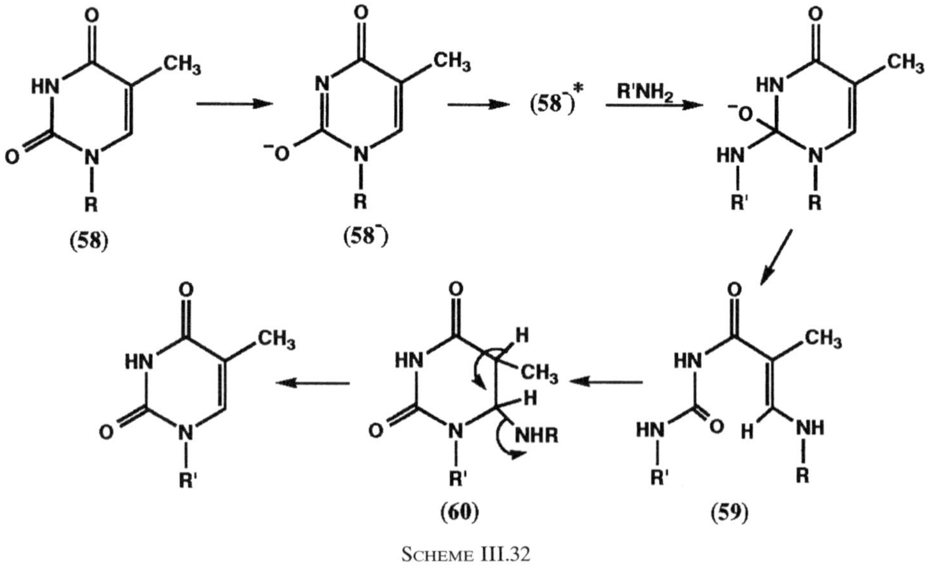

SCHEME III.32

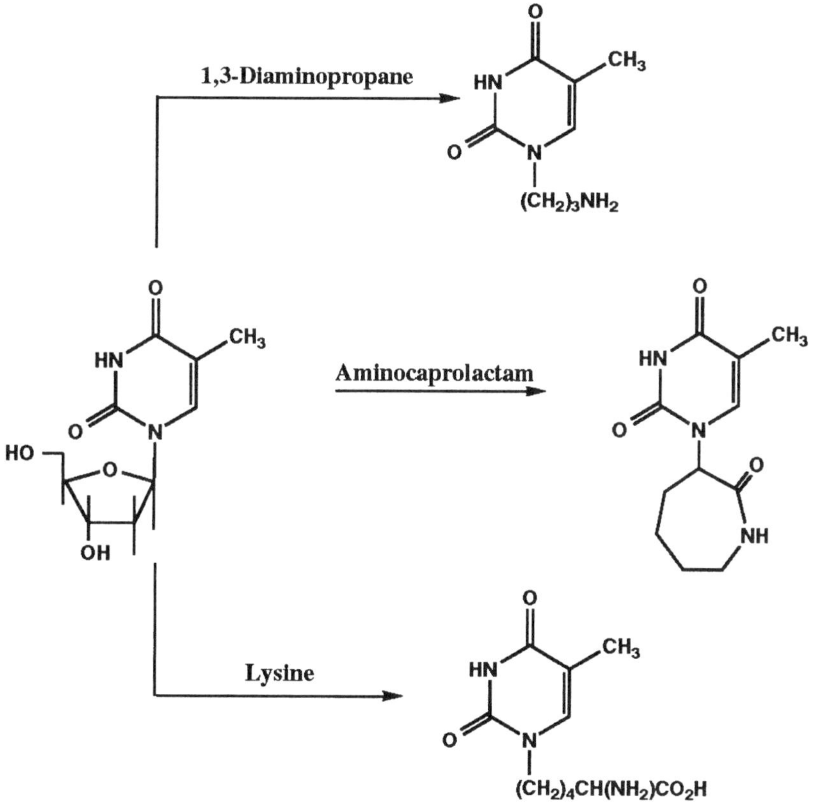

SCHEME III.33

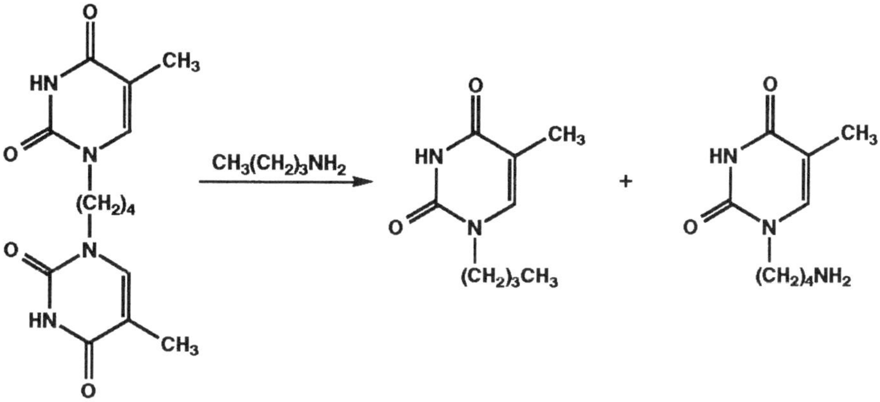

SCHEME III.34

introduction of the thymyl group at position 3 in azepin-2-one, when thymidine reacts with aminocaprolactam (81TL2365).

Another interesting reaction is the conversion of 1,4-di(1-thymyl)butane into a mixture of 1-butylthymine and 1-(4-aminobutyl)thymine when reacted with *n*-butylamine (Scheme III.34).

A further exploration of these photoconversion reactions has shown that dAdo, dGuo, and dCyt do not give any detectable photoadduct (81MI1). Irradiation of a mixture of thymidine and dAdo in *n*-butylamine only gave 1-*n*-butylthymine, suggesting that the photoreaction preferentially occurs with thymidine (thymine). In the light of these results it is not surprising that

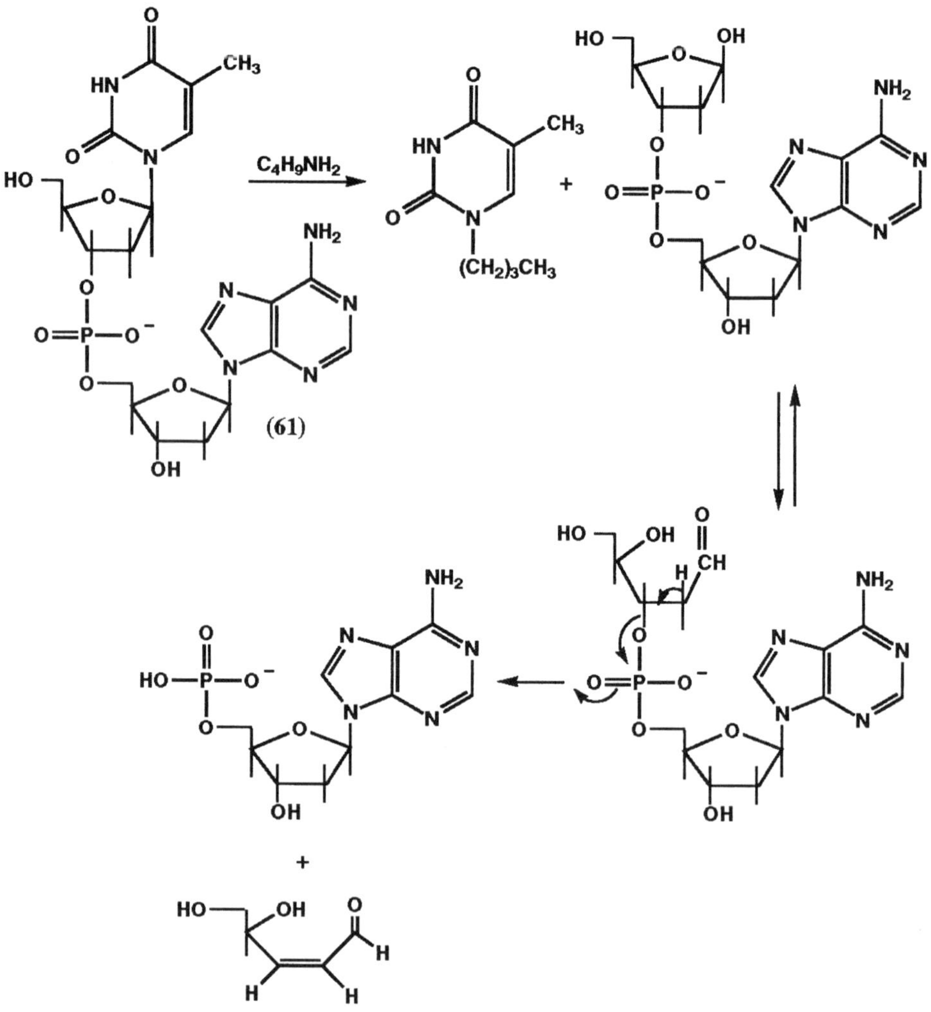

SCHEME III.35

a low-temperature irradiation (0°C) of an aqueous solution of the nucleotide TpdA (**61**), featuring both the thymine and the adenosine ring system, in the presence of *n*-butylamine gave a photoproduct that, after subsequent heating at 90°C, gives 1-*n*-butylthymine and dAMP. The formation of both products can be explained according to the route given in Scheme III.35.

These results seem to suggest that by irradiation in alkylamines DNA (single stranded) can undergo a selective release of thymine, accompanied by strand scission. This has indeed experimentally been verified. Irradiation of calf thymus DNA in 5% aqueous *n*-butylamine gave a photoproduct that on heating yields 1-*n*-butylthymine (81MI1).

3. *N*-Aryl-1,2,4-Triazinones

Hydrazine-induced degenerate ring transformation reactions are recorded (80JHC1733) with 3-thiooxo- or 3-thiomethyl-4-aryl-5-oxo-4,5-

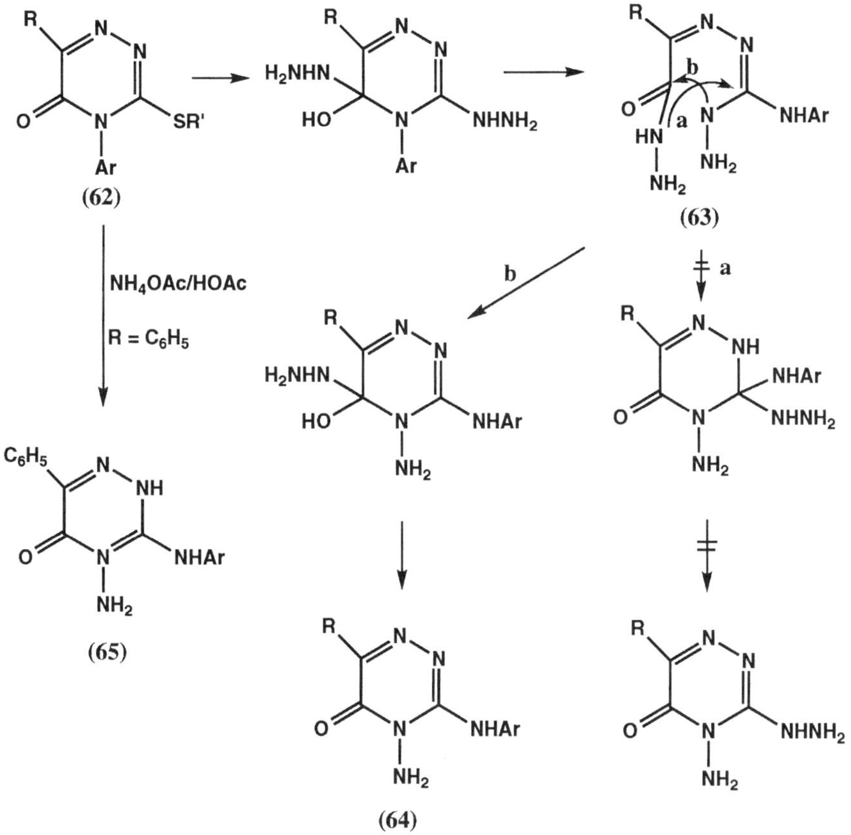

Scheme III.36

dihydro-1,2,4-triazine (**62**), giving the 3-arylamino-4-amino-5-oxo-4,5-dihydro-1,2,4-triazines (**64,** Scheme III.36). The reaction is assumed to proceed via the intermediacy of the hydrazino covalent adduct at C-5 and the ring-opened product (**63**). As illustrated in Scheme III.36, the ring closure of the open-chain compound can in principle take place either by pathway (a) or by pathway (b). Pathway (a) can, however, be excluded, since it could be expected that the 3-hydrazino compound, too, should have been formed, which in fact has not been observed. A similar rearrangement has also been found in the reaction of **62** with ammonium acetate/acetic acid; it affords 4-amino-3-arylamino-4,5-dihydro-5-oxo-1,2,4-triazine (**65**) (83H51). This *N*-aryl–amino "exchange" follows the same reaction pathway as described for the *N*-aryl–hydrazino replacement.

Hydrazination of 4-aryl-3,5-dioxo-2,3,4,5-tetrahydro-1,2,4-triazine (**66**) also affords a rearranged product, 4-amino-3,5-dioxo-2,3,4,5-tetrahydro-1,2,4-triazine (81JHC953). This replacement of the *N*-aryl group by the *N*-amino group requires the intermediacy of the ring-opened compound **67**. This intermediate could be isolated under mild conditions and its structure was proven by an independent synthesis. From this structure it can be concluded unequivocally that the ring transformation involves initial addition of the hydrazine at C-5 and not at C-3 (Scheme III.37).

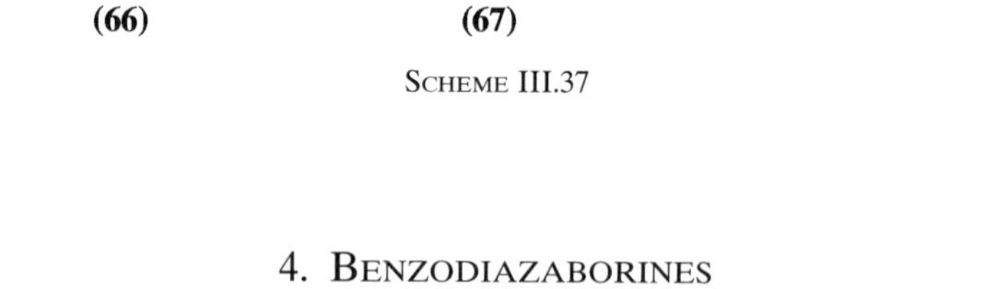

SCHEME III.37

4. BENZODIAZABORINES

It has been suggested (97JA7817) that the formation of 1,2-dihydro-1-hydroxy-2,3,1-benzodiazaborine (**70**) from its 2-(4-methylbenzenesulfonyl) derivative **68** with hydrazine may occur via a S_N(ANRORC) process. This degenerate transformation may involve as intermediate the benzene derivative **69** (Scheme III.38).

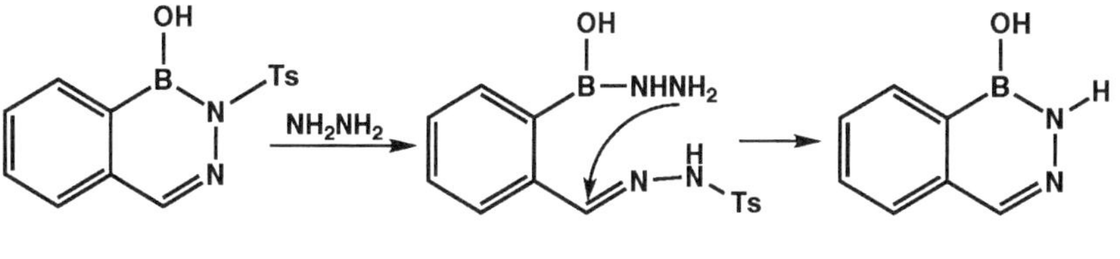

Scheme III.38

(68) → (69) → (70)

5. IMIDAZOLES

The occurrence of ANRORC-type reactions in nonactivated aromatic five-membered heterocycles has not been observed. The π-excessive character of these compounds makes a facile addition of a nucleophilic reagent, the initial step in this ANRORC-process, highly unlikely. However, when the π-excessive character is diminished by the presence of electron-attracting substituents such as nitro and arylsulfonyl groups, nucleophilic addition and substitution become possible. 1,4-Dinitroimidazole, for example, is highly susceptible to nucleophilic attack, and several sites are available: the nitrogen atom at position 1, and both positions C-4 and C-5 of the imidazole ring. The course of the reaction strongly depends on the kind of nucleophile and the reaction conditions applied (94H(37)1511). No experimental evidence is available for a nucleophilic attack at position 2.

In basic media, nucleophilic reagents (ZH), such as water, alcohols, and secondary amines, may react with 1,4-dinitroimidazole into 5-substituted 4-nitroimidazole. Thus, a base-catalyzed *cine*-substitution, initiated by addition of a nucleophile at C-5, takes place and nitrite is released (Scheme III.39) (91PJC323). With primary amines R–NH₂ or very weak alkaline aqueous–methanol solution, 1-R-4-nitroimidazoles are obtained with release of nitrous oxide (Scheme III.39) (90AP949, 90PJC813; 91PJC515; 91PJC1071; 92AP317; 95MI1). Used as primary amines were aminosulfonic acids and esters, α-aminoalkane carboxylic acids (at pH 9) and their esters, *m*-aminophenol, aminobenzoic acids and esters, and aminosugars.

The interesting question of whether we deal with a direct replacement of the nitro group by group R of the amine or by a two-atom exchange of the *N*-nitro moiety by an alkyl-amino group was solved by reacting each of the *S*- and *R*-enantiomers of an α-aminoalkane carboxylic acid (ester) with 1,4-dinitroimidazole and observing that the product, α-(4-nitro-1-imidazolyl)alkanecarboxylic acid (ester), is chiral and exhibits equal but opposite optical rotation (Scheme III.40) (91TA941).

This result proves unequivocally that the products were enantiomeric

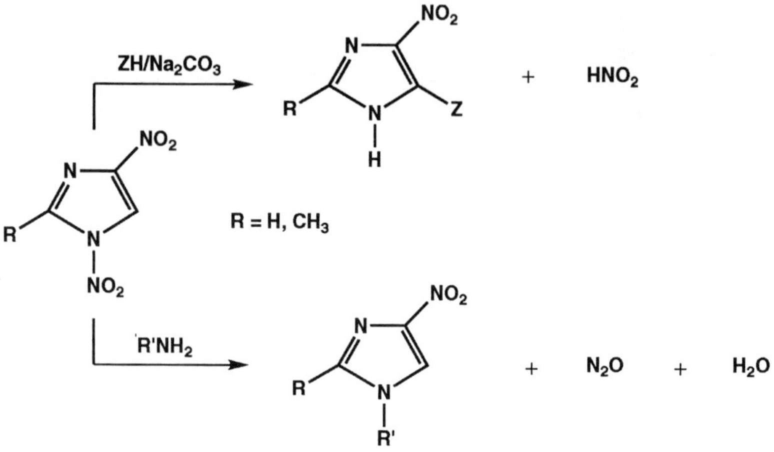

SCHEME III.39

R' NH$_2$ = ⁻SO$_3$CH$_2$CH$_2$NH$_2$; RCH(NH$_2$)CO$_2$H (Ester); m-NH$_2$C$_6$H$_4$OH; NH$_2$C$_6$H$_4$CO$_2$H(Ester);

and that the bond between amino-nitrogen and chiral carbon, present in the amino substrate, is not subjected to cleavage during the replacement reaction.

These results are further substantiated by ^{15}N-labeling studies. Reaction of 1,4-dinitro-2-methyimidazole with [^{15}N]glycine gives (2-methyl-4-nitro-

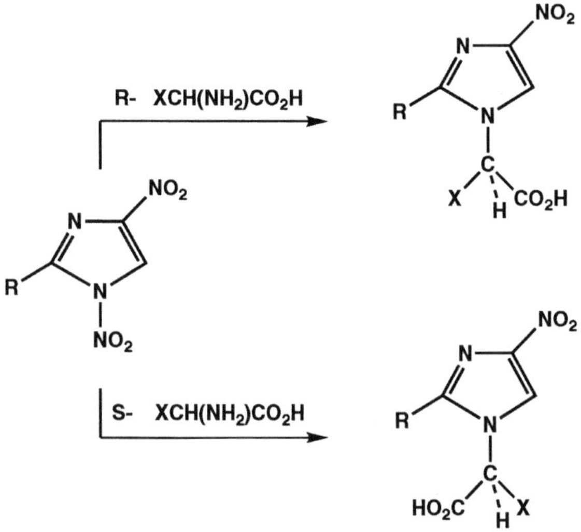

SCHEME III.40

1-imidazolyl)acetic acid (**71,** 70%), position 1 of which carries the ^{15}N label (as proven by mass spectrometry) (91MI3; 96MI1). This compound proved to be an appropiate starting material for the preparation of 2-methyl-4-nitro-[1-^{15}N]imidazole (**72**), a compound that could be utilized in investigations of the metabolism of 4-nitroimidazole drugs (Scheme III.41) (75MI2).

Attempts to obtain **72** directly by reaction with ^{15}N-labeled ammonia were only partly successful, as besides incorporation of the label into position 1 of the imidazole ring, i.e., **72,** a considerable part of the label was found to be present in the nitrous oxide, being formed by hydrolysis of the ^{15}N-nitroamide (96MI1). It means that during the reaction, as well as a nucleophilic addition of ammonia at position 5, attack has also taken place on the nitrogen of the 1-nitro group. Based on MNDO calculations (77JA4899), the four-center structure **74** was suggested as intermediate in this aminodenitration, explaining the partial formation of the unlabeled compound **73** (Scheme III.41).

Spectroscopic and kinetic studies were undertaken to get a better understanding of the mechanism of this degenerate rearrangement. From studies

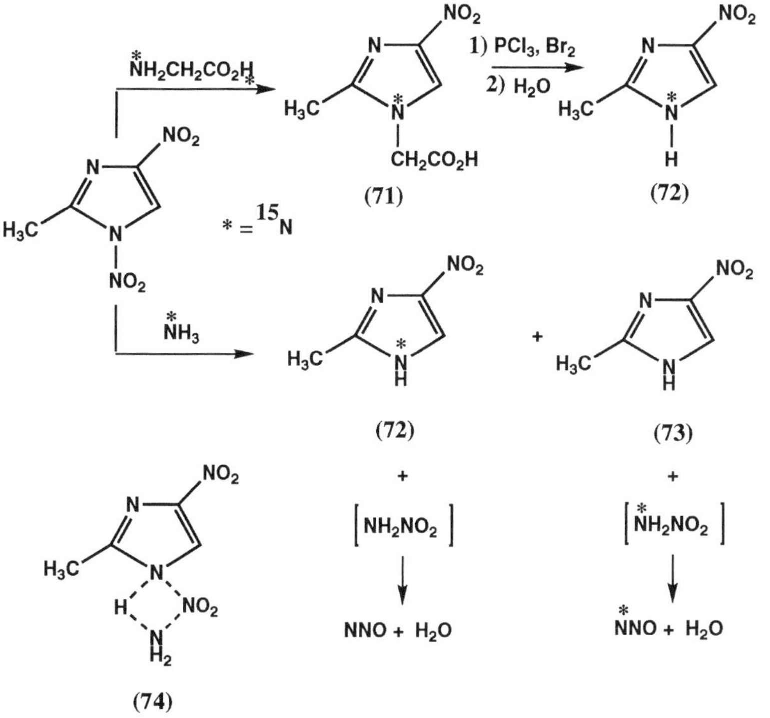

Scheme III.41

between 1,4-dinitroimidazole (and some of its methyl derivatives) and aromatic amines, the following results were obtained (92PJC943).

Reaction of 1,4-dinitro-2-methylimidazole with arylamines (X-$C_6H_4NH_2$; X = H, p-Cl, m-Cl, p-OCH$_3$) in organic solvents leads to the formation of relatively pure adducts, which absorb strongly in the spectral region beyond 370 nm. This long-wavelength band fades quickly in time, and a new band appears at 300–310 nm, characteristic of 1-aryl-2-methyl-4-nitroimidazoles. The UV-vis spectra run at certain time intervals show sharp isosbestic points at 280–320 nm; the changes in absorbance are in agreement with first-order reaction kinetics. The adduct formed between 1,4-dinitroimidazole and aniline showed in its ^{1}H NMR spectrum a multiplet at 7.0–7.7 ppm, which could not be ascribed to imidazole hydrogen signals (they appear over 8.0 ppm). With time the multiplet changes into two doublets at 8.43 ppm (J = 1.6 Hz) and 8.89 ppm (J = 1.6 Hz), characteristic of 1-phenyl-4-nitroimidazole.

Results from kinetic studies on the reaction of 1,4-dinitro-2-methylimidazole with aromatic amines (X-$C_6H_4NH_2$; X = H, p-Cl, m-Cl, p-OCH$_3$) are summarized in Table III.1. These kinetic studies lead to the reaction in Scheme III.42. It is assumed that $k - 1 = 0$, indicating that the intermediate product C is irreversibly formed; it has an acyclic structure. 1-Aryl-4-nitroimidazole is formed from this acyclic compound by a relatively slow intramolecular nucleophilic addition, followed by a fast elimination of nitroamide. Considering that k_1 and k_2 are characteristic for the two consecutive stages of the nucleophilic addition, they are consistent with those predicted by the Hammett relation: electron-withdrawing groups (m-Cl and p-Cl) retard the rate of addition (k_1) and cyclization (k_2), while for the p-methoxy group the opposite effect is observed.

All these results are consistent with the occurrence of an S_N(ANRORC) mechanism for the degenerate ring transformation involving the replace-

TABLE III.1

ESTIMATED EXPERIMENTAL REACTION RATE CONSTANTS FOR THE REACTION OF 1,4-DINITRO-2-METHYLIMIDAZOLE (A) WITH ARYLAMINES (B, X–$C_6H_4NH_2$) IN AQUEOUS METHANOL (1:1) AT 25°C

Substituent X	k_1 (dm^3/mol.s)	k_2 (s^{-1})
p-OCH$_3$	1.67 × 10^{-7}	1.28 × 10^{-2}
H	5.20 × 10^{-2}	6.40 × 10^{-3}
p-Cl	2.05 × 10^{-3}	5.25 × 10^{-3}
m-Cl	6.30 × 10^{-3}	3.50 × 10^{-3}

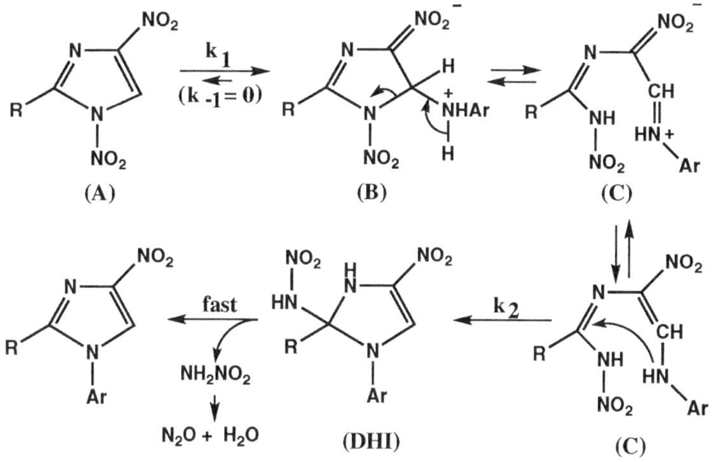

SCHEME III.42

ment of the *N*-nitro moiety at position 1 in 1,4-dinitroimidazoles by an *N*-aryl group.

ANRORC-type displacements have also been observed in reaction of 1-arylsulfonyl-4-nitroimidazoles with aniline (92PJC1623; 94T5741). On heating a methanol–water solution of 1-(*p*-toluene sulfonyl)-4-nitroimidazole (**75**) and aniline at 65–70°C, 1-phenyl-4-nitroimidazole (**76**, 42%) was obtained, together with a nearly equimolar amount of *p*-toluenesulfonamide (46%) (Scheme III.43). As by-product, 4-nitroimidazole was isolated, due to hydrolysis of the starting material. *p*-Toluenesulfonanilide was not formed, indicating that no nucleophilic attack of aniline on the sulfur of the sulfonyl group had taken place. However, when the *p*-methyl group in the phenyl ring was replaced by an electron-withdrawing group at the *para* position (*p*-NO$_2$, *p*-Cl), the formation of arylsulfonanilides was considerably favored, although the 1-aryl-4-nitroimidazoles are still the main products of the reaction.

Attempts to introduce an amino group at position 1 by replacing the *N*-nitro moiety in 1,4-dinitro-2-methylimidazole with hydrazine were not successful; ring expansion into 1,2-di-(3-methyl-1,2,4-triazin-5-yl)azine (**77**) was observed (Scheme III.44) (96T14905). It is beyond the scope of this book to discuss in detail the course of this quite interesting rearrangement.

Reaction of 1,4-dinitroimidazole with 1,1-disubstituted hydrazine, i.e., *N*-aminomorpholine, appeared to be more successful; it yields 1-(*N*-

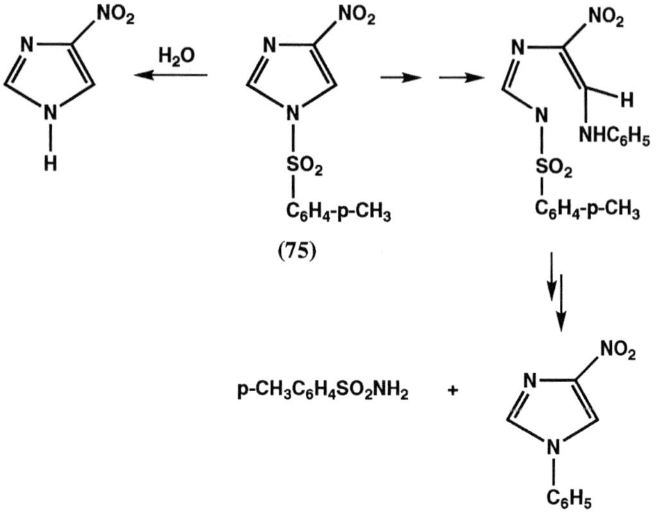

SCHEME III.43

morpholino)-4-nitroimidazole (**78**, 30%) (Scheme III.45). The mechanism follows the same reaction pattern as described before, involving addition at C-5 of the imidazole ring, ring opening, and ring closure [S_N(ANRORC mechanism)].

Hydrazine-induced rearrangements have been observed with imidazoles, which contain an *N*-arylcarboxamide function as part of the ring system. 3-Aryl-5-arylidene-2-methylthioimidazolin-4-one (**79**) reacts with hydrazine to produce 5-arylidene-3-amino-2-arylimino-imidazolidin-4-one

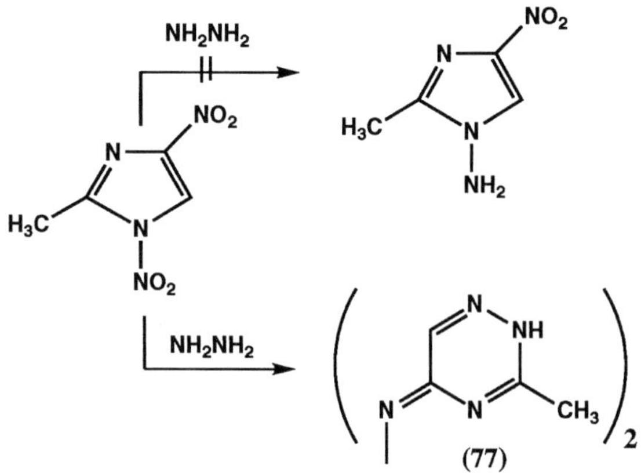

SCHEME III.44

(80) (Scheme III.46) (82JHC41). In case a hydrogen or methyl group is present on the nitrogen at N-3, i.e., **82,** this rearrangement does not occur, only replacement of the methylthio group by the hydrazino group being found.

An *N*-aryl/NH exchange was observed when compound **79** is fused with ammonium acetate, forming the imidazolidin-4-one **81** (85JHC155). This hydrazine-induced *N*-aryl/*N*-amino exchange follows the same reaction pathway as described with 3-phenylquinazolin-4-one (see Section III,A,2,c), phenyl-1,2,4-triazinones, and phenyl-1,2,4-triazinediones (see Section III,A,3). All these ring systems feature the presence of an *N*-arylcarboxamide functionality. If the aryl group is absent, rearrangement does not take place, or it requires severe reaction conditions. The essential role of the aryl group is ascribed to its electron-withdrawing character, which enhances the electrophilicity of the carbon of the adjacent carbonyl group.

B. Degenerate Ring Transformations Involving the Replacement of the C–N Fragment of the Ring by a C–N Moiety of the Reagent

A new variant of the S_N(ANRORC) substitutions was found in reactions of *N*-methylpyrimidinium salts with bifunctional nucleophiles, such as *S*-methylisourea, *O*-methylisourea, and cyanamide.

When 1-methyl-4-phenylpyrimidiniun iodide reacts with *S*-methylisourea, it is converted in the corresponding 2-amino-4-phenylpyrimidine (yield 70%). A similar conversion has been reported for 1-methyl-5-phenylpyrimidinium iodide, 2-amino-5-phenylpyrimidine being obtained in 40% yield (Scheme III.47) (76RTC209). In both reactions, no detectable amounts of 2-methylthio-4-phenyl- or 2-methylthio-5-phenylpyrimidines are ob-

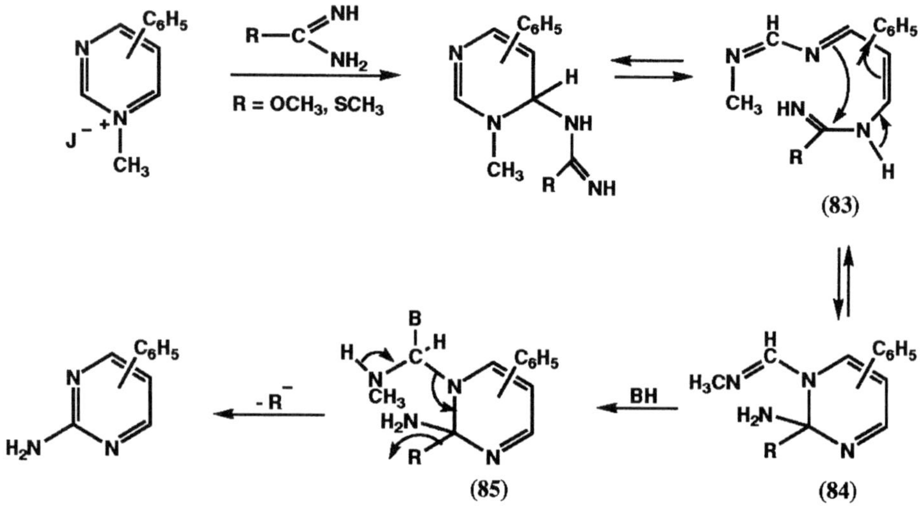

SCHEME III.47

served. When O-methylisourea was used as reagent, the same degenerate transformation took place. 2-Amino-4-phenylpyrimidine and 2-amino-5-phenylpyrimidine, respectively, were obtained, although the yields are considerably lower than with S-methylisourea (35 and 15%, respectively). No detectable amounts of 2-methoxy-4-phenyl- or 2-methoxy-5-phenylpyrimidine were found. It is evident that in all these conversions the C(2)–N(1) fragment of the pyrimidine ring is replaced by the C–N fragment present in both isourea reagents.

Concerning the mechanism of the rearrangement, it is proposed to involve the addition of the nucleophilic nitrogen of the 1,3-ambident nucleophiles to C-6 (see Section III,A,2,a). This covalent adduct is in equilibrium with the open-chain diamidino compound **83** (R = OCH_3, SCH_3). Cyclization into **84**, followed by a base-catalyzed fragmentation of the nitrogen–carbon bond at N-3 and expulsion of the thiomethyl or methoxy anion, yields the 2-amino-4(5)-phenylpyrimidines (Scheme III.47).

Based on this mechanism, one can expect that in reactions of both phenylpyrimidinium salts with cyanamide, the 2-amino compound would also be obtained. This has indeed been found (76RTC209). Reaction of both 4-phenyl- and 5-phenylpyrimidinium iodides with cyanamide gave the corresponding 2-amino compounds (60 and 35%, respectively). With guanidine no reaction occurs; no explanation was offered.

C. Degenerate Ring Transformations Involving Replacement of a Three-Atom Fragment of the Heterocyclic Ring by Three-Atom Reagent Moiety

In the reactions described in Section III,C, a three-atom fragment in the heterocycle is replaced by the same atoms of the reagent and with the same sequence of atoms. This type of reaction has been referred to as an *intermolecular transfragment* reaction (67JA6911) and therefore can be clearly distinguished from molecular rearrangements, such as the Dimroth reactions (68MI1; see also Chapter IV) and photochemical transposition reactions.

1. CCN AND CCC FRAGMENT REPLACEMENT

a. *Pyridines*

In pyridine chemistry interesting examples are reported of degenerate ring transformations occurring by replacement of either the C–C–N or the C–C–C part of the pyridine ring.

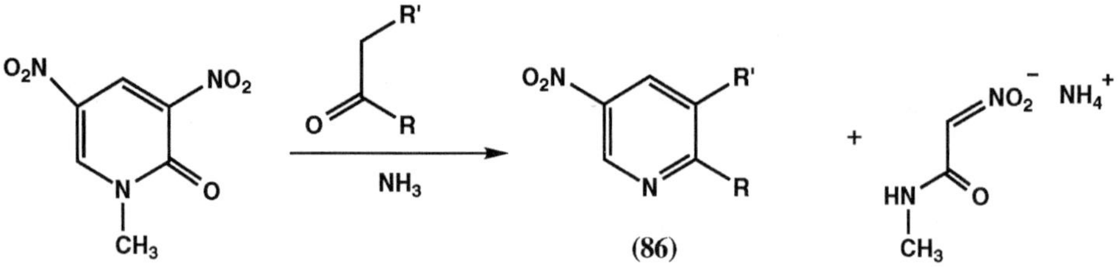

SCHEME III.48

An elegant method for the preparation of 3-nitropyridines (**86**) with alkyl- and/or aryl substituents at either the 5- or 6-position is based on the replacement of the C(3)–C(2)–N part of 3,5-dinitro-1-methylpyridin-2(1*H*)-one (90BCJ2830). Reagents are a ketone, providing two carbon atoms, and ammonia, the source for nitrogen. This degenerate transformation was performed on heating a methanolic solution of 3,5-dinitro-1-methylpyridin-2(1*H*)-one with ketones, having α-methyl or α-methylene hydrogens, in the presence of ammonia. As a by-product in this reaction is formed the ammonium salt of *N*-methyl α-nitroacetamide (Scheme III.48).

A variety of aliphatic and aromatic ketones have been investigated (cyclohexanone, cyclopentanone, acetophenone, *p*-nitroacetophenone) and they all usually lead to 2-R'-3-R"-5-nitropyridine in moderate to good yields. Better yields were recorded when enamines of the respective ketones were used. With aldehydes the yields are in general lower than with ketones.

In order to clarify the reaction mechanism, 6-deuterio-labeled 3,5-dinitro-1-methyl-2-pyridone (**87**, 1 mmol) was prepared and heated with cyclohexanone (2 mmol) and ammonia (10 mmol). A mixture of 2-deuterio- (**88**) and 4-deuterio-3-nitro-5,6,7,8-tetrahydroquinoline (**89**) was obtained in the ratio 58:42 (Scheme III.49).

It has been proposed that two isomeric *meta*-bridged intermediates are involved. They are formed by the consecutive addition of the ammonia at C-6 and the α-carbon of the ketone at the C-4 (path A, Scheme III.50)

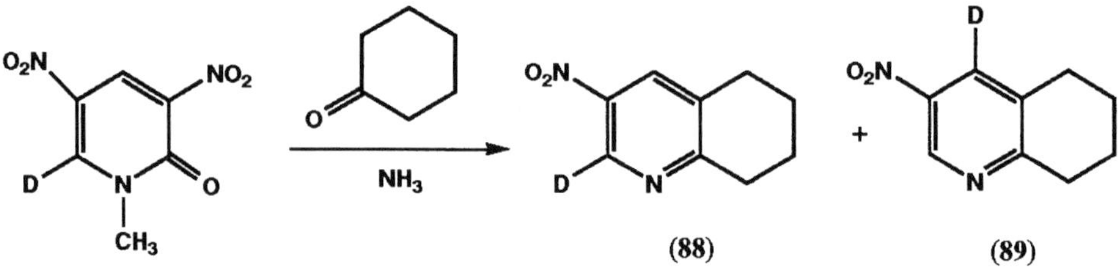

SCHEME III.49

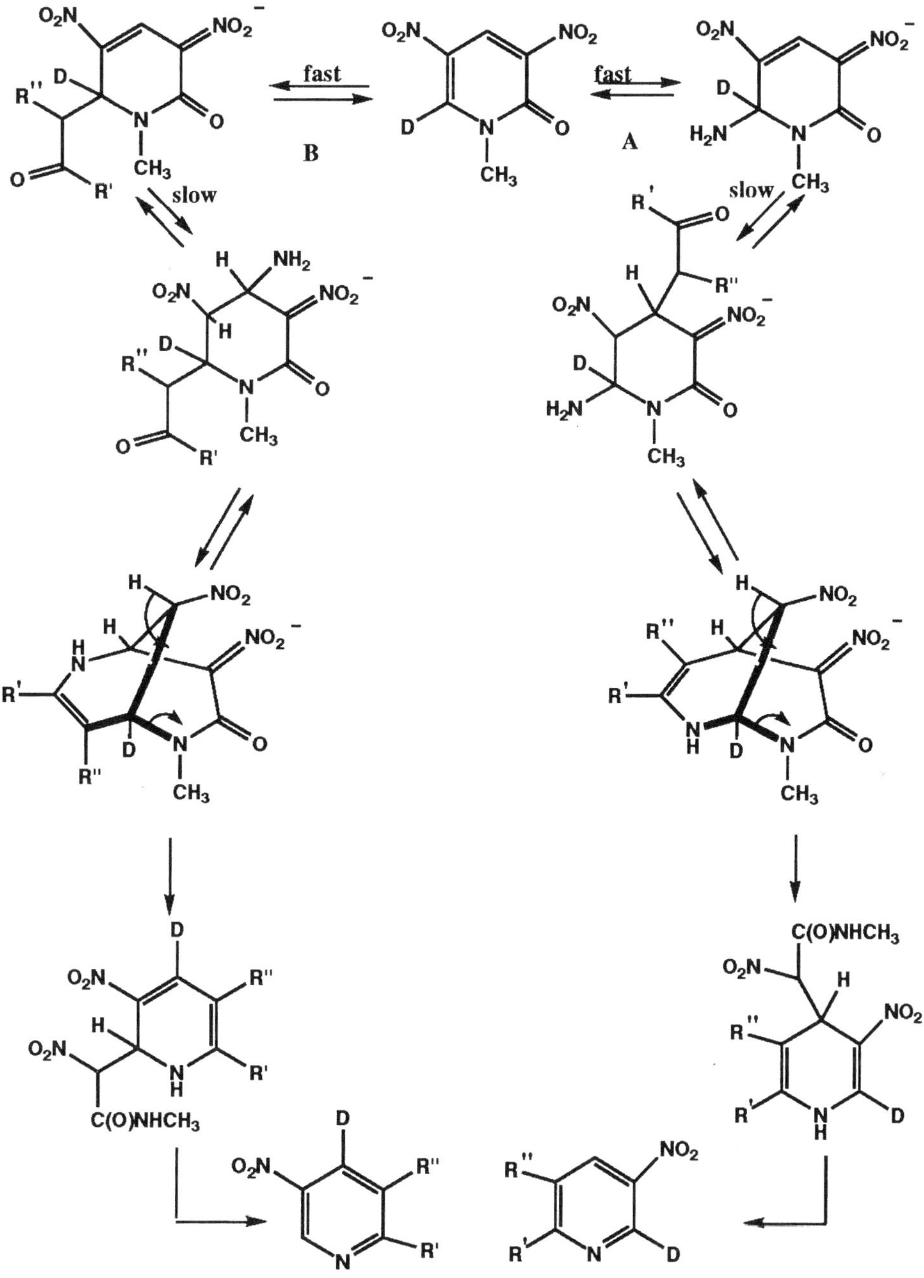

S_N(ANRORC) REACTIONS IN AZAHETEROCYCLES   133

Scheme III.50

or alternatively by the addition of the α-carbon of the ketone at C-6 and ammonia at C-4 (path B). Based on the preferred structure of the Meisenheimer adducts, formed between C-, N-, and O- nucleophiles and electron-deficient aromatics (83CL715; 84MI3), it can reasonably be expected that the first addition takes place at C-6 and that this first addition is certainly faster than the second addition, since the formation constants of the *meta*-bridged adducts are generally smaller than those of the Meisenheimer adducts [69JCS(B)330; 70CRV667]. The predominant formation of the 2-deuterio-3-nitro compound indicates that the rate-determining second addition step with the C-nucleophile is faster that with ammonia (Scheme III.50).

A somewhat similar degenerate ring transformation involving the replacement of the three-atom C(3)- C(2)-N fragment of the pyridine ring is the rearrangement of 3-nitro-4-methylpyridinium salt (**90**) into 1,2,3,4-tetrahydro-2,2-dialkyl-4-nitromethylenepyridine (**94**) when it undergoes a reaction with dialkyl ketones in the presence of methylamine (87KGS488; 92KGS792). The reaction pathway has been proposed to occur according to

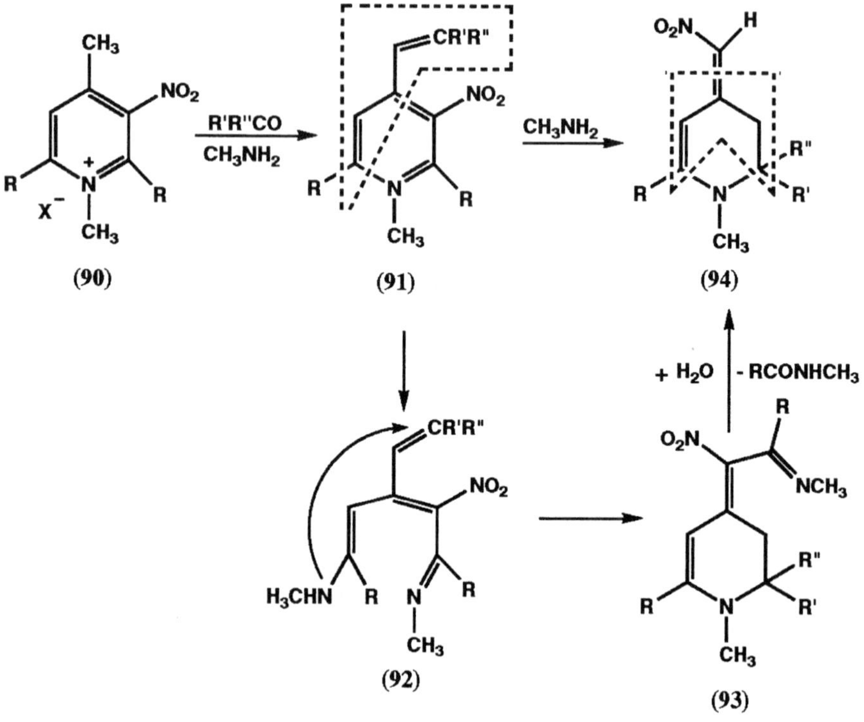

SCHEME III.51

Scheme III.51, although a more detailed investigation is required to establish this mechanism more firmly. It initially involves condensation of the carbonyl group in the ketone with the activated methyl group at C-4, i.e., formation of **91,** and is followed by an ANRORC process, in which after addition of the nucleophilic methylamine and ring opening into **92,** ring closure occurs into the pyridine derivative **93,** containing three atoms of the starting pyridinium salt, the original two carbons of the side chain at position 4 in **91** and the nitrogen of the methylamine group. Carbon atom C-3, being connected to the nitro group, is built into the exocyclic nitromethylene side chain; during the workup the N-methylcarbonamide of the corresponding carboxylic acid is split off.

The course of the reaction is determined to a considerable extent by the structure of the starting material and the conditions under which the reaction takes place. So, for example, reaction of **90** (R = H) with acetone (R' = R" = H) gives instead of the pyridine derivative **94** (R = R' = R" = H) a benzene derivative (88KGS1213). In general N-methyl-3-nitropyridinium salts show great versatility in their reaction behavior. Ring transformation into benzene derivatives is a well-studied subject and the reader is referred to the original literature (92KGS792).

b. *Pyrimidines*

Smooth replacements of the N(3)–C(4)–C(5) fragment of the pyrimidine ring by the N–C–C moiety of a 1,3-ambident nucleophile occurs when a 1-R', 3-R"-5-nitrouracil (**95,** R' = R' = CH_3; R' = c-C_6H_{11}, R" = CH_3; R' = CH_3, R" = c-C_6H_{11}) reacts with malonamide in ethanolic sodium ethoxide solution; it affords 1-R'-5-carbamoyluracil (Scheme III.52) [81TL2409; 84JCS(PI)1859]. This intermolecular transfragment reaction can be described as an ANRORC process, in which the carbanion of malonamide attacks position 6, yielding adduct **97.** Abstraction of the exocyclic α-proton from the malonamide part in **97** and a retro-Michael reaction involving scission of the C(5)–C(6) bond gives the open-chain intermediate **98.** Cyclization affords 5-carbamoyluracil (**96**) and N-methyl-α-nitroacetamide.

The reaction was found to fail with 1-methyl-5-nitrouracil and 5-nitrouracil. This is because both uracils, containing dissociable protons, exist in the basic solution in the anionic forms, which inhibit the addition of the nucleophiles at the C-6 position. Attempts to bring about these pyrimidine-to-pyrimidine ring transformations with cyanoacetamide, acetoacetamide, and phenylacetamide were not successful. A substituent at C-6 of 5-nitrouracil suppressed the reaction: 1,3,6-Trimethyl-5-nitrouracil was recovered almost quantitatively.

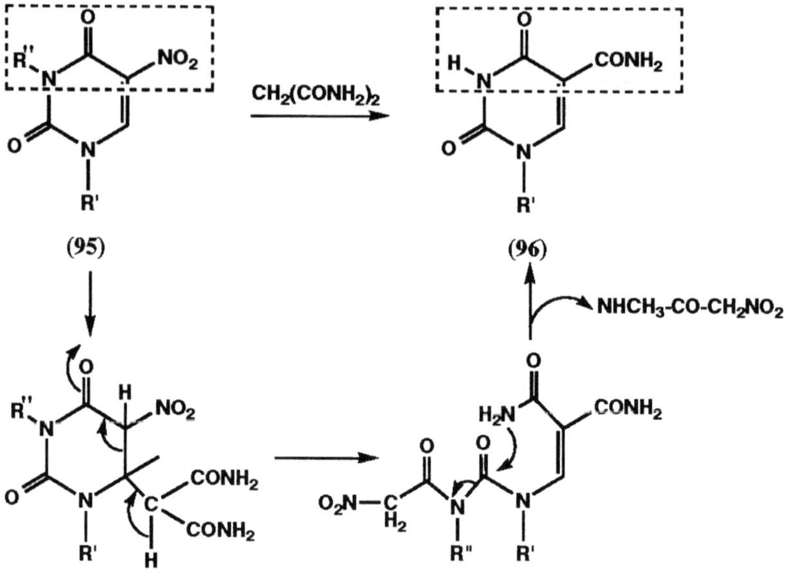

136   REPLACEMENT OF A THREE-ATOM FRAGMENT

SCHEME III.52

2. REPLACEMENT OF A CCC FRAGMENT

In only a few cases are ring transformation reactions found in which a intermolecular CCC transfragment reaction occurs. An example of this rearrangement has been observed in the reaction of 3,5-dinitropyridin-4(1H)-one (**99**) with diethyl sodio-3-oxopentanedioate. It provided in good yield 3,5-di(ethoxycarbonyl)pyridin-4-(1H)-one (**102**) and 1,3-dinitroacetone as second product (78H108). Its structure was proven by the formation of 1-phenyl-3-nitro-4-hydroxy-5-phenylazopyrazole (**103**), obtained on heating of the diphenylhydrazone of 1,3-dinitroacetone (Scheme III.53).

The ring transformation of **99** into **102** can be described by the ANRORC process involving first the formation of the C-2 adduct **100**, subsequent ring opening into **101**, recyclization into **102** by carbanionic attack on the carbon adjacent to the NR group, and elimination of 1,3-dinitroacetone. The suggestion can be made that, instead of the open-chain intermediate **101**, the bicyclic adduct **104** acts as a transient intermediate, being formed by addition of both nucleophilic carbon centers of the 3-oxopentanedioate to the electron-deficient positions C-2 and C-6 in **99**. Bicyclic adduct formation with nitroarenes is not unprecedented (70JOC383; 71TL2349; 82JOC1081, 82T1405).

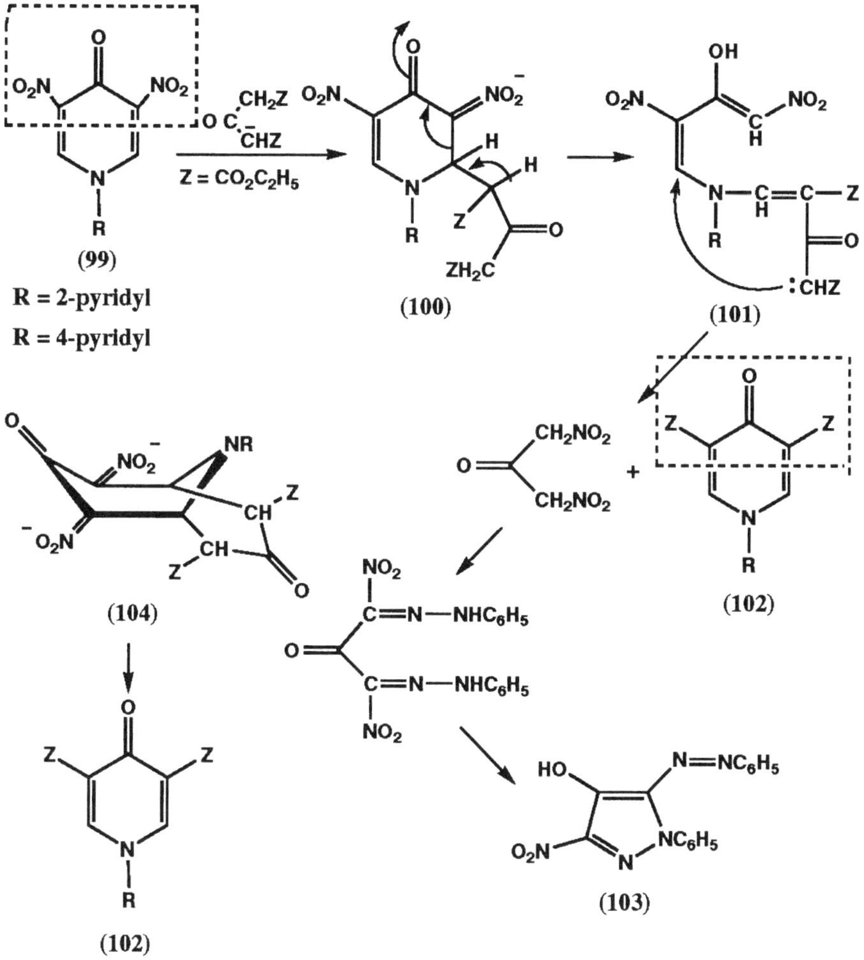

SCHEME III.53

3. REPLACEMENT OF THE NCN OR CNC FRAGMENT

a. *Pyrimidines*

Intermolecular NCN or CNC transfragment reactions have been extensively studied and many examples of these transformations are reported in the literature.

N-Methylpyrimidinium iodide reacts with benzamidine and pivalamidine to give 2-phenylpyrimidine and 2-*t*-butylpyrimidine, respectively (Scheme

III.54) (76RTC209). It is assumed that the addition reaction and the ring opening follow the same mechanistic pattern as decribed in Scheme III.47. However, in contrast to the one-way cyclization of the open-chain intermediate **83** in Scheme III.47 (due to the presence of the methylthio or methoxy leaving group), now two alternative routes can be formulated for cyclization of the ring opened diamidino intermediate **105**: route A, in which the cyclization takes place in such a way that in the product the C–N of the amidine is built into the pyrimidine ring (the amidine acts as a C–N donor); and route B, in which the amidine acts as an N–C–N donor (Scheme III.54).

In order to provide more evidence for one of these possible mechanisms, the phenylation reaction was studied with the *double*-labeled N-methyl[1,3-^{15}N]pyrimidinium salt as substrate. It was found that the 2-phenylpyrimidine obtained did not show any ^{15}N enrichment, proving that in the ring transformation the benzamidine acts exclusively as an N–C–N donor and has replaced the N(1)–C(2)–N(3) moiety of the pyrimidine ring. This result rules out the occurrence of a cyclization according to route A (Scheme III.54) (76RTC209; 95H(40)441].

Besides quaternary pyrimidinium salts, pyrimidines with a strong electron-withdrawing substituent on one of the carbons of the pyrimidine ring show enhanced reactivity of the heterocyclic ring toward nucleophiles. In light of the results mentioned previously with pyrimidinium salts, it is not too surprising that reflux of an ethanolic solution of 5-nitropyrimidine with benzamidine hydrochloride or pivalamidine hydrochloride in the presence of triethylamine afforded 5-nitro-2-phenylpyrimidine (**108**, 84%

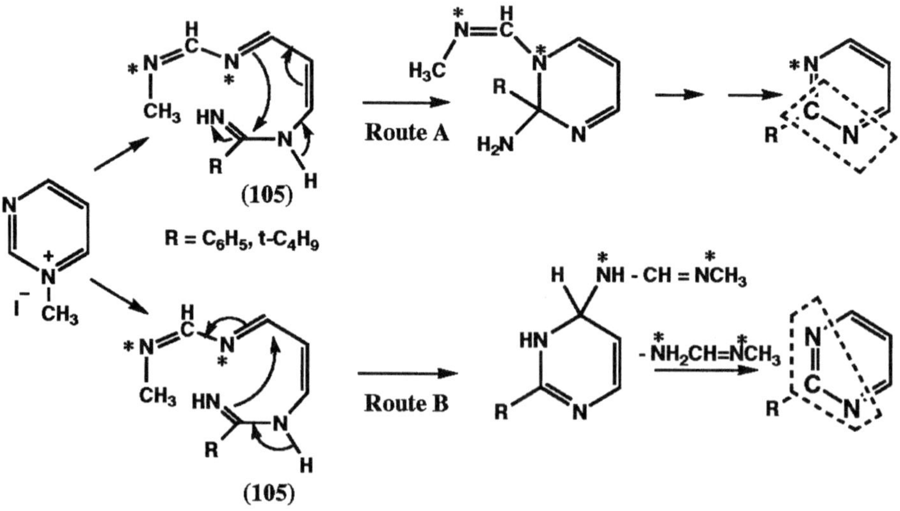

SCHEME III.54

yield). Using pivalamidine or acetamidine gives similar results, 2-*t*-butyl-5-nitropyrimidine (**108**, R = C_4H_9) and 2-methyl-5-nitropyrimidine (**108**, R = CH_3) being obtained, although in the last-mentioned reaction a pyridine derivative is also formed.

Experiments with *mono*-^{15}N-labeled amidines (the label is equally scrambled over both nitrogen atoms) have shown that in the 2-R-5-nitropyrimidines the excess of the ^{15}N enrichment is lower than that in the amidine (see Table III.2) (86JOC71). These measurements clearly indicate that during the reaction some ^{15}N labeling is lost. This result leads to the conclusion that besides formation of a *mono*-^{15}N-labeled 2-phenyl-5-nitropyrimidine (**108**), in which the ^{15}N-label is scrambled over *both* nitrogens, a *mono*-^{15}N-labeled 2-phenyl-5-nitropyrimidine (**109**) is also obtained in which only *one* nitrogen is ^{15}N-labeled (the benzamidine acts in this reaction as N–C donor) (Scheme III.55).

These results justify the conclusion that two routes of cyclization, i.e., route A (NCN donation) and route B (NC donation), are operative in this degenerate intermolecular transfragment reaction. Both reactions can be described as occurring according to the ANRORC process. Both involve an initial attack of the nucleophile amidine nitrogen atom on the 6-position of the 5-nitropyrimidine, forming the covalent σ-adduct **106**. Ring opening leads to **107**. Ring closure according to route A provides **108**, and the N–C–N fragment of the amidine is built into the ring. Ring closure B gives **109**; only the N–C fragment is built into the ring system. It is peculiar that in the reactions of the *N*-methylpyrimidinium salt with benzamidine, no indication of C–N donation has been found. It cannot be excluded that the formation of **107** can also be visualised by expulsion of formamidine from the bicyclic species **110**. This species can be formed by a 1,3-cycloaddition of the vicinal amino/imino groups across the C-4 and C-6 position of 5-nitropyrimidine.

The challenging question can be raised as to whether the formation of **109** can be explained in an alternative way, i.e., by a Diels–Alder reaction with an inverse electron demand, in which the C=N group of the electron-rich benzamidine forms a regiospecific cycloadduct across the 1,4-position

TABLE III.2

^{15}N Content in Amidines and in Products Obtained in the Reaction of 5-Nitropyrimidines with Amidines

Substituent	Amidines	Products (**108/109**)	Ratio (**108/109**)
R = C_6H_5	% ^{15}N = 8.9	% ^{15}N = 6.4	44/56
R = *t*-C_4H_9	% ^{15}N = 8.3	% ^{15}N = 5.6	20/80
R = CH_3	% ^{15}N = 8.7	% ^{15}N = 5.5	26/74

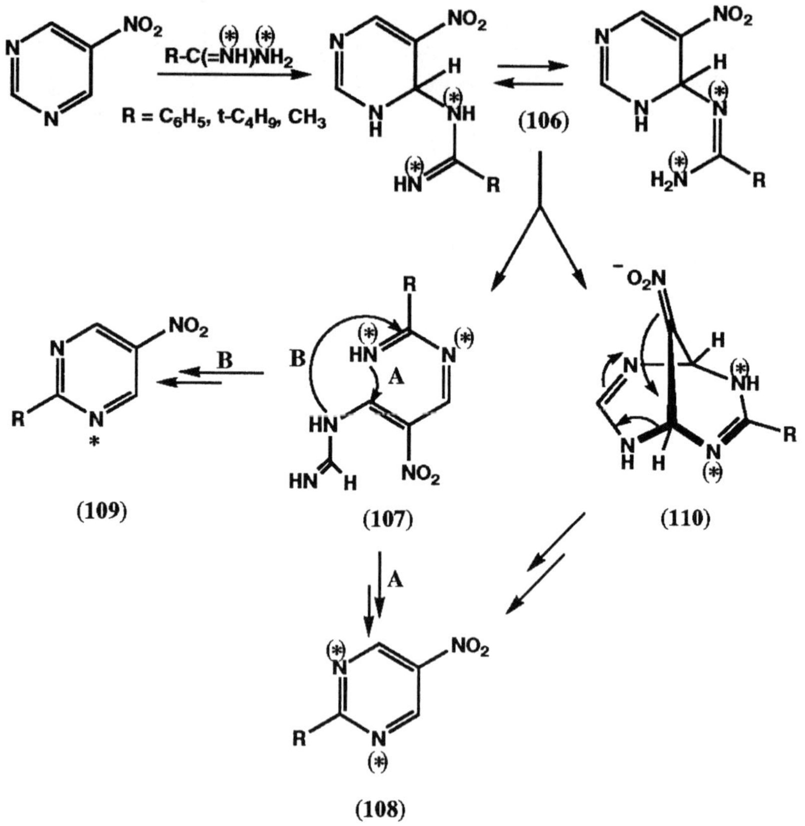

140 REPLACEMENT OF A THREE-ATOM FRAGMENT

SCHEME III.55

in the electrondeficient 5-nitropyrimidine. If this reaction occurs the bicyclic species **111** is formed, which on release of ammonia and hydrogen cyanide yields the product **109** (Scheme III.56).

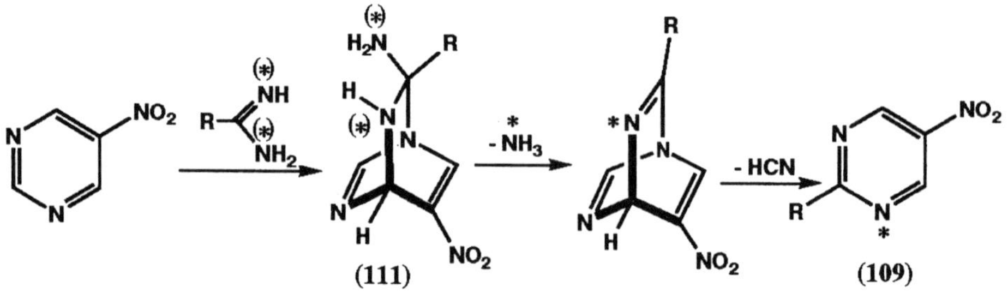

SCHEME III.56

There is interesting supporting evidence for this view (82TL3965; 83RTC373; 89T2693; 91MI4). Convincing proof has been presented that 5-nitropyrimidine is indeed able to undergo an inverse Diels–Alder 1,4-cycloaddition reaction with electron-rich olefins. Thus, for example, 5-nitropyrimidine with 1,1-dimorpholinoethene gives 2-morpholino-5-nitropyridine, the bicyclic 1,4-cycloadduct **112** being suggested as intermediate (Scheme III.57). The preference of enamines to add across N-1/C-4 in 5-nitropyrimidines was correctly predicted by FMO perturbation theory (86JOC4070). The fact that **109** is obtained does not contradict the view that a bicyclic adduct such as **111** is a real intermediate species in this N–C transfragment reaction.

Further support for the occurrence of this 1,4-cycloaddition reaction came from detailed studies of the reaction of 5-nitropyrimidine with ^{15}N-enriched pivalamidine and acetamidine (86JOC71). From measurements of the ^{15}N enrichment in the products (**108/109**), it was observed that the ratio (**108/109**) was considerably *increased* when compared with the data found for benzamidine (see Table III.1). This result clearly indicates that with pivalamidine and acetamidine the contribution of the C–N donation is enhanced. These results are certainly not in agreement with the C–N donation mechanism as presented by route B in **107** (see Scheme III.55). The electron-donating properties of the *t*-butyl as well as the methyl group should disfavor instead of promoting the ring closure according to route B in **107**. From this result it was prudently concluded that C–N incorporation in the degenerate ring transformation of 5-nitropyrimidine into **109** occurs not only via an open-chain intermediate such as **107,** but certainly also via a mechanism for which an inverse 1,4-cycloaddition might be suggested. An intermediate such as **111** can certainly not be excluded.

Another supporting piece of evidence was obtained in a study of the reaction of 5-nitropyrimidine with benzamidine hydrochloride in the *absence* of triethyl amine (which usually was added to liberate the free benzamidine). In amidinium salts the C–N bonds are about equal in length [1.314 Å

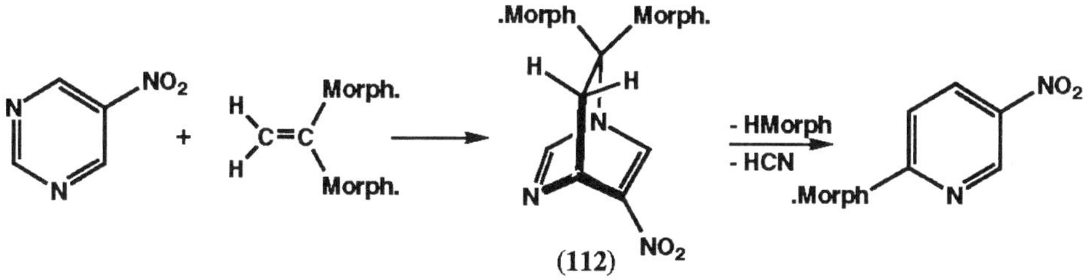

SCHEME III.57

(75MI4)], have less double-bond character, and are less electron-rich than the free bases. All these features do suggest that amidinium salts are less favored than amidines to react in an inverse Diels–Alder cycloaddition reaction with electron-poor substrates. One could argue that if in the reaction of 5-nitropyrimidine with ^{15}N-labeled benzamidine no cycloaddition takes place with the free amidine, it will certainly not take place with the amidinium salt. Therefore, one can expect that, if with ^{15}N-labeled free benzamidine no cycloaddition takes place, the ^{15}N distribution in 2-phenyl-5-nitropyrimidine, obtained with the benzamidinium salt, would then be more or less the same as with benzamidine. This has not been found: Reaction with ^{15}N-benzamidine hydrochloride (8.9% ^{15}N) gives 2-phenyl-5-nitropyrimidine in which 8.4% ^{15}N label has been incorporated into the product. This almost complete incorporation of the ^{15}N label indicates that more than 90% of **107** undergoes ring closure and supports the view that with free amidines the conversion does involve a regiospecific inverse 1,4-cycloaddition reaction, as shown in Scheme III.56

Extension of studies on the occurrence of ring transformations with amidines RC(=NH)NH$_2$, having in group R an acidic hydrogen, showed that no degenerate ring transformations but transformation into pyridines occur as the main reaction [83RTC373; 95H(40)441]. So, for example, 5-nitropyrimidine and phenylacetamidine give, not 2-benzyl-5-nitropyrimidine, but 2-amino-3-phenyl-5-nitropyridine (Scheme III.55). Acetamidine is a borderline case: Besides formation of 2-methyl-5-nitropyrimidine (**108**, R = CH$_3$/**109**, R = CH$_3$), 2-amino-5-nitropyridine is also obtained (Scheme III.45). These ring transformations are beyond the scope of this book and will not be discussed further. The author refers the reader to the original literature [83RTC373; 95H(40)441].

Other interesting examples of intermolecular N–C–N transfragment replacement are the ones being found when 1,3-dimethyluracil (**113**, R' = R'' = H) and several of its C-5/C-6 *mono*-substituted or C-5,6 *di*-substituted derivatives react with different 1,3-ambident nucleophiles (77JHC537; 84H(2)89). Reaction of (**113**, $R' = R'' = H$) with guanidine gives isocytosine **115** ($R' = R'' = H$) in reasonable-to-good yields.

The ease of the reaction depends on the electronic nature of the substituent at C-5 or C-6 and the steric environment at C-6. For example, the 5-fluoro compound reacts more easily than the 5-methyl derivative. A similar difference in reactivity is observed between 1,3,6-trimethyluracil and its 5-bromo derivative. Whereas 1,3,6-trimethyluracil requires fusion with guanidine to be converted into 6-methylisocytosine, 5-bromo-1,3,6-trimethyluracil can readily be converted into 5-bromo-6-methylisocytosine on treatment with guanidine in refluxing ethanol (77JHC537; 78JOC1193). The ring transformation can easily be explained by the ANRORC process involving the intermediacy of the C-6 adduct (**114**) (Scheme III.58). This

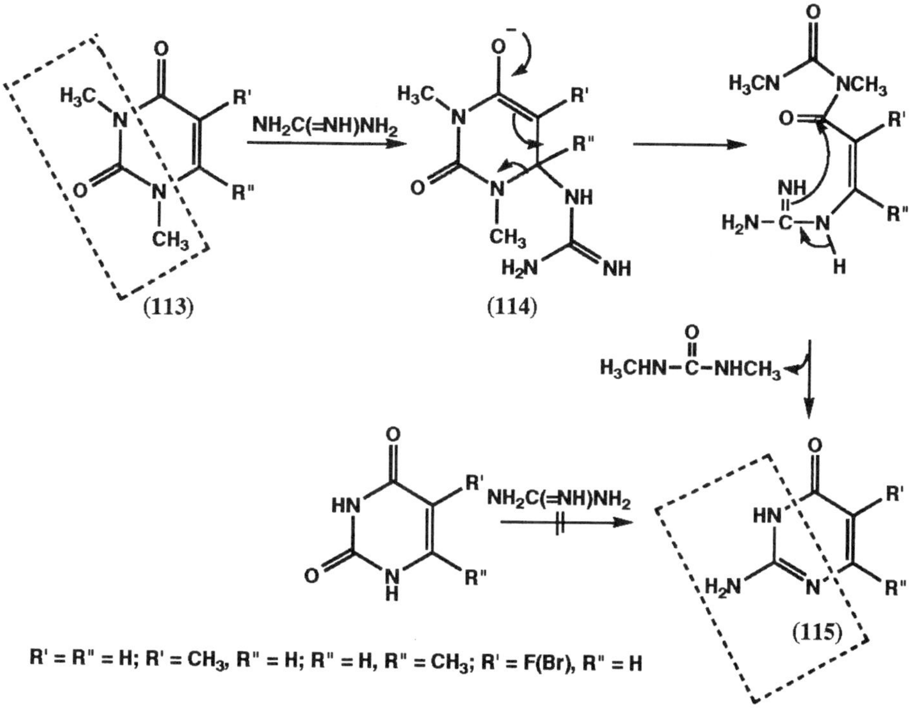

S_N(ANRORC) REACTIONS IN AZAHETEROCYCLES 143

SCHEME III.58

adduct undergoes scission of the N(1)–C(2) bond to the ring-opened product, which undergoes a subsequent ring closure with concomitant expulsion of 1,3-dimethylurea.

Transformation of uracil into isocytosine by treatment with guanidine did not occur, because of deprotonation of the NH group by the base guanidine. A uracil anion is generated in which an attack of the nucleophile is strongly prohibited because of electrostatic repulsion with the negatively charged uracil. The synthetic usefulness of these pyrimidine-to-pyrimidine transformations can also be demonstrated by conversions of 1,3-dimethyluracil with methylguanidine and cyanoguanidine. It is interesting that in case of methylguanidine, besides 2-(methylamino)pyrimidin-4(3H)-one, 1-methylisocytosine is also obtained, ratio 3:1 (Scheme III.59).

The formation of both isomers is due to competition for attack on C-6 between the stronger nucleophilic CH_3NH part in methylguanidine and the sterically less hindered NH_2 group.

The reaction of **113** with urea and thiourea (both compounds are weaker bases than guanidine) requires the presence of sodium ethoxide. With urea, uracil and with the thiourea, 2-thiouracil are obtained, respectively.

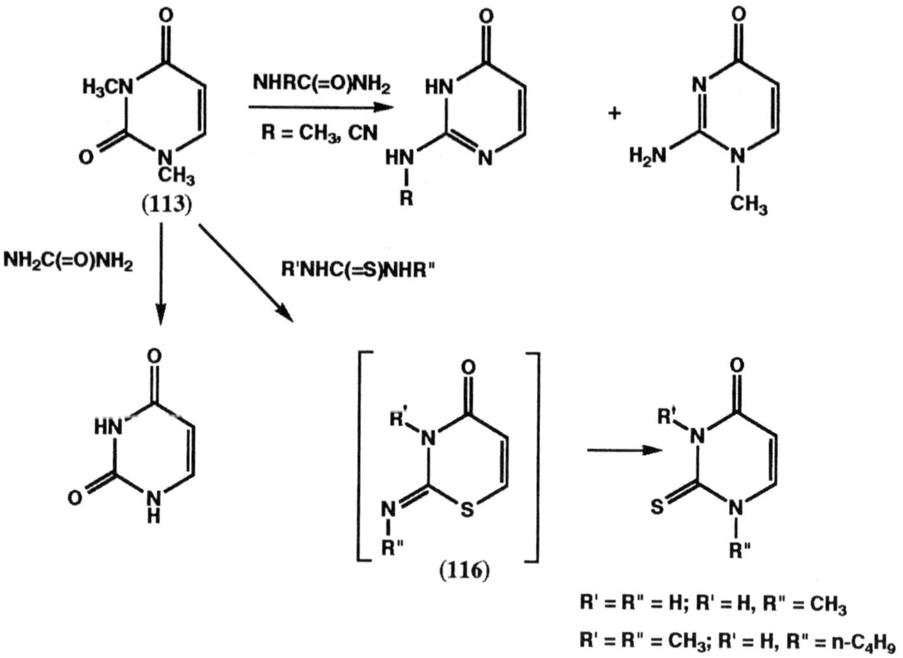

SCHEME III.59

Reaction with *N*-methyl- and *N*-butylthiourea gave *N*-methyl(butyl)-2-thiouracil. It is possible that the conversions with the thiourea derivatives take place via the intermediacy of 1,3-thiazine (**116**), which undergoes an internal sulfur–amino rearrangement (Scheme III.59) (78JOC1193). Alkali-induced rearrangement of amino-1,3-thiazines into thiopyrimidines is a well-established phenomenon (67CB3671; 70AJC51).

An interesting and useful application of these intermolecular N–C–N transfragment reaction has also been found in preparing the anti-leukemic *C*-nucleoside 5-(β-D-ribofuranosyl)isocytosine (**118**, R = H, pseudoisocytidine) from 1,3-dimethylpseudouridine (**117**) and guanidine (77JHC537; 78JOC1193). In a similar way, *N*-methylpseudoisocytidine (**118**, R = CH$_3$) and 2-thiopseudouridine (**119**) have been obtained (Scheme III.60).

Intermolecular N–C–C transfragment replacement occurs less widely than the N–C–N replacement. A synthesis of alkyl(aryl)pyrimidines has been published that was based on the replacement of the N–C–C fragment of the pyrimidine ring by an identical sequence of atoms derived from a ketone in the presence of ammonia [94H(38)249]. Heating a solution of 3-methyl-5-nitropyrimidin-4(3*H*)-one (**120**) in acetonitrile with cyclohexanone and ammonia gas in a sealed tube at 100°C for 3 hours gave tetrahy-

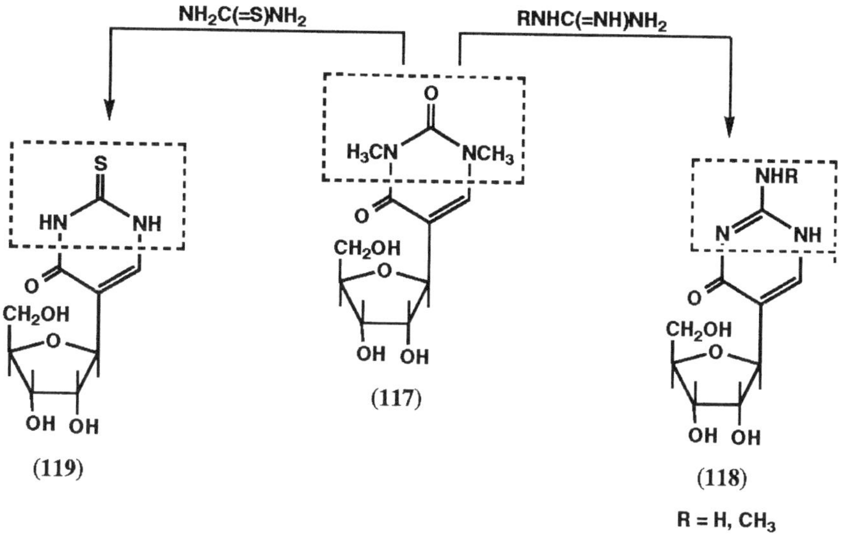

SCHEME III.60

droquinazoline (**121**) in reasonable yield (Scheme III.61). When the reaction was carried out in methanol, the yields were considerably lower. The same product could also be obtained with 1-morpholino-1-cyclohexene. The general applicability of this ring transformation was demonstrated by the conversion of **120** into cyclopentapyrimidine (**122**), 4-phenylpyrimidine

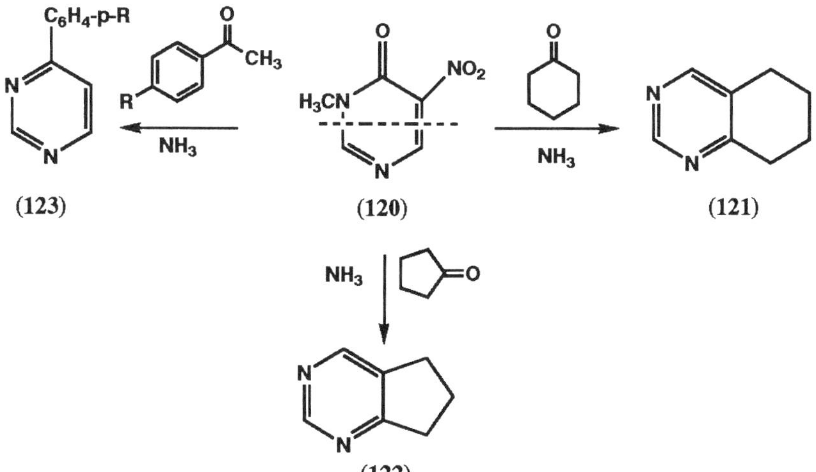

SCHEME III.61

(**123**, R = H), and 4-*p*-nitrophenylpyrimidine (**123**, R = NO₂) by reaction with cyclopentanone, acetophenone, and *p*-nitroacetophenone, respectively. The yields are, however, quite low (15–30%) and the reaction is found to be applicable to restricted substrates. In some of these reactions *N*-methyl-α-nitroacetamide could be isolated as the other reaction product.

The mechanism of this conversion was formulated to occur by an initial addition of the ammonia at position 2 and of the anion of the keto compound (or the enamine) at position 6, i.e., formation of **124**. It is of course possible that this addition pattern can be reversed: addition of the ammonia at position 6 and of the anion at position 2. In both addition products an internal cyclization occurs by attack of the nitrogen of the amino group on the keto function, yielding the tricyclic intermediate **125**. Aromatization occurs by loss of *N*-methyl-α-nitroacetamide (Scheme III.62).

When ammonium salts were used instead of ammonia as nitrogen source, the reaction took a very different course. Reaction of **120** with acetophenone and ammonium acetate in methanol for 1 day gave, besides formation of the degenerate ring transformation product **123** (R = H) (47%), the rearranged product 3-nitro-6-phenyl-2-pyridone (**126**) (48%) (Scheme III.63) [97JCS(PI)2261, 97S1277]. The present reaction was also applied to other ketones. Acyclic or cyclic ketones also afforded pyrimidines and/or pyridines. A relationship between the ratio of these products and the reaction conditions or the structure of the substrates could not be observed. It requires further investigations to control the selectivity of this process. It has been argued that the nitropyrimidone **120** behaves as an activated diformylamine in case of the conversion into **121**, but as an synthetic equivalent of nitrodiformylmethane in the nitropyrimidine–nitropyridine ring transformation (**120** to **126**) (Scheme III.63).

The observed difference of behavior of **120** toward the ammonium ion and ammonia is ascribed to the coordinating power of the ammonium ion,

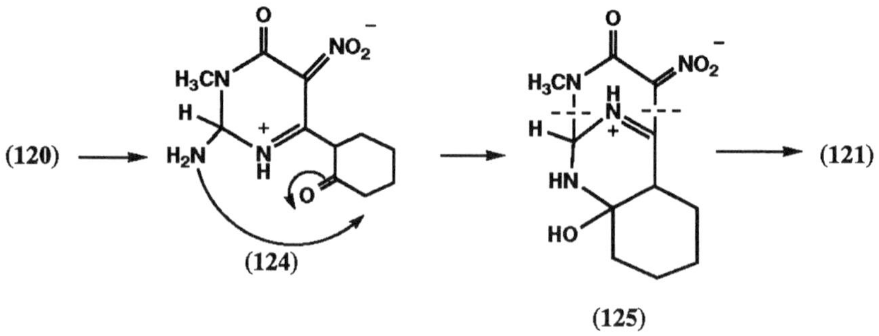

SCHEME III.62

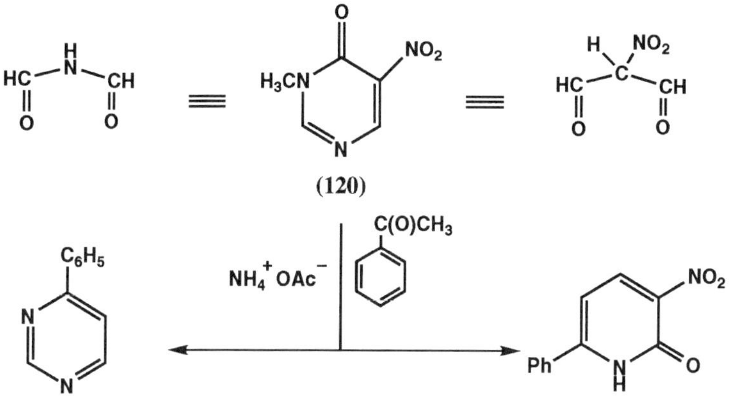

SCHEME III.63

in contrast to ammonia, with the carbonyl group in position 4, activating it for a nucleophilic attack [97JCS(P)2261]. Addition of the ammonia at C-4 and addition of the enol of the ketone at position 6 forms adduct **127**. Intramolecular nucleophilic addition in **127** yields a bicyclic adduct, which is converted into **128** by elimination of *N*-methylformamidine (Scheme III.64).

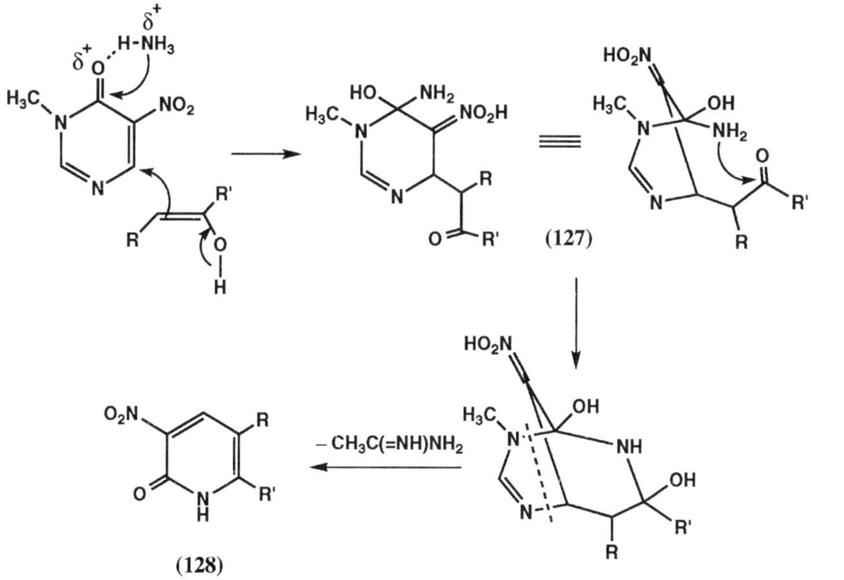

SCHEME III.64

b. *1,3,5-* and *1,2,4-Triazines*

As can be expected, the reactions discussed in this section show strong similarities with those presented in the previous section on pyrimidines.

Reaction of 1,3,5-triazine with various substituted amidines provides a very elegant and efficient method for synthesizing *mono*-substituted 1,3,5-triazines in reasonable-to-good yields (79JA3982). As by-product, formamidine is formed. The reaction involves an overall replacement of the N–C–N portion of the 1,3,5-triazine ring by the N–C–N part of the amidine and can be described as an S_N(ANRORC) process, in which an initial addition of the nucleophilic amidine nitrogen takes place at the ring carbon, followed by ring opening and ring closure (Scheme III.65). An interesting extension of this N–C–N intermolecular trans-fragmentation reaction is the formation of *para*-1,4-di(1,3,5-triazin-1-yl)benzene from 1,3,5-triazine and the bisamidine terephthalamidine (Scheme III.65) (79JA3982). 1,3-Dimethyl-5-azauracil was found to be extremely susceptible to ring transformation. Treatment with guanidine yields 5-azacytosine; with urea, 5-azauracil is obtained (Scheme III.66) (79JOC3982). Both conversions show a strong similarity with the ring degenerate transformation reported for 1,3-dimethyluracil with the same reagents (see Section III,B,3).

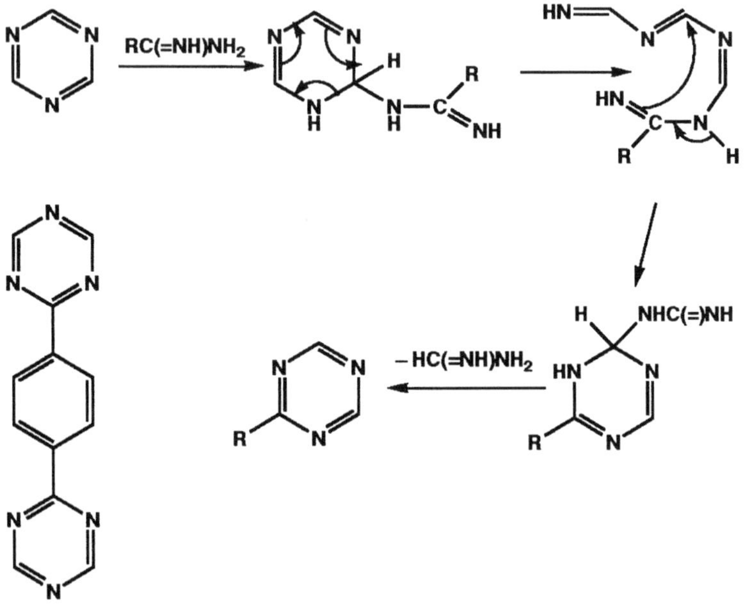

SCHEME III.65

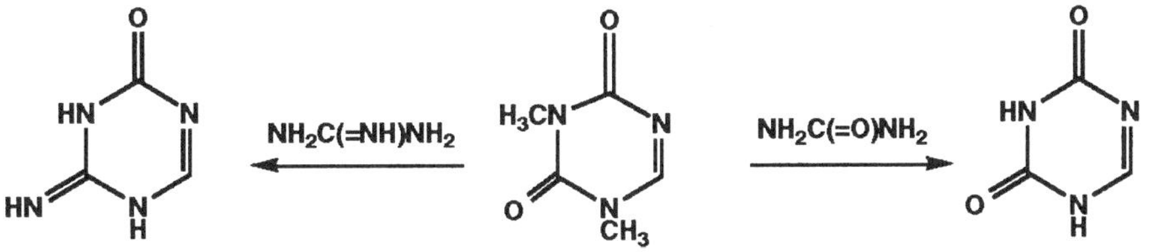

SCHEME III.66

An interesting degenerate ring transformation in the 1,2,4-triazine series has already extensively been discussed in Section II,D,1,b. It concerned the formation of 3,5-diphenyl-1,2,4-triazine in the reaction of 3-halogeno-5-phenyl-1,2,4-triazine with potassium amide in liquid ammonia (80JOC881). The presence of the second phenyl group in the 1,2,4-triazine ring is due to a reaction with benzamidine, which is formed by decomposition of 3-halogeno-5-phenyl-1,2,4-triazine (Scheme III.67). For more detailed information the reader is referred to Section II,D,1,b, in which is explained how ^{15}N-labeling studies were used to unravel the mechanism of this complex degenerate ring transformation.

D. Degenerate Ring Transformation Involving the Replacement of a Carbon Atom of the Heterocyclic Ring by a Carbon Atom of a Nucleophilic Reagent

This type of degenerate ring transformation, involving the replacement of a carbon atom of the heterocyclic ring by the carbon atom of a reagent, has been observed in only a few rare cases.

In attempts to formylate the bicyclic compound 5-methyl-9-phenylhydrazono-6,7,8,9-tetrahydro-4-oxo-4H-pyrido[1,2-a]pyrimidine 3-carboxylate (**129**) with dimethylformamide–phosphoroxychloride at 90–100°C, a degenerate ring rearrangement took place, resulting in the formation of 7-(α-chloroethyl)-8-chloro-9-(N,N-dimethylaminomethylene)amino-6,7,8,9-tetrahydro-4-oxo-4H-pyrido[1,2-a]pyrimidine-3-carboxylate (**131**)

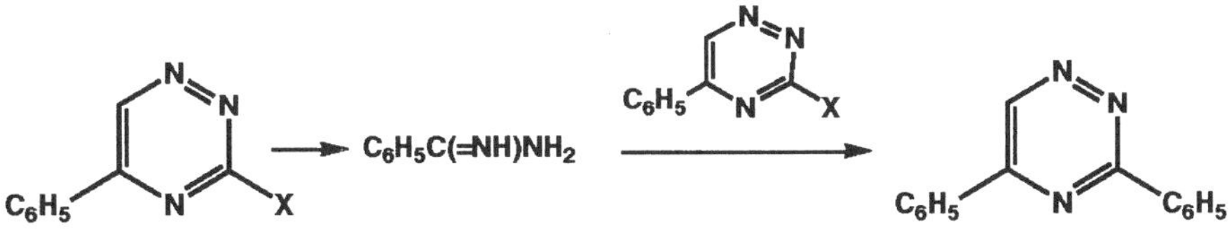

SCHEME III.67

(91H(32)1455, 91JHC783). When **129** reacts with dimethylformamide–phosphoryl chloride at lower temperature (40–45°C), the 9-(*N,N*-dimethlaminomethylenamino)-8-chloro-7-(*N,N*-dimethlaminomethylene)-6,7-dihydro-4*H*-pyrido[1,2-*a*]pyrimidine-3-carboxylate (**130**) was formed. Further heating of **130** at around 100°C gave **131** (Scheme III.68).

It was suggested that the ene hydrazino compound **132** and not the hydrazone **129** is the active species that plays an essential role in the degenerate ring transformation. This is based on the experimental finding that compound **133** as well as compound **134** (both compounds lack a double bond between C-8 and C-9 in the ring) are not able to undergo this rearrangement reaction under the Vilsmeyer–Haak conditions (Scheme III.68).

The reaction mechanism, taking into account these different experimental facts, is tentatively proposed as follows (Scheme III.69). (91JHC783;

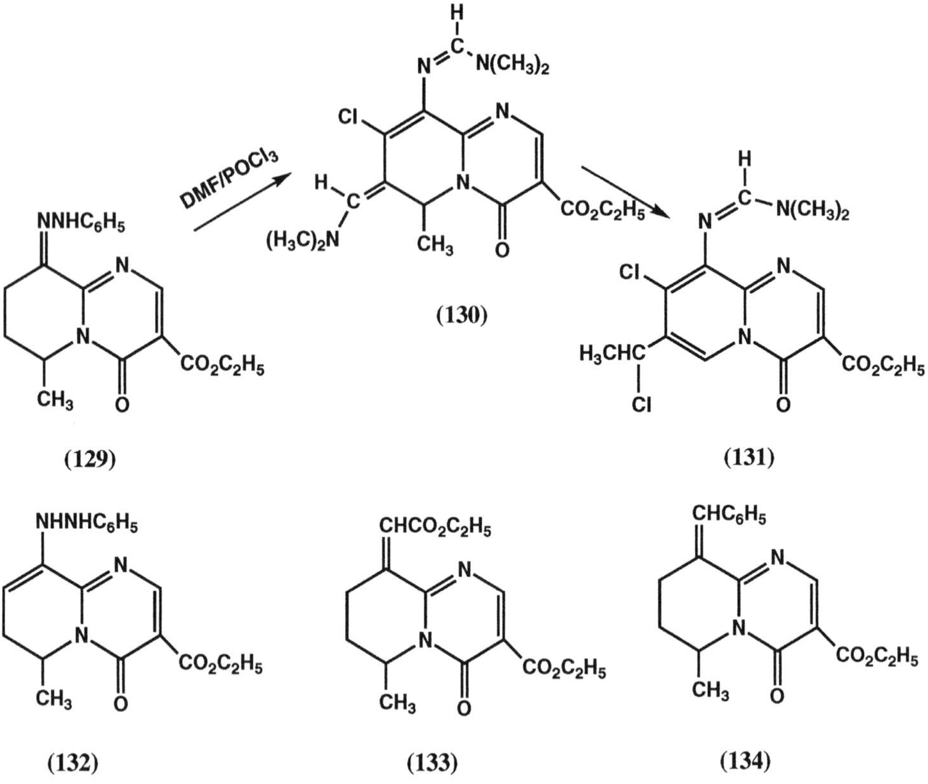

SCHEME III.68

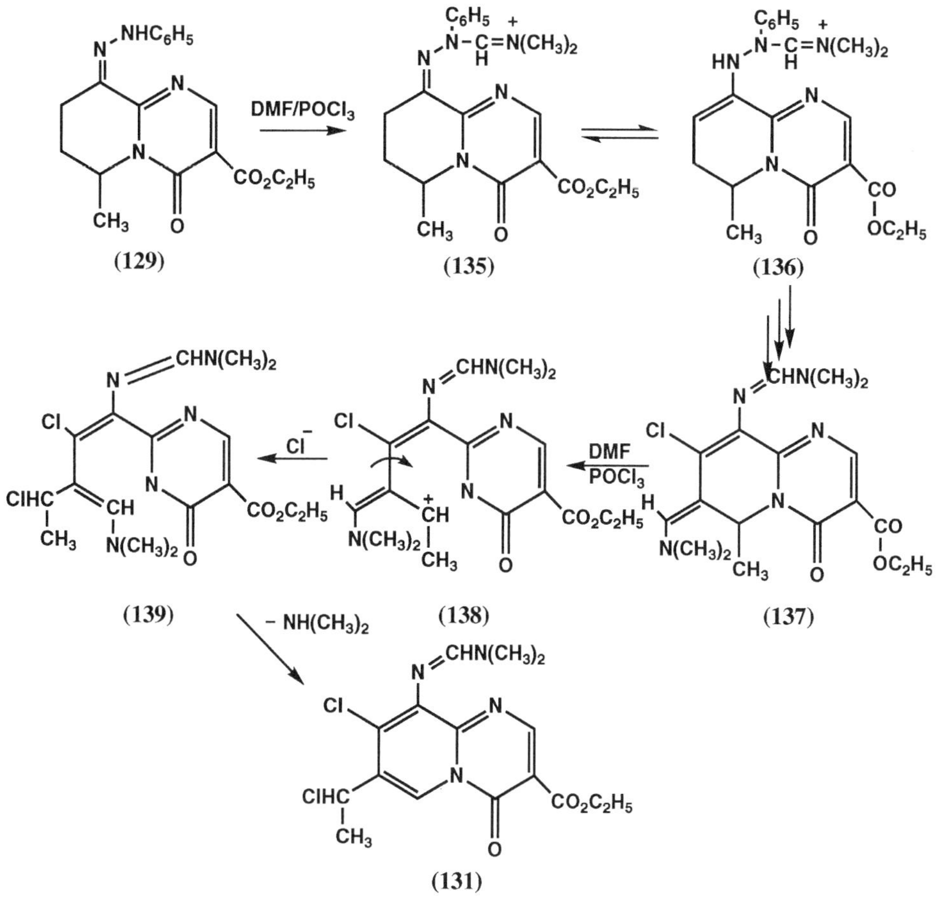

SCHEME III.69

97MI3). First, the amino group of the 9-phenylhydrazone moiety in **129** is acylated with the Vilsmeyer–Haak reagent. Since the nitrogen of the 9-hydrazono group in **135** contains an electron-withdrawing substituent, the equilibrium between the hydrazono tautomer **135** and the enehydrazino tautomer **136** is shifted to **136** (91JHC783). By a series of reaction steps **137** is formed, which after N-1 protonation undergoes heterolytic scission into the relatively stable carbocation **138**. Reaction with the chloride ion and cyclization yield **131**.

Chapter IV

Degenerate Ring Transformations Involving Side-Chain Participation

A. Introduction

In this chapter degenerate ring transformations are discussed of five-membered and six-membered aromatic and nonaromatic heterocycles, involving the participation of a side chain. These degenerate ring transformations form an interesting class of reactions among the many nondegenerate rearrangements with participation of a saturated or unsaturated side chain in mononuclear and benzo-fused systems. They have been extensively discussed and reviewed [67JCS(C)2005; 79JCR(S)64; 81AHC1, 81AHC141, 81AHC251; 82JCS(PI)759; 84JHC627; 86JST215; 90DOK1127; 94AHC49, 94H(37)2051]. Degenerate ring rearrangements with side chain participation can occur by thermal induction or photostimulation or can be initiated by acids or bases. Depending on the number of atoms in the side chain, the following classification can be made:

a. Rearrangements, which involve the participation of one atom in the side chain. These rearrangements can formally be described as a 1,3-exoannular interchange of two heteroatoms X, which are bound to a common ring atom R, located in positions between the hetero atom X. For the occurrence of a *degenerate* ring transformation, it is evident that in these 1,3-exoannular rearrangements the two atoms have to be identical. A well-known representative in this category of reactions is the Dimroth rearrangement (R is carbon; X is nitrogen) (09LA183; 27LA39; 63JCS1276; 74AHC33; 84MI2). These rearrangements are found to occur in both five- and six-membered heterocycles ($n = 3,4$), see Section IV,B (68MI1). Schematically they are represented in Scheme IV.1. However, several rearrangements are reported in the literature, in which it is not the ring heteroatom, but a carbon atom located between the heteroatom X and the ring atom R, that undergoes an exchange with a carbon atom present in the side chain as part of a functional group, for example CN, CO_2Alk(Ar), and being linked to the ring atom R. In order to distinguish these two different types of 1,3-exoannular rearrangements, they are earmarked in this book as 1,3-*exo*-annular heteroatom (HA) exchange and 1,3-exoannular carbon–carbon (CC) exchange (Scheme IV.1). Both types are extensively discussed in Section IV,B.

INTRODUCTION

1,3-exoannular HA exchange

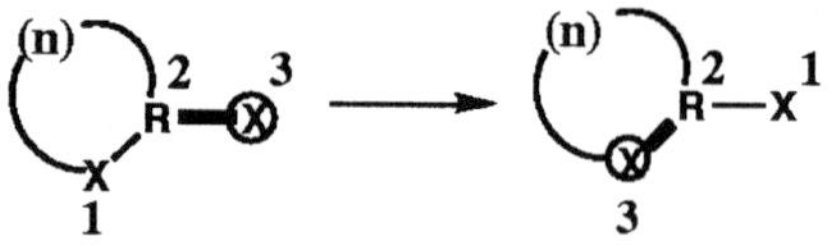

n = 3, 4

1,3-exoannular CC exchange

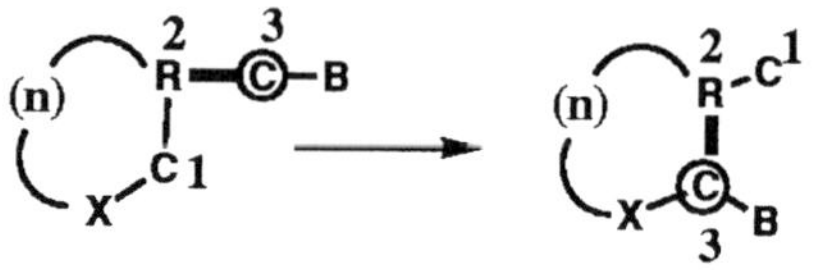

n = 2, 3

SCHEME IV.1

b. Degenerate ring transformations in which formally interchange occurs between a two-atom side-chain and an identical two-atom fragment of the heterocyclic ring, both being bound to the same ring atom R. These (1,2)–(4,5) exo-annular rearrangements are schematically represented in Scheme IV.2; they are extensively discussed in Section IV,C. The atoms X and Y can be heteroatoms and/or carbon atoms. It is evident that the rearrangement, characterized in Scheme IV.1 as CC exchange, takes an intermediary position between the HA rearrangement (Scheme IV.1) and the two-atom rearrangement indicated in Scheme IV.2.

c. Degenerate ring transformations, which occur by interchange of a three-atom side chain with a three-atom ring fragment. For these (1,2,3)–(5,6,7) exoannular degenerate rearrangements, it is required not only that both three-atom fragments have identical atoms, but, in addition, that they also have the same *sequence* of atoms and be bound to the same ring atom R

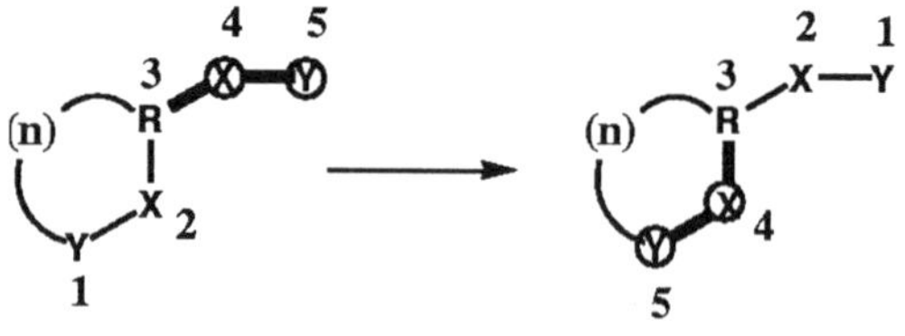

n = 2, 3

SCHEME IV.2

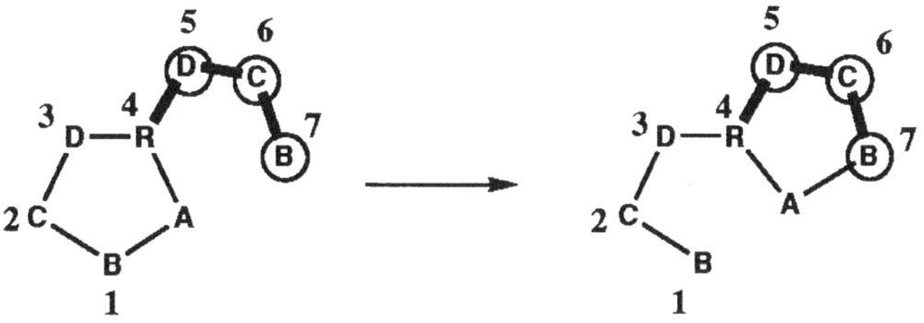

SCHEME IV.3

(Scheme IV.3). The general scheme of this type of rearrangements in aromatic five-membered heterocycles with an unsaturated side chain is known as the Boulton–Katritzky rearrangement [67JCS(C)2005].

Extension of this scheme to nonaromatic heterocycles and saturated side chains has also been described [94H(37)2051]. See Section IV,D.

B. Degenerate Ring Transformations Involving Participation of One Atom of a Side Chain

1,3-Exoannular HA exchange. In most of the rearrangements being reported, permutation occurs of a ring nitrogen atom with an exocyclic nitrogen atom being bound to the same carbon atom. These rearrangements involve ring fission and a subsequent recyclization and are now recognized as a quite general phenomenon in nitrogeneous containing heterocycles, both in five-membered and in six-membered compounds. The highlights are discussed in Sections IV,B,1 and IV,B,2.

1,3-Exoannular CC exchange. The chemistry of this type of ring conversion is extensively discussed in the last part of Section IV,B (see Schemes IV.34–39).

1. Five-Membered Heterocycles

a. *1,2,3-Triazoles and Tetrazoles*

Very early in the history of heterocyclic chemistry it was discovered by Dimroth (09LA183) that 1-aryl-5-amino-1,2,3-triazole (**1**, R = H) undergoes a facile, reversible isomerization to 5-(arylamino)-1,2,3-triazole (**2**, R = H) (Scheme IV.4).

A more recent example of a 1,3-exoannular rearrangement in the 1,2,3-triazole series is the isomerization of 5-amino-4-carboxamido-1-

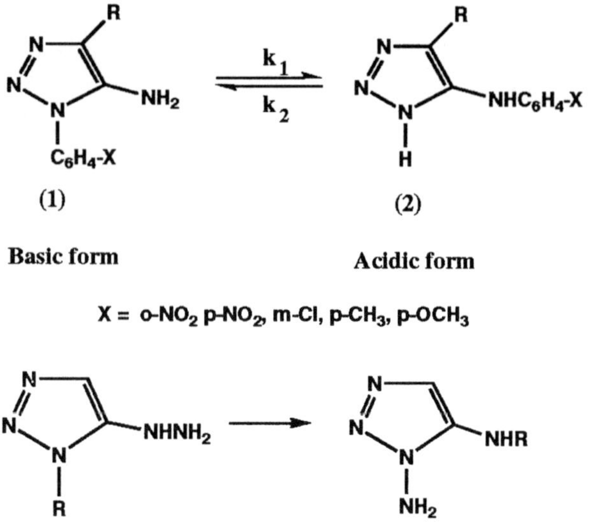

SCHEME IV.4

(2-nitrophenyl)-1H-1,2,3-triazole (**1**, R = C(O)NH$_2$, X = 2-NO$_2$) into 4-carboxamido-5-(2-nitroanilino)-1H-1,2,3-triazole (**2**, R = C(O)NH$_2$, X = 2-NO$_2$) in boiling ethanol (96JHC1847). A similar thermal rearrangement has been reported with 1-R-5-hydrazino-1H-1,2,3-triazole. It leads to the formation of 1-amino-5-(substituted amino)-1H-1,2,3-triazole (Scheme IV.4) (89BSB343).

More detailed investigations of the conversion of **1** into **2** have shown that in undisturbed melts an equilibrium exists between the basic form **1** and the acidic form **2** (56JOC654) and that in the presence of a base (pyridine, 4-picoline) nonequilibrium conditions are created, favoring the formation of the acidic form **2** (56JOC654). The presence of substituents in the phenyl ring at position 1 influences the equilibrium between the basic and acidic form (Scheme IV.4).

The kinetics of the equilibrium involved in the isomerization of 1-aryl-5-amino-4-phenyl-1,2,3-triazoles (**1**, R = C$_6$H$_5$) and of 5-(arylamino)-4-phenyl-1,2,3-triazoles (**2**, R = C$_6$H$_5$) shows that the initial stages of the isomerization follow the first-order law (57JA5962). The rate k_1 increases with increasing electronegativity of the aryl substituent X and the rate of the reversed reaction k_2 decreases with increasing electronegativity of substituent R. A good correlation was found between the logarithm of the rates of isomerization and the Hammett's σ-values of substituent X (57JA962).

It is known that 1H-1,2,3-triazoles can exist in equilibrium with diazoimines. In the conversion of **1** into **2** as intermediate the diazo-imine **3** can be proposed. In several cases these intermediary diazo-imines could be isolated (53JOC1283; 72CB2963, 72S571; 74CB2513; 81BSB615). In the formation of **2** Z/E isomerization and a 1,3-prototropic hydrogen shift in the intermediate stages of the reaction are involved (Scheme IV.5). Recent calculations show that in the cyclization step an *E*-configuration of the imino group is required; the lone pair of the electrons of the imino nitrogen is then oriented to the diazo functionality. A low barrier of a *non*rotatory transition state of ca. 10 kcal/mol was calculated (98JOC5801). From *ab initio* calculations of the Dimroth rearrangement in 5-amino-1,2,3,4-tetrazoles, it was concluded that the rate-determining step is not the ring opening, but either the Z/E isomerization around the C=N bond or the 1,3-sigmatropic shift (82JHC943).

A very similar rearrangement has been reported for the conversion of 5-amino-1,2,3,4-tetrazole (**4**) into the 5-(substituted amino)-1,2,3,4-tetrazole (**5**) and vice versa (Scheme IV.6) [53JOC779; 53JOC1283; 54JA88; 71JCS(CC)674]. The reaction proceeds via the intermediacy of an azido-

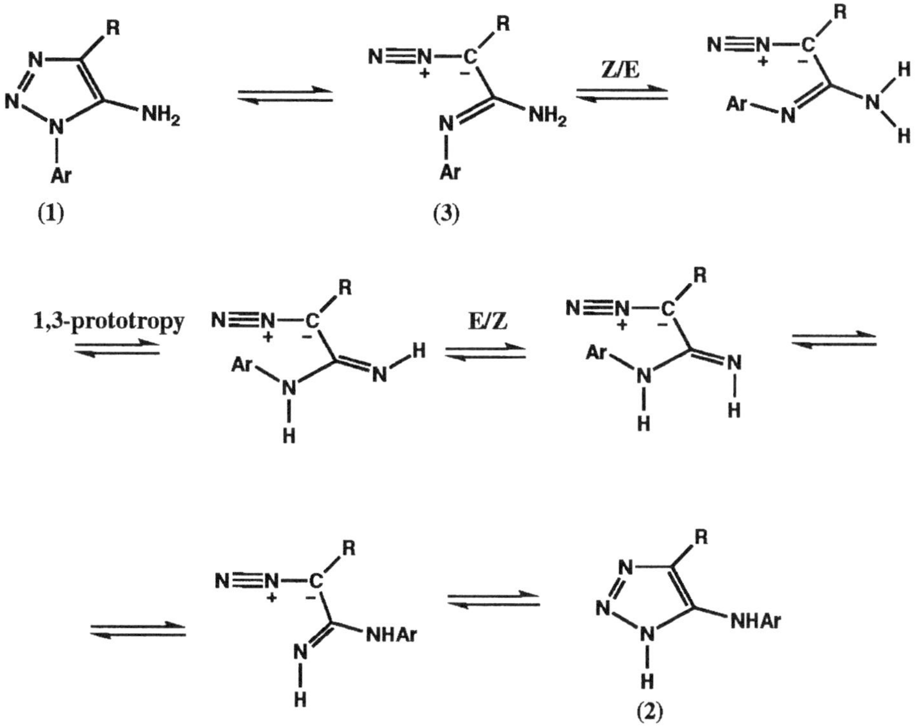

SCHEME IV.5

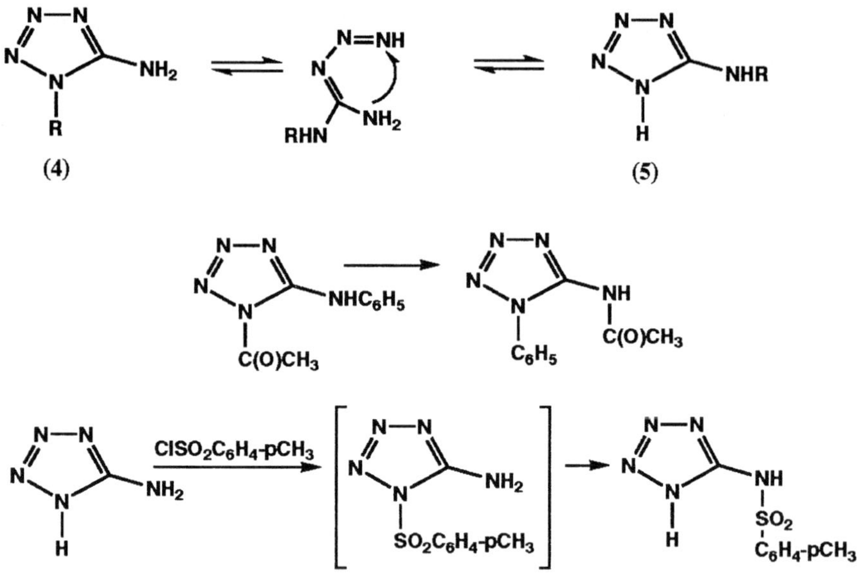

SCHEME IV.6

imine (53JOC1283; 54JA88). Similarly, as observed in the 1,2,3-triazole series, the acidic form **5** is favored when electron-withdrawing groups are present on N-1. Nice illustrations of the preference of the acidic form **5**, when an electron-withdrawing group is present on N-1, are the formation of 1-phenyl-5-acetamido-1,2,3,4-tetrazole on heating of 1-acetyl-5-anilino-1,2,3,4-tetrazole (Scheme IV.6) (53JOC1283; 58JOC1912) and the formation of 5-(*p*-toluenesulfonylamino)-1,2,3,4-tetrazole from 5-amino-1,2,3,4-tetrazole and *p*-toluenesulfonyl chloride (43RTC207; 53JOC1283; 58JOC1912; 60JA1609; 66AG548; 61ACS991). Although the formation of 5-(*p*-toluenesulfonylamino)-1,2,3,4-tetrazole seems to involve a direct substitution at the exocyclic amino group at C-5, in fact one deals here with a degenerate ring transformation, in which first a substitution occurs at nitrogen N-1 and subsequently a Dimroth rearrangement, which is fast because of the presence of the electron-withdrawing group at N-1.

b. *1,2,4-Triazoles*

An amidine-type interconversion has also been found to occur with the mesoionic 1,2,4-triazolium-3-aminides **6** [74JCS(PI)627, 74JCS(PI)638]. On heating of an ethanolic solution of compound **6** a (partial) conversion into the aminides **8** took place and led to an equilibrium between **6** and **8** (Scheme IV.7). As can be expected, this equilibrium is dependent of sub-

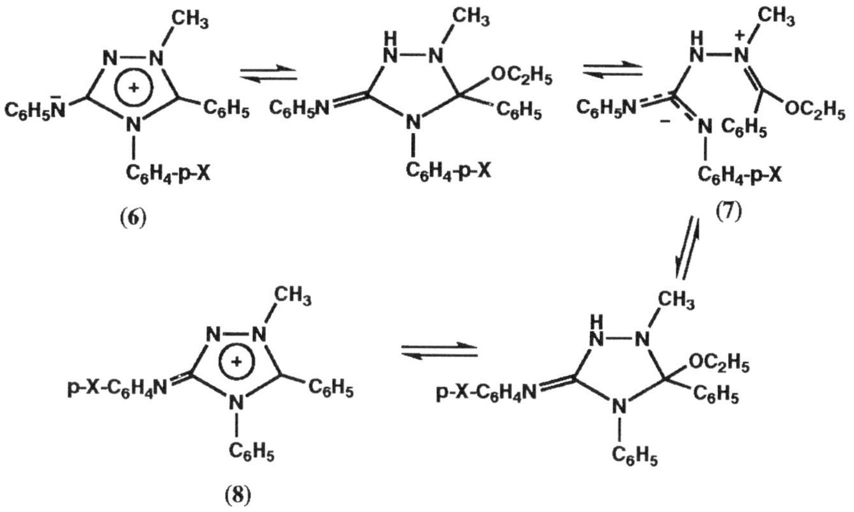

SCHEME IV.7

stituent X, as is convincingly demonstrated by the quantitative rearrangement of the p-chloro compound **6** ($X = p$-Cl) into compound **8** ($X = p$-Cl) on heating, and the great stability of compound **6** ($X = p$-OCH$_3$) on thermolysis. Both p-tolyl isomers **6** ($X = p$-CH$_3$) and **8** ($X = p$-CH$_3$) give on heating a 1:1 equilibrium mixture. An acceptable interpretation of the results in terms of a different electronic influence of the substituents on the relative thermodynamic stabilities of the isomers **6** and **8** was not offered.

The reaction can be described as occurring by an initial addition of ethanol and ring opening into the bipolair intermediate **7** (or its cationoid equivalent), which on recyclization and aromatization gives **8** (ANRORC mechanism).

c. *1,2,4-Thiadiazolines and 1,2,4-Thiadiazolidines*

Several Dimroth rearrangements were reported with 1,2,4-thiadiazolines and 1,2,4-thiadiazolidines. On dry heating of 4-*R*-5-imino-1,2,4-thiadiazoline (**9** $R = $ CH$_3$, C$_2$H$_5$), a color change from yellow to colorless was observed. This was attributed to a rearrangement, involving an interchange of the ring nitrogen of the N-*R* group at N-4 with the exocyclic imino group at C-5, leading to 5-alkylamino-1,2,4-thiadiazole (**10**) (Scheme IV.8) (54CB68; 63CB534). As ring-opened intermediate was postulated the bipolar species **9A**.

A somewhat similar conversion was found when the 3-anilino-4-phenyl-5-imino-1,2,4-thiadiazoline (Hector's base) was heated with ammonia,

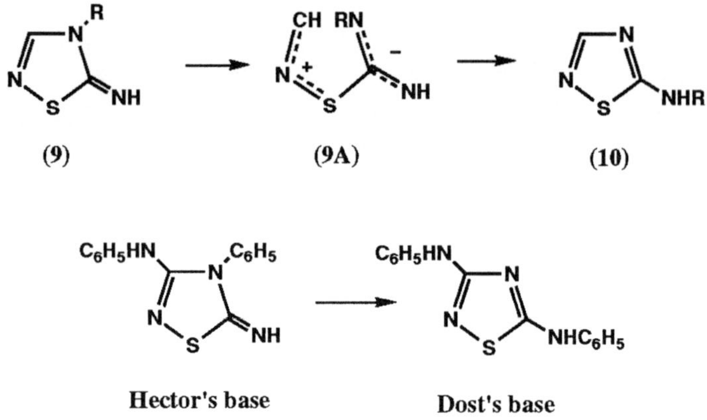

160   PARTICIPATION OF ONE ATOM OF SIDE CHAIN

SCHEME IV.8

3,5-bis(anilino)-1,2,4-thiadiazole (Dost's base) being obtained [06CB863; 78JCS(CC)652; 79JCR(S)316]. The structure of the Dost's base was for some time a matter of controversy, but by detailed ^{13}C and ^{15}N NMR spectroscopic investigations that problem was elegantly solved [80JCR(S)114]. More complicated ring degenerate rearrangements were reported for the 2-alkyl-5-alkylimino-4-aryl-3-arylimino-1,2,4-thiadiazolidines (**11**) (Scheme IV.9) (79BCJ1225).

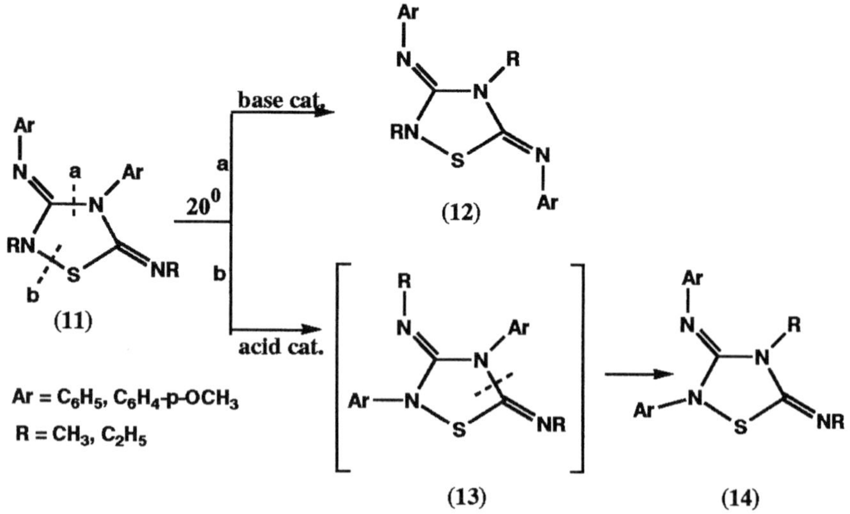

SCHEME IV.9

Depending on conditions (use of a base or an acid) 2,4-dialkyl-3,5-bis(arylimino)-1,2,4-thiadiazolidine (**12**) and/or 2-aryl-4-alkyl-3-arylimino-5-alkylimino-1,2,4-thiadiazolidine (**14**) are formed. The conversion of **11** into **14** is accelerated by acid and involves *two* amidine rearrangements: first breaking of the S–N bond in **11** (bond breaking b) and formation of intermediate **13**, in which subsequently bond breaking occurs between C-5 and N-4, followed by recyclization into **14**. The conversion of **11** into **12** is promoted by base and involves bond breaking between C-3 and N-4 (bond breaking a) (Scheme IV.9).

d. *1,2,4- and 1,3,4-Dithiazolidines*

In the 1,2,4-dithiazolidine series, 4-phenyl-3-aroylimino-5-methylimino-1,2,4-dithiazolidine (**15**) rearranges under the influence of benzoyl isothiocyanate into the 3-aroylimino-4-methyl-5-phenylimino-1,2,4-dithiazolidine (**16**) (78JOC4951). The Dimroth rearrangement is suggested to occur via a betaine as intermediate (Scheme IV.10). It is of interest to mention that the isomerization of **15** into **16** was not observed with the Lewis acids aluminum trichloride or benzoyl chloride.

A somewhat similar isomerization was observed with the sultam 5-methylimino-4-phenyl-1,3,4-dithiazolidine 1-dioxide (**17** $R = H$). On heating at 60°C in the presence of benzoyl chloride, rearrangement into the isomeric 5-phenylimino-4-methyl-1,3,4-dithiazolidine 1-dioxide (**19**, $R = H$) was found (Scheme IV.11). As intermediate can be proposed the amidinium salt **18** (78JOC4951). Furthermore, NMR-controlled test tube experiments revealed that this rearrangement also occurs under influence of aluminum trichloride and methanesulfonyl chloride. Also, the 2-phenyl derivative **17** ($R = C_6H_5$) could be isomerized; it required heating in acetone with *m*-dichlorobenzoic acid as catalyst (Scheme IV.11).

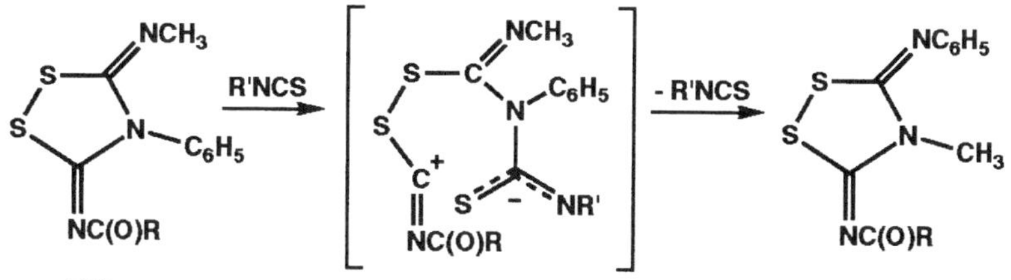

$R = C_6H_5, p\text{-}ClC_6H_4, OCH_3C_6H_4$

SCHEME IV.10

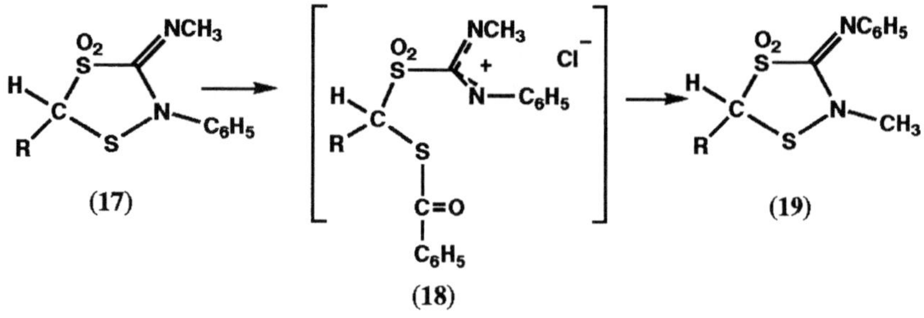

SCHEME IV.11

e. Thiazolines and Imidazolidines

Similarly to the rearrangement observed with 1-aryl-5-hydrazino-1,2,3-triazole (see Section IV,B,1,a), 3-phenyl-2-hydrazonothiazoline (**20**) is rearranged by acid at room temperature into 3-amino-2-(phenylimino)thiazoline (**21**) (Scheme IV.12) (61LA66).

Also, 2-iminoimidazolidines undergo amidine rearrangements, as shown by the conversion of 1-R-4,4'-diphenyl-2-iminoimidazolidin-5-one (**22**, R = CH_3, $CH_2C_6H_5$) into 4,4'-diphenyl-2-(R-imino)-imidazolidin-5-one (**23**) on heating with base or ammonium acetate (59E412; 60E107). Labeling stud-

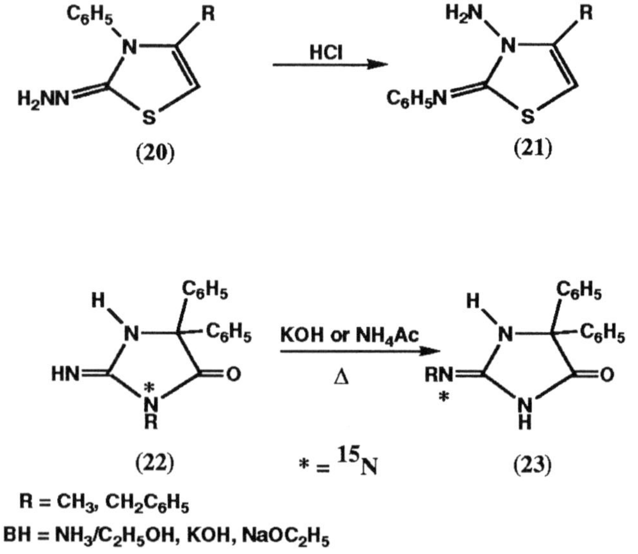

$R = CH_3, CH_2C_6H_5$
$BH = NH_3/C_2H_5OH, KOH, NaOC_2H_5$

SCHEME IV.12

ies with 2-imino[3-^{15}N]imidazolidin-5-ones (**22***) showed that after base treatment the exocyclic 2-(*R*-imino) group is ^{15}N-labeled, unequivocally proving ring opening during the degenerate rearrangement (Scheme IV.12) (63MI1).

2. SIX-MEMBERED HETEROCYCLES

a. *Pyridines*

An early observation of nitrogen interchange between the nitrogen of an amino group attached to the pyridine ring and the ring nitrogen has been described to occur when 2-[^{15}N-amino]pyridine (**24**) reacts with ammonia at 200°C for 50–100 hours (66ZC181). A partial 1,3-exoannular exchange took place and compound **24** is converted into the ring-labeled isomer **27**. Its formation can be formulated to involve a covalent σ-adduct **25** and its ring-opened amidino species **26**. ^{15}N-Scrambling in the amidine function and subsequent ring closure leads to thr formation of 2-amino[^{15}N]pyridine **27** (Scheme IV.13). Interchange between an exocyclic nitrogen and ring nitrogen has also been observed in the Dimroth rearrangement of 2-imino-1-methylpyridine (and its 5-chloro-, 3,5-dichloro-, and 5-cyano derivatives) into the corresponding 2-(methylamino)pyridines (28CB1223; 63CB534; 65JCS5542). The reaction occurs with great difficulty, in contrast to the 5-nitro derivative, which reacts very easily and provides in high yield the corresponding 2-(methylamino)pyridine derivative.

1,3-Exoannular nitrogen exchange has also been found in 2-pyridylnitrene. Heating, for example, 4-methyltetrazolo[1,5-a]pyridine (**28**) at 380°C/ 0.05 mm gives in small yield a 1:1 mixture of 2-amino-4-methylpyridine and 2-amino-5-methylpyridine (76JA1259; 80JA6159). Similar observations were also recorded with 5-methyltetrazolo[1,5-*a*]pyridine. This result was explained by the intermediate formation of (4-methylpyrid-2-yl)nitrene

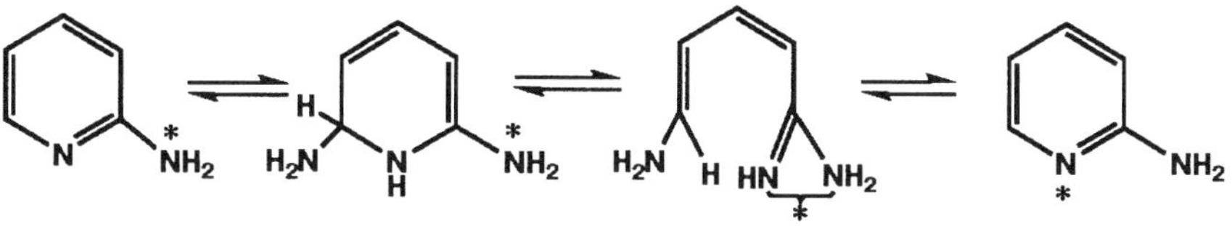

SCHEME IV.13

and (5-methylpyrid-2-yl)nitrene, being in equilibrium with an intermediate tentatively identified as 2,7-diazatropylidene (**29**) [69JCS(CC)1387]. More recent work, based on IR observations in cryogenic matrices (80JA6159), proved that this intermediate is the cyclic heterocumulene 1,3-diazacyclohepta-1,2,4,6-tetraene (**30**) (Scheme IV.14) (80JA6159). More recently, the trifluoromethyl derivative of this diazacycloheptatetraene has been isolated (96JA4009).

In order to establish more firmly the degenerate character of this rearrangement, 1(3)-[^{15}N]tetrazolo[1,5-a]pyridine (**31**) was prepared and subjected to pyrolysis. It can be expected that when unsubstituted **32** is intermediate, both nitrogen atoms are equivalent and consequently a 50:50 scrambling of the ^{15}N label over the ring nitrogen and amino nitrogen in the nitrene and thus in the 2-aminopyridine must be observed. This has indeed been found (Scheme IV.14) [69JCS(CC)1387].

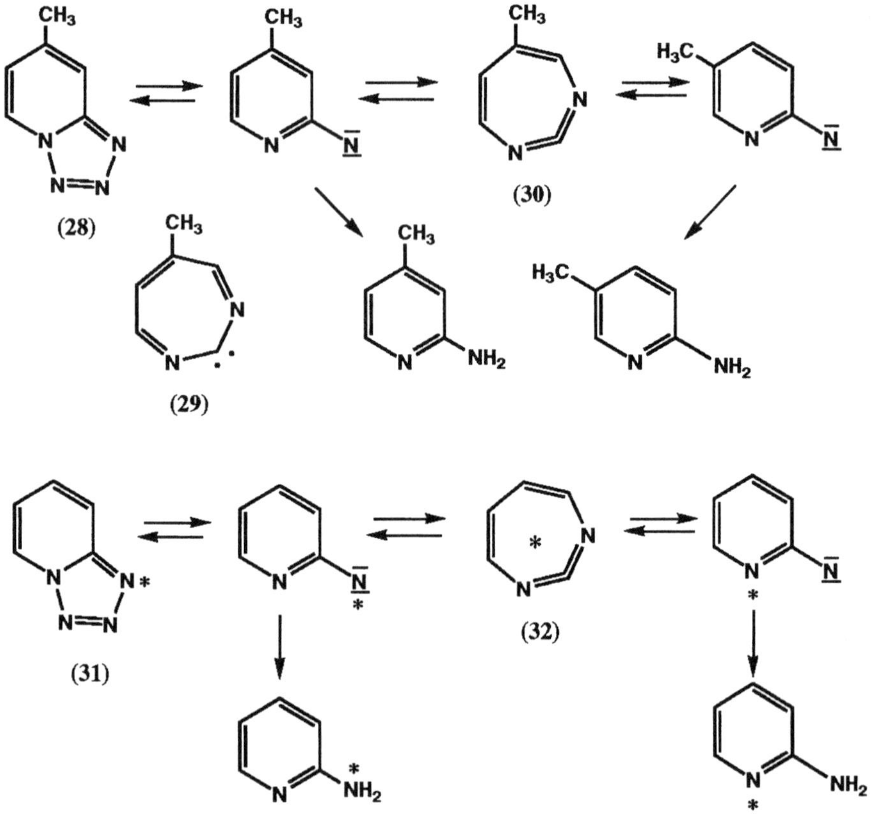

SCHEME IV.14

b. *Pyrimidines*

As already mentioned in Section IV,A, amidine (Dimroth) rearrangements are well-known representatives of exocyclic-nitrogen ring-nitrogen exchange. They have been studied in great detail in pyrimidine systems. An example of a well-studied rearrangement is the alkaline-induced conversion of 1,2-dihydro-2-imino-1-methylpyrimidine into 2-(methylamino) pyrimidine. When the reaction was studied with a substrate that was isotopically labeled with ^{15}N at the imino group, it was found that after the rearrangement the ^{15}N label was present on the ring nitrogen of 2-(methylamino)pyrimidine, showing that an 1,3-exoannular exchange between the exocyclic amino nitrogen and the ring nitrogen has occurred. This incorporation of the ^{15}N label leads to the unequivocal conclusion that the rearrangement must involve the intermediacy of an open-chain compound **33** (Scheme IV.15) (61N828; 63CB534, 63JCS1276; 68MI1). When the course of the reaction was followed by UV spectroscopy, it was found that the disappearance of the 1,2-dihydro-2-imino-1-methylpyrimidine is a first-order reaction with time of half disappearance $t_{1/2}$ = 114 min (25°C, pH 14). For the formation of the 2-(methylamino)pyrimidine, $t_{1/2}$ of 108 min was found (64MI2). From these data it can be concluded that during the rearrangement a buildup of the intermediate **33** can be precluded. Moreover, the data suggest that the ring fission is somewhat slower than the recyclization to the formally aromatic product. However, supporting proof for the existence of an open-chain intermediate was obtained by carrying out the isomerization in the presence of hydroxylamine; the intermediate could be trapped as its oxime **34** (65JCS7071).

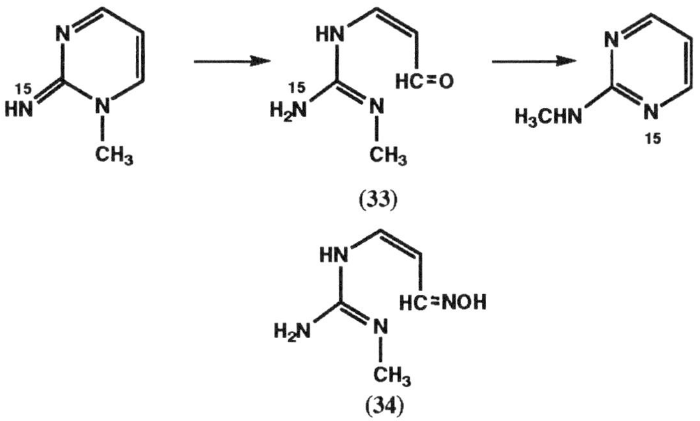

Scheme IV.15

Several reports deal with studies of rearrangement rates related to various electron-donating and electron-withdrawing substituents on the ring carbons and/or ring nitrogen in 1,2-dihydro-2-imino-1-methylpyrimidine.

Electron-donating substituents at position 5 were found to retard or even to stop the rearrangement [$t_{1/2}$ (min), pH 14, 25°C: H 114; 5-C_2H_5 196; 4,6-di-CH_3 166; 4-$(CH_3)_2N$ ≈2000; 4,6-di-NH_2: ∞] (64MI2). In the case of 4-dimethylamino-1,2-dihydro-2-imino-1-methylpyrimidine, the rearrangement is so slow that the competing hydrolysis into the corresponding pyrimidone-2 is the main reaction. It became also evident that introduction of electron-withdrawing groups in the parent system accelerates the rearrangement [$t_{1/2}$ (min), 25°C, pH 14: H 114; $t_{1/2}$ (min), 20°C, pH 12.2: 5-Br 39; 5-Cl 49; 5-I 31, 5-C(O)NH_2 7.5] [66JCS(C)164; 67JCS(C)903]. The rate of acceleration is in accordance with the degree of electron-withdrawing character of the substituents as is shown by their pK_a values [63JCS12/6, 63JCS1284; 64MI1; 65AJC471, 65JCS5542, 65JCS7071; 66JCS(C)164; 67JCS(C)903, 67JCS(C)1922]. Other examples, illustrating the striking rate-accelerating effect of an electron-withdrawing substituent on the amidine rearrangement, are the $t_{1/2}$ values for the 4-amino-5-nitro-6-imino- and the 4-dimethylamino-5-nitro-2-imino-1-methylpyrimidines, **37** and **38**, respectively. When compared with the $t_{1/2}$ values found for the corresponding iminopyrimidine derivatives **35** and **36**, respectively, they were found to be quite low (Scheme IV.16).

Variation of the N-1 substituent, i.e., change of methyl to ethyl, propyl, and butyl, has shown that the methyl derivative has a lower rearrangement rate than the higher homologs [$t_{1/2}$(min for N-R: R = CH_3 114; R = C_2H_5 63; R = C_3H_7 55(?); R = C_4H_9 58] [67JCS(C)903, 67JCS(C)1922]. Since there is only a small difference in electron-donating properties of the alkyl groups (the pK_a values of these compounds are very close) the appreciable difference in $t_{1/2}$ may be attributed to a steric factor.

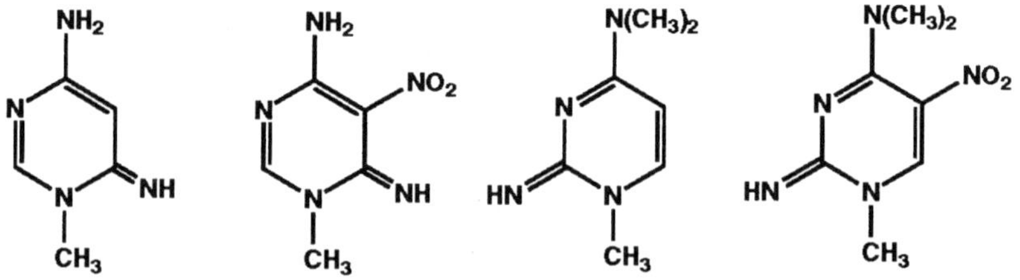

(35) τ$_{1/2}$ (min.) 15
(37) τ$_{1/2}$ (min.) << 0.1
(36) τ$_{1/2}$ (min.) 2000
(38) τ$_{1/2}$ (min.) < 0.1

SCHEME IV.16

It has been established that in the Dimroth rearrangement of 2-iminopyrimidines, water plays an essential role (65JCS7071). In water the imine is, in fact, in equilibrium with its hydrate, the carbinolamine **39**. That participation of the hydrate is important is shown by the experimental fact that in dry tetrahydrofuran, acetone, dioxane, or ether, the imine is quite stable and is not inclined to undergo rearrangement. However, on addition of a little water, rearrangement occurs; its rate is proportional to the concentration of the water (65JCS7071).

Based on the results of those experiments, the mechanism can be described as the initial formation of **39**, being in equilibrium with the ring-opened acyclic tautomeric intermediate **40a/40b**. Several of these intermediates have been characterized [65AJC471, 65JCS7071; 66JCS(C)164]. Recyclization into the formally aromatic compound is an irreversible process (Scheme IV.17). It is evident that in cases where the imino group carries a substituent, i.e., **41**, the recyclization product **42** is in equilibrium with the starting substrate [66JCS(C)1163]. The steric and/or electronic properties of the substituent determine on which side the equilibrium exists (Scheme IV.18). Based on a PMR study of differently substituted 1-R-1,2-dihydro-2-(R'-imino)pyrimidine (**41**, R = C_2H_5, n-C_4H_9, $CH_2C_6H_5$, and R' = CH_3, n-C_3H_7, i-C_3H_7), it was established that for R = C_2H_5, n-C_4H_9, $CH_2C_6H_5$ and R' = CH_3, the equilibrium lies on the side of the tautomer in which the larger group is located on the exocyclic nitrogen. For the ethyl and the butyl group, this effect is steric, but it has been argued that for the

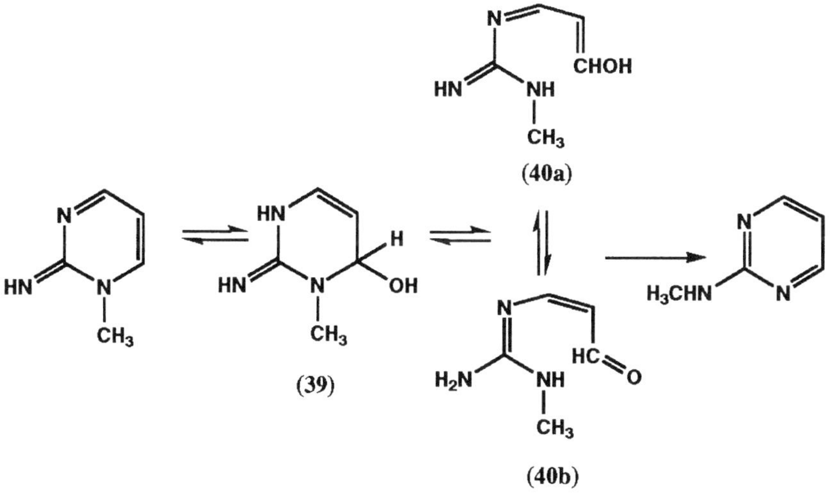

SCHEME IV.17

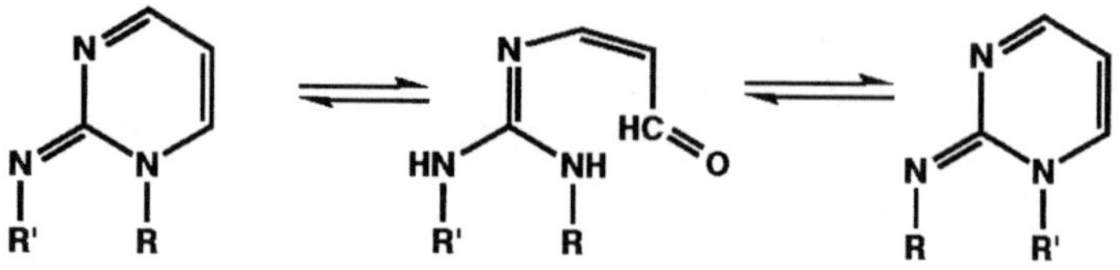

SCHEME IV.18

benzyl group its electron-withdrawing character partly also plays a role in establishing the equilibrium [66JCS(C)1163].

1,2-Dihydro-2-imino-1,4-dimethylpyrimidine (**43**) on treatment with alkali at room temperature showed a somewhat peculiar behavior. Although the expected rearranged product, i.e., 2-methylamino-4-methylpyrimidine (**44**), was formed, the yield is surprisingly low (10%) and its rate of formation is low ($t_{1/2}$ about 2000 min) [67JCS(C)1928]. The main product is 1,4-dimethylpyrimidin-2(1*H*)-one (80%), directly formed by hydrolysis of the imine. Reactions carried out at higher temperatures and/or introduction of electron-withdrawing substituents in the pyrimidine ring increased the yields of the rearranged product, but never leads to a clean product, because of the omnipresence of the pyrimidone-2 in the reaction mixture (Scheme IV.19).

Interesting Dimroth rearrangements in cytosine and its derivatives occur when they are allowed to react with acetic anhydride–acetic acid. Cyto-

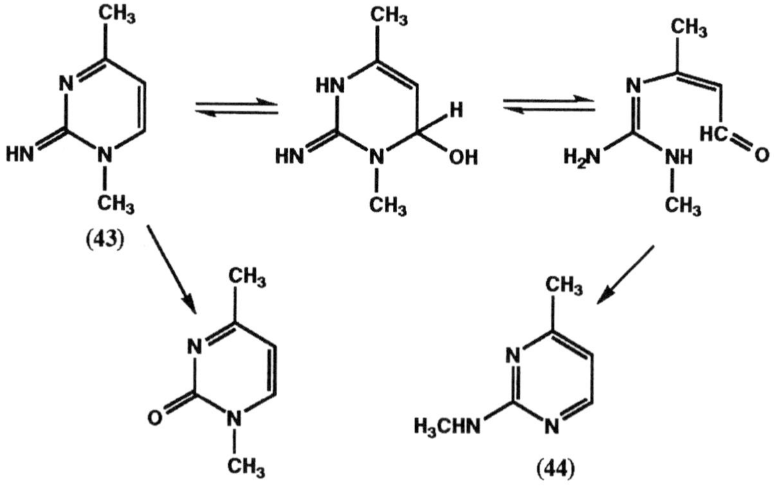

SCHEME IV.19

sine,[15]N-labeled on the exocyclic nitrogen, when being refluxed with acetic anhydride–acetic acid, yields N⁴-acetylcytosine, in which the label is scrambled over the exocyclic nitrogen, i.e., **46,** and the ring nitrogen, i.e., **47,** (ratio 1 : 1) (Scheme IV.20) (65B54). Acid hydrolysis provides a *mixture* of amino-labeled and ring-labeled cytosine.

Similar degenerate rearrangements were also observed with N⁴-methylcytosine (**48**) and 3-methylcytosine (**49**) (64JOC1770). Reaction of **48** with acetic anhydride–acetic acid gave an acetylated product that, after hydrolysis, yields 3-methylcytosine (**49**) alongside the starting substance **48.** When **49** was used as starting substance, the rearrangement went equally well in the reverse direction (Scheme IV.21). Both rearrangements have to proceed through an acyclic intermediate. Another interesting Dimroth rearrangement reported in the cytosine chemistry is the acetic anhydride induced conversion of [N⁴-β-carboxyethyl]cytosine (**50**) into 1,2,3,4-tetrahydro-2,6-dioxopyrimido[1,2-c]pyrimidine (**52**) (Scheme IV.21) (64JOC1762, 64JOC1770). The mechanism of this degenerate ring rearrangement is envisaged as follows. First the carboxylic side chain forms a mixed anhydride, offering the possibility for a cleavage of the C(2)–N(3) linkage of the pyrimidine ring with expulsion of the acetate anion. Another (degenerate) pyrimidine ring is formed with a side chain at position 2, featuring an isocyanate group, i.e., **51.** Recyclization by an attack on the carbon of the isocyanate leads to the required product **52.** That the N(1) hydrogen plays a crucial role in the rearrangement is experimentally established by the fact that the N(1) methyl analogue of **50** does not undergo this conversion.

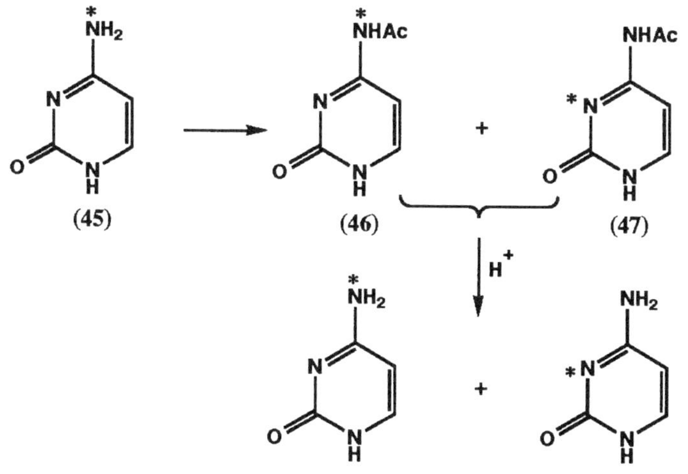

SCHEME IV.20

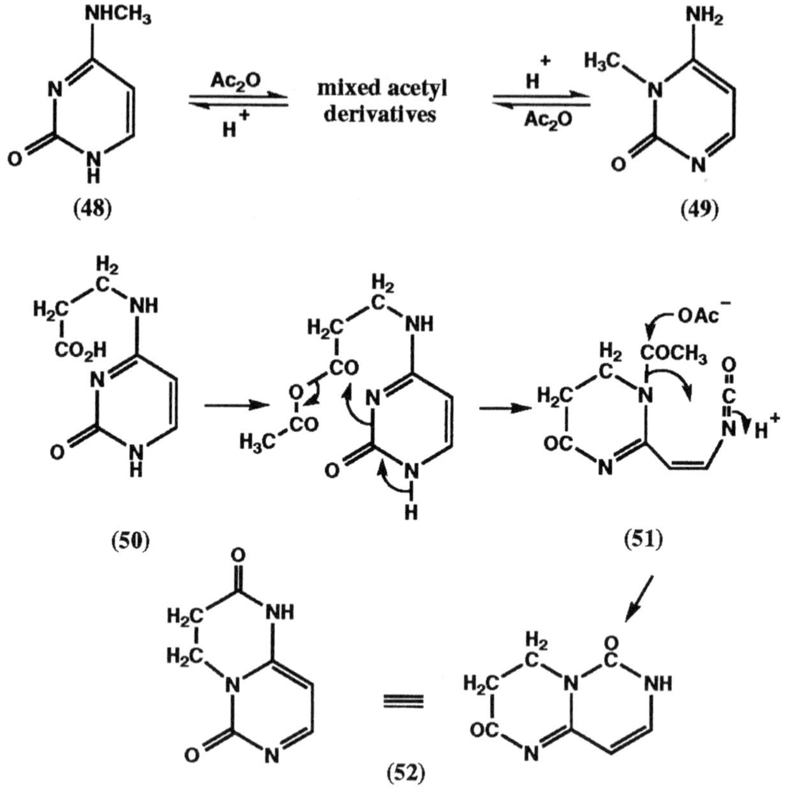

SCHEME IV.21

Several reports deal with Dimroth rearrangements of 4-aminopyrimidine 3-oxides and its tetrahydro derivatives. Examples are the rearrangement of cytosine 3-oxide (**53**) into N^4-acetoxycytosine (**54**) by treatment with acetic anhydride (65JOC2765) and the conversion of 1,2-di-R-4-amino-1,2,5,6-tetrahydropyrimidine 3-oxides (**55**, $R = C_6H_5$, 2-ClC_6H_4, 4-ClC_6H_4, 4-$CH_3OC_6H_4$) into the corresponding 1,2-di-R-4-(hydroxyamino)hexahydropyrimidines (**57**) (Scheme IV.22). The last-mentioned reaction occurs if **55** is dissolved in an aprotic solvent (chloroform, benzene, toluene) and allowed to stand for some time (Scheme IV.22) (94LA19). It is evident that in this strictly aprotic medium the rearrangement of the cyclic N-oxide **55** into **57** containing the exocyclic hydroxyimino function cannot occur according to the usual addition–elimination Dimroth mechanism. Therefore, it is assumed that a ring-opened dipolar species **56** is involved as interme-

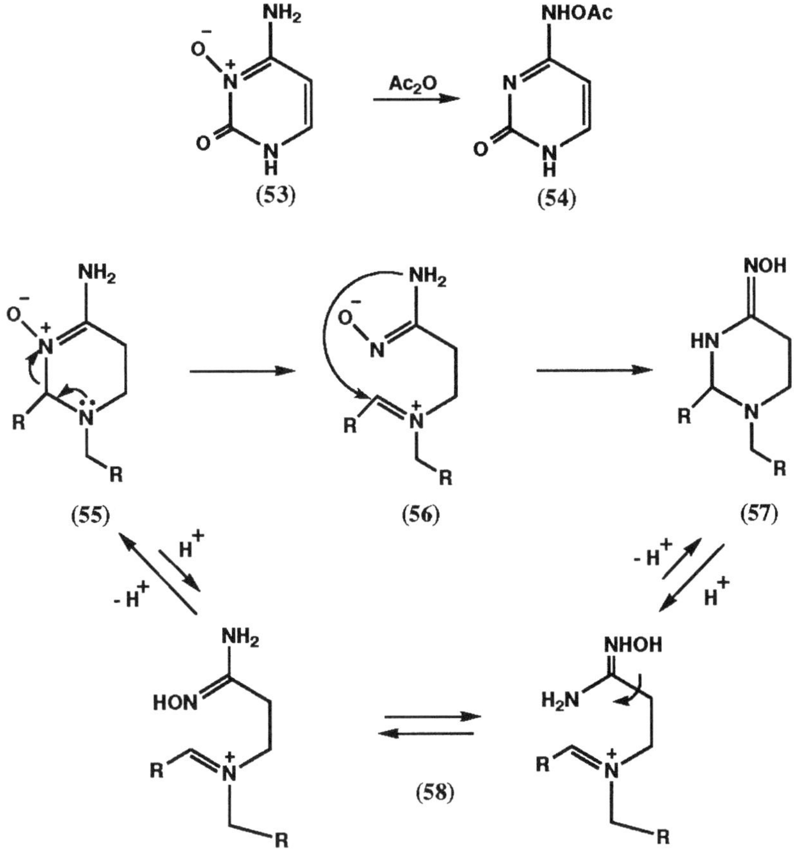

SCHEME IV.22

diate (94LA19). Since this species is hardly stable in the aprotic solvent, it will undergo a fast ring closure into **57**. The ring opening of **55** into **56** is the rate-determining slow step in the rearrangement. The electronic character of the N-1 substituent is decisive for the occurrence of the Dimroth rearrangement of **55** into **57**.

When instead of a benzyl group a phenyl group is present at position 1, no *N*-oxide–hydroxyimino rearrangement occurs, but a ring contraction into a 1,2,4-oxadiazoline (83TL5763; 86JCS(PI)2163; 94LA19), probably formed via a seven-membered oxadiazepine derivative. The reversed Dimroth reaction, i.e., conversion of **57** into **55**, takes place when a methanolic solution of **57** is allowed to stand for 1 week at room temperature. This hydroxylamino–cyclic *N*-oxide degenerate rearrangement can be rationalized by formation of the tautomeric open-chain iminium ion **58,** being in an acid-base equilibrium with **57** and **55** (Scheme IV. 22).

c. *Triazines*

The first report of a degenerate ring transformation, later recognized as a Dimroth rearrangement, appeared in 1888 and described the conversion of 4,6-dianilino-1,2-dihydro-2-imino-1-phenyl-1,3,5-triazine (**59**) into 2,4,6 tri(anilino)-1,3,5-triazine by heating in alcoholic ammonia. More recent examples are the conversions of 2,4-diamino-1-aryl-6,6-(mono(di) alkyl-1,6-dihydro-1.3.5-triazine (**60**) into 2-anilino-1,6-dihydro-1,3,5-triazines in boiling water or in hot alkali (54JCS1017) and the complete rearrangement of hexahydro-2,4,6-tris(imino)-1,3,5-trimethyl-1,3,5-triazine (**60a**) into 2,4,6-tri(methylamino)-1,3,5-triazine by dry heating (Scheme IV.23) (60JOC1043).

An amidine-type rearrangement, involving the degenerate transformation of the 1,2,4-triazine ring being annulated to the quinazoline ring system, was reported to occur on heating a mixture of 4,6-dioxo-6H,11H-11-methyl-3-phenyl[1,2,4]-triazino[3,4-b]quinazoline (**61**) and sodium acetate in acetic anhydride, the [1,2,4]-triazino[3,2-b]quinazoline **62** being obtained (Scheme IV.24) (94PJC1115). These degenerate rearrangements involve

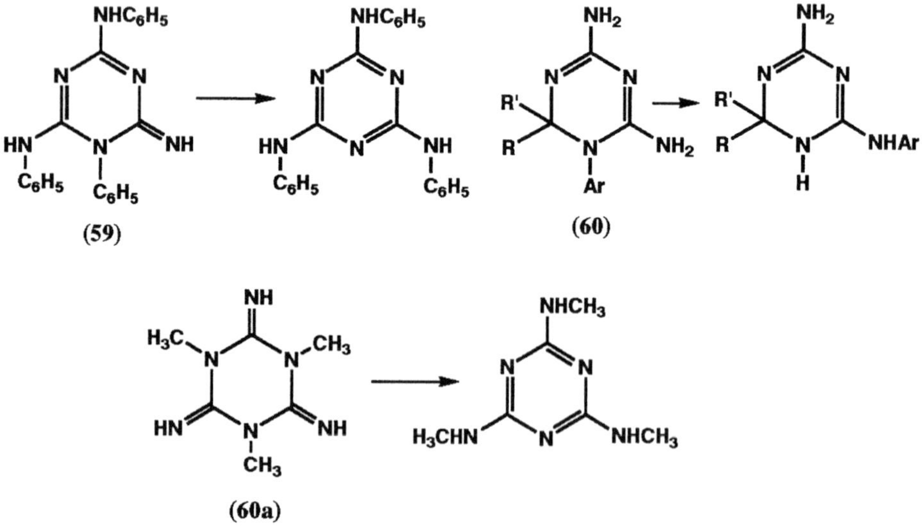

SCHEME IV.23

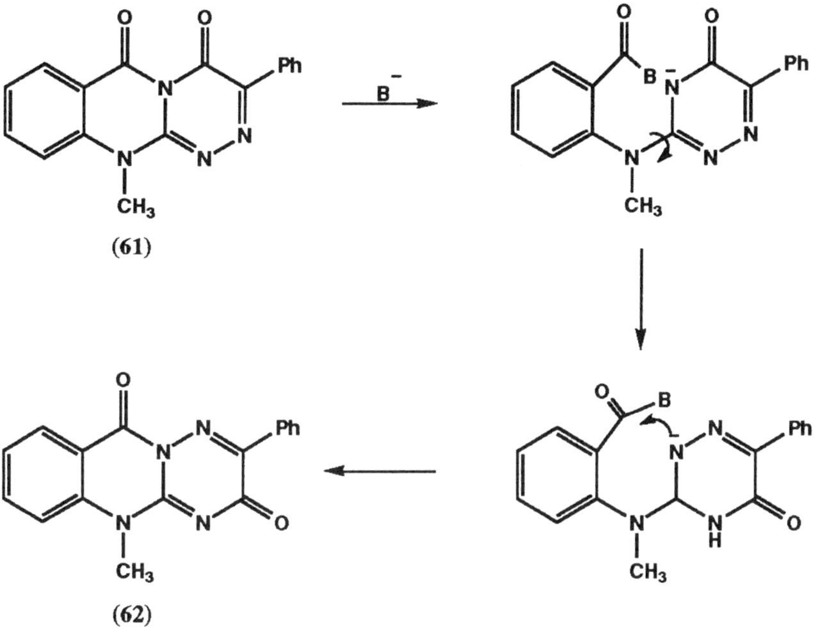

DEGENERATE RING TRANSFORMATIONS 173

SCHEME IV.24

the addition of the nucleophilic acetate ($B^- = OAc^-$), ring opening, and recyclization (ANRORC mechanism).

They also have the special interesting feature that it is one of the few cases of a degenerate involvement of a *ring*-nitrogen/*ring*-nitrogen exchange, since in almost all Dimroth rearrangements one deals with an exocyclic-nitrogen/*ring*-nitrogen exchange (or vice versa). This degenerate isomerization has also been found for pyrimido[1,2-*b*][1,2,4]-triazines.

A similar case is that of the conversion of the angular phenyl[1,2,4]-triazino[4,3-*a*]quinazoline (**63**) into the linear [1,2,4]-triazino[3,2-*b*]quinazoline (**65**) on heating with sodium acetate in acetic anhydride (Scheme IV.25) (94HAC97).

The conversion of **63** into **65** involves a double ANRORC-type rearrangement reaction: first a rearrangement of **63** into the [1,2,4]-triazino[3,4-*b*]quinazoline **64** [by bond breaking in **63**, rotation and recyclization], followed by a second rearrangement of **64** into **65**; both rearrangements occur according to the same mechanism as that presented in Scheme IV.24. It has been recorded that surprisingly the *N*-methyl derivative of **63** is unaffected when heated in acetic anhydride/sodium acetate; however, the *N*-

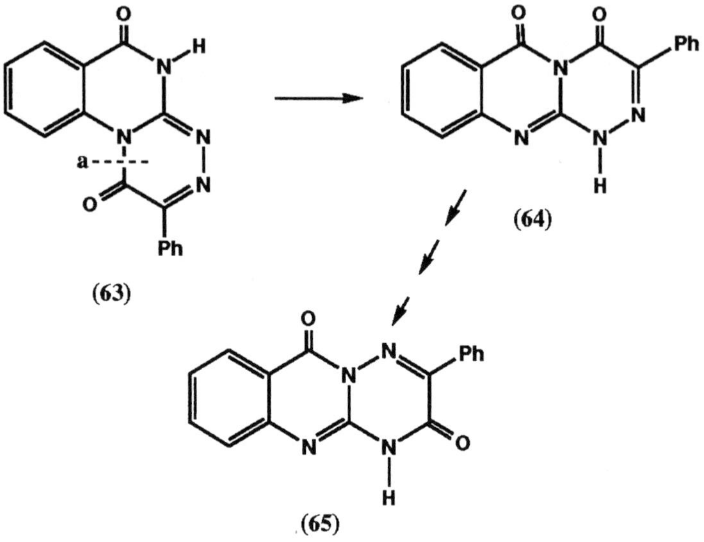

174   PARTICIPATION OF ONE ATOM OF SIDE CHAIN

SCHEME IV.25

amino compound easily rearranges under these conditions (Scheme IV.26) [94IJC(B)881].

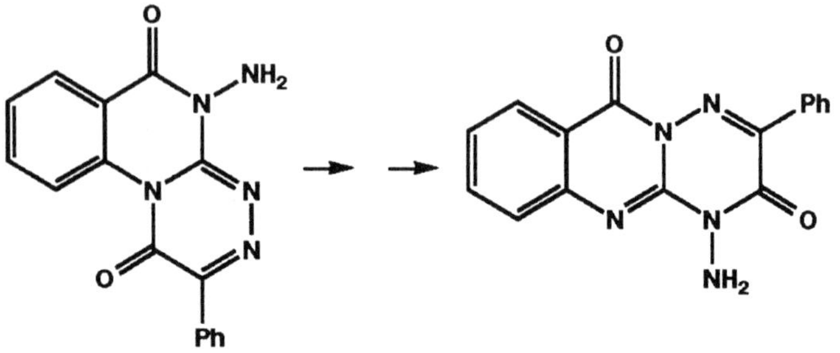

SCHEME IV.26

d. *Pteridines, Purines, Quinazolines, and Azolopyrimidines*

Detailed studies were made of the alkali-induced Dimroth rearrangement of 2-amino-3,4-dihydro-3-methyl-4-oxopteridine (**66**, $R = H$) into 2-methylamino-3,4-dihydro-4-oxopteridine (Scheme IV.27). Rapid-flow UV spectroscopy shows that (i) the ring opening of the anionic pteridine as well as the cyclization of the acyclic intermediate are first-order reactions and that (ii) the ring-opening step is faster than the ring-closure reaction (at 20°C, pH 14.7, $t_{1/2}$ ring opening 11 sec, $t_{1/2}$ ring closure 81 sec) (63JCS1284).

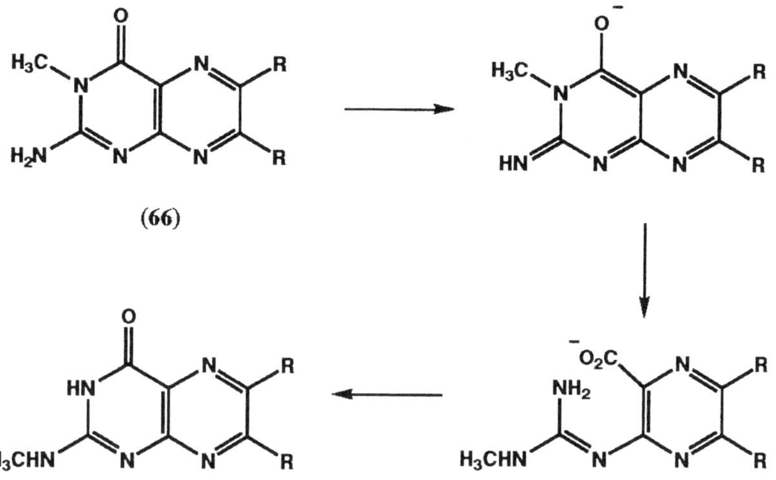

SCHEME IV.27

The 6,7-dimethyl derivative (**66,** R = CH$_3$) followed similar lines, although the rate of reaction in the different stages occurred more slowly ($t_{1/2}$ ring opening: 47 sec; $t_{1/2}$ ring closure; 960 sec) (Scheme IV.27).

As can be expected from the results mentioned in Section IV,B,2,b, the Dimroth rearrangements in pteridines are facilitated by the presence of electron-withdrawing groups. This can be exemplified by the observed lower rate of the rearrangement of the dianion of 2-imino-3-β-carboxyethyl-2,3-dihydro-4-oxopteridine compared to that of the compound containing a carbalkoxy substituent in position 6 (9 hours vs 4 hours for complete conversion) (Scheme IV.28) (60JCS539).

Since an electron-donating substituent retards (or even can stop) the Dimroth rearrangement, one can expect that purines react more slowly than pteridines, as in alkaline media the purine ring is negatively charged, which leads to deactivation of the pyrimidine ring for addition and ring opening. This explains why the ammonia-induced rearrangement of 1-methyladenine into 6-methylaminopurine (60JCS539) and the formation of 6-methylaminopurine from 1,6-dihydro-1-methyl-6-thiopurine and ammo-

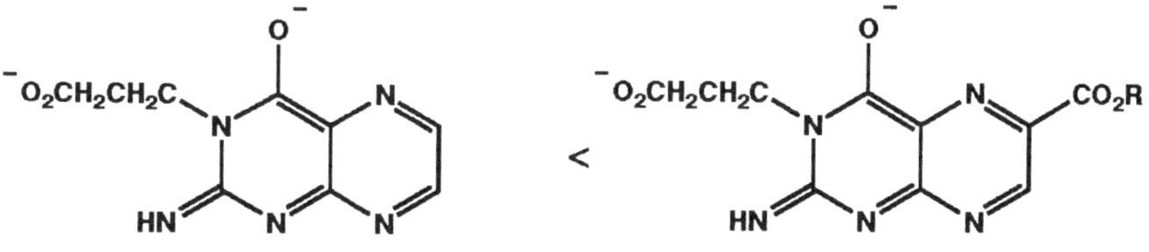

SCHEME IV.28

nia (60JA3147; 62JOC2478) require rather drastic conditions (Scheme IV.29) and why the 7-methyl derivative rearranges easily into the corresponding 6-methylamino-7-methylpurine just by boiling in water, even without using a base (Scheme IV.29) (62JOC2622). Also, the rearrangement of 1-amino-9-benzyl-6-iminopurine into 9-benzyl-6-hydrazinopurine by hydrazine occurs under mild conditions (Scheme IV.29) (88JOC382). All these degenerate rearrangements involve bond breaking between N(1) and C(2).

Similar base-induced Dimroth rearrangements have also been reported with the related iminopyrazolo[3,4-*d*]pyrimidines (60JA3147) and 6-imino-8-azapurines [73JCS(PI)2659].

Dimroth rearrangement under acidic conditions has also been reported; acid hydrolysis of 1-hydroxyisoguanine provides 6-hydroxylamino-2,3-dihydro-2-oxopurine (Scheme IV.29) (67JOC1151).

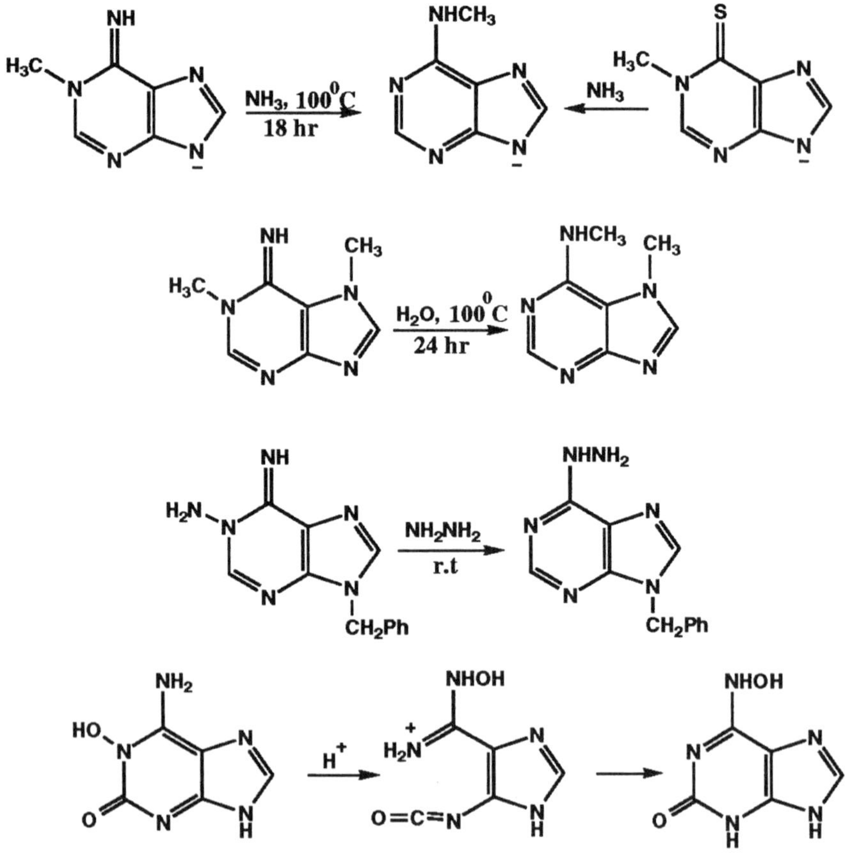

SCHEME IV.29

The Dimroth rearrangement was also succesfully applied to produce N^6-functionalized adenosines from their N(1) isomers. The conversion of N(1)-methyladenosine into N^6-methyladenosine is very slow at pH 6.0 and 7.0 and requires harsh alkaline conditions (pH 10–11, 60–70°C, 2 hours) to achieve an acceptable rearrangement rate (68B3453; 73JOC2247; 77CB373). However it was surprisingly found that milder rearrangement conditions could be applied (pH 6.0–7.0, 50°C, 7 hours) when at N-1 of adenosine a β-aminoethyl group is present, i.e., **67** [95H(41)1399]. It is quite interesting that the rearrangement not only leads to the formation of the N^6-(β-aminoethyl)adenosine (**68**) (65%), but also to a new rearrangement product, namely N^6-ethanoadenosine (**69**) (32%) (Scheme IV.30). These percentages do not change even after prolonged heating, clearly indicating that

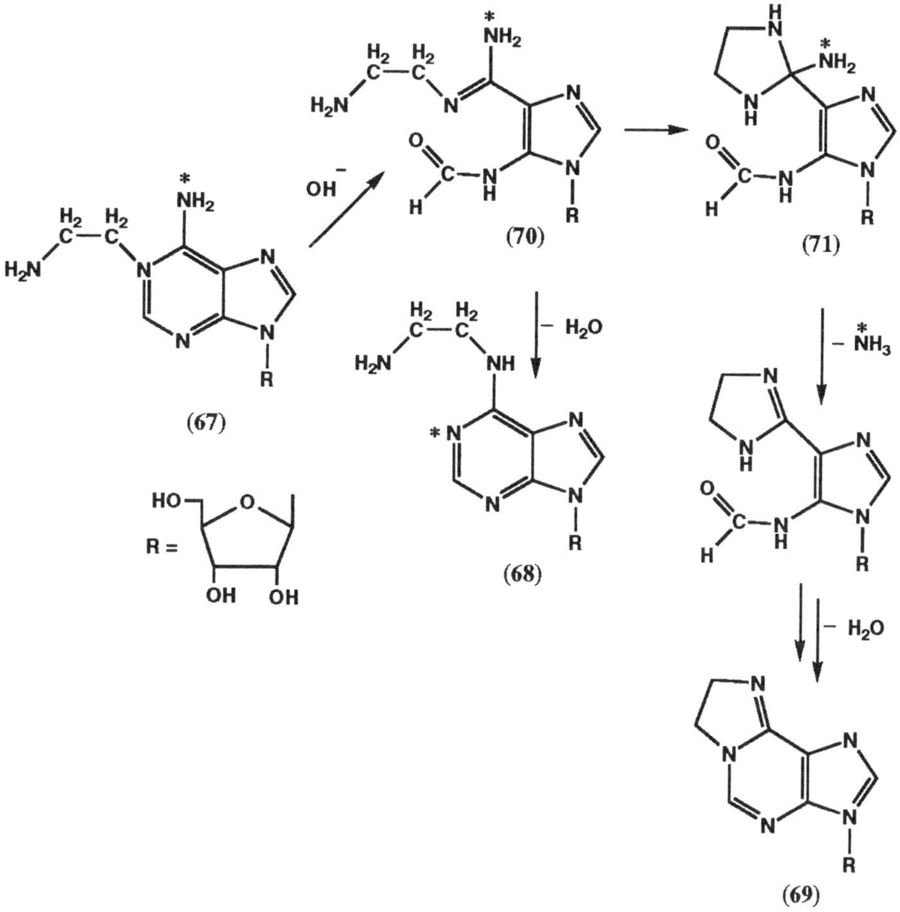

SCHEME IV.30

68 and **69** are formed in two parallel, independent reactions. Therefore, the conclusion seems justified that the ethano derivative **69** cannot serve as the precursor of **68** [95H(41)1399].

It seems reasonable to assume that the open-chain compound 5-(formylamino)-4-[(N-β-aminoethyl)formamidino]imidazole (**70**) is intermediate for the formation of both products **68** and **69**. Recyclization can occur to either the N^6-derivative **68** or to 5-(formylamino)-4-(2-aminoimidazolin-2-yl)imidazole (**71**). A subsequent internal cyclization between the NH group and the carbonyl functionality gives the ethano derivative **69** (Scheme IV.30). Support for this mechanism comes from the observation that adenosine **67**, being ^{15}N-labeled on the 6-amino group, yields unlabeled **69** [95H(41)1399].

It is of interest to mention that in the presence of small amounts of sodium phosphate or arsenate the formation of **69** becomes the dominant reaction; at pH 6.0 and 50°C the ratio **69/68** is 90/10. The strong catalytic effect of the phosphate and arsenate anions on the rate of formation of the ethano derivative **69** bears some analogy to the semicarbazone formation between ketone and amine, which is also found to be effectively catalyzed by these anions (68JA4319).

Based on the experience mentioned previously, the Dimroth rearrangement was succesfully applied for the preparation of water-soluble macromolecular adenosine derivatives of the redox enzymes **72** (NAD(H), NADP(H), and FAD) (Scheme IV.31) (86MI1; 87MI3, 87MI2; 88H1623; 89MI1; 90MI1).

Several studies have been performed on the formation of N^6-methyladenosine during heating of milk (94MI2). This compound, not being present in raw milk, has proved to be formed by a thermally induced Dimroth rearrangement of 1-methyladenosine, one of the compounds in raw milk. Study of the kinetics of the rearrangement revealed that these data are suitable as chemical process parameters in heat processing of milk (94MI2).

A Dimroth rearrangement has been used as a chemical tool to determine whether the product obtained by oxidation of N^6-benzyladenine with *m*-perchloroperoxybenzoic acid was an N-1 or N-3 oxide (96CPB967). After conversion of the N-oxide into the $^+N\text{-}OCH_3$ derivative, it was found that this methoxy derivative can undergo a rearrangement into 1-benzyl-N^6-methoxyadenine, which after a nonreductive removal of the benzyl group yields N^6-methoxyadenine (**73**) (Scheme IV.32). From this result it may be concluded that the oxidation has taken place at position N-1 and not at N-3. It is of interest to mention that oxidation of N^6,N^6-dimethyladenine gives the N-3 oxide (96CPB967). A similar rearrangement was also observed with 1-alkoxy-7-alkyladenines, affording N^6-alkoxy-7-alkyladenines (97CPB832).

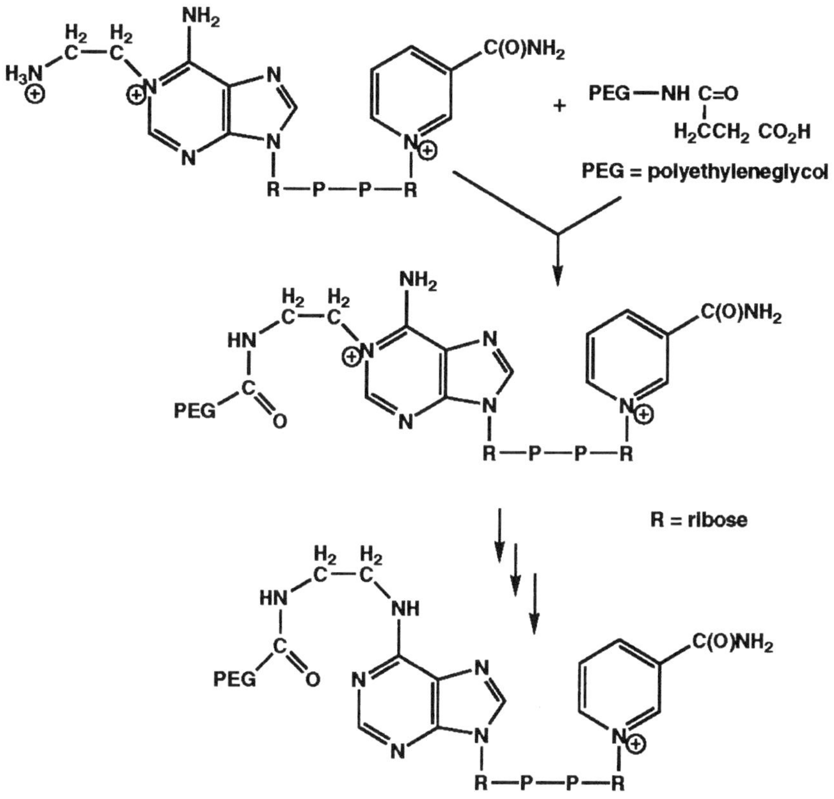

DEGENERATE RING TRANSFORMATIONS 179

SCHEME IV.31

Another interesting rearrangement, involving a pyrimidine–pyridine ring transformation combined with a Dimroth rearrangement, is observed when 6-nitro[1,2,4-triazolo][1,5-a]pyrimidine (**74**) reacts with ethyl cyanoacetate,

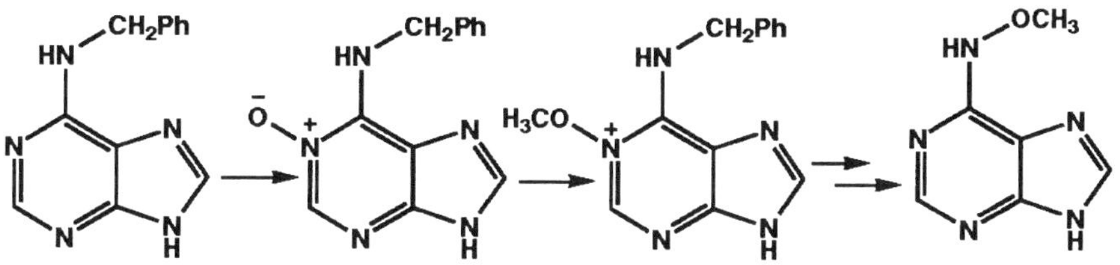

SCHEME IV.32

9-oxo-7-nitro-4,9-dihydrotriazolo[1,5-*a*]pyrido[2,3-*d*]pyrimidine (**79,** $X =$ O) being formed (Scheme IV.33) [87KGS857; 90KGS256, 90KGS1632, 90S713; 91KGS101; 92KGS116, 92TL3695; 95H(40)441]. As an intermediate 1-[1,2,4-triazolyl]-2-imino-5-nitropyridine-3-carboxylate (**77**) was isolated; it undergoes a further 1,3-exoannular rearrangement into 2-[1,2,4-triazol-5-ylamino]-5-nitropyridine-3-carboxylate (**78,** $R' = CO_2C_2H_5$). NMR spectroscopic investigation of the reaction of dimethylamino-6-nitro-1,2,4-triazolo[1,5-*a*]pyrimidine (**74,** $R = N(CH_3)_2$ with ethyl cyanoacetate shows that first addition takes place of the nitrile at position 7, i.e., formation of **75** ($R = N(CH_3)_2, R' = CO_2C_2H_5$). Ring opening and ring closure leads to **77**. It has been suggested that the formation of **77** could also occur by an initial intramolecular cyclization between the cyano functionality and the NH group in **75**, yielding the tricyclic compound **76**, followed by ring fission at bond *a*.

That the nitropyrimidine–nitropyridine ring transformation is not unprecedented and in fact well documented, and many examples of this kind of ring transformation by action of substituted acetonitriles have been described [95H(40)441]. Intermediate **77** undergoes an exocyclic nitrogen/ring nitrogen exchange (Dimroth rearrangement) into **78.** Support for this mechanism was provided by the results of a study of the reaction with ethyl cyanoacetate, containing two isotopic labels (^{13}C and ^{15}N) in the nitrile group (93JOK78G). Both isotopes were shown to be present at N-1 and C-2 in the pyridine ring containing the nitro group, providing evidence that both atoms originate from the cyano function of the reagent (Scheme IV.33).

With ethyl cyanoacetate, the 9-oxo compound **79** ($X = O$) is obtained. When the reaction is carried with malononitrile a similar rearrangement into the 9-imino compound **79** ($X = N$) takes place.

Also in the quinazoline series Dimroth rearrangements have been reported. 2,3-Dihydro-2-imino-3-methylquinazoline (**81**) rearranges very easily (alkali, room temperature) into 2-methylaminoquinazoline (68AJC2813) (Scheme IV.33a).

As expected, the conversion of 2-amino-3,4-dihydro-3-alkyl(aryl)-4-oxoquinazoline (**82**) into 3,4-dihydro-2-alkyl(aryl)amino-4-oxoquinazoline is quite slow; it requires boiling in 10 N sodium hydroxide solution for 8 hours to get complete conversion (Scheme IV.33a) (60JCS3540).

Similarly, the 1,2,3,4-tetrahydro-4-imino-1-methyl-2-oxo-3-phenylquinazoline (**83**) rearranges into the 4-anilinoquinazoline (Scheme IV.33a) (62JOC2622). Heating of 2-thio-4-iminotetrahydroquinazoline **84** in DMF leads to the rearranged product **85** (62JOC2622). In this 1,3-*exo*annular rearrangement the imine **84** acts as basic catalyst. Attempts to effect the rearrangement with dilute sodium hydroxide resulted only in hydrolysis of the imino group. The mechanism can be described as an ANRORC mechanism (Scheme IV.33a).

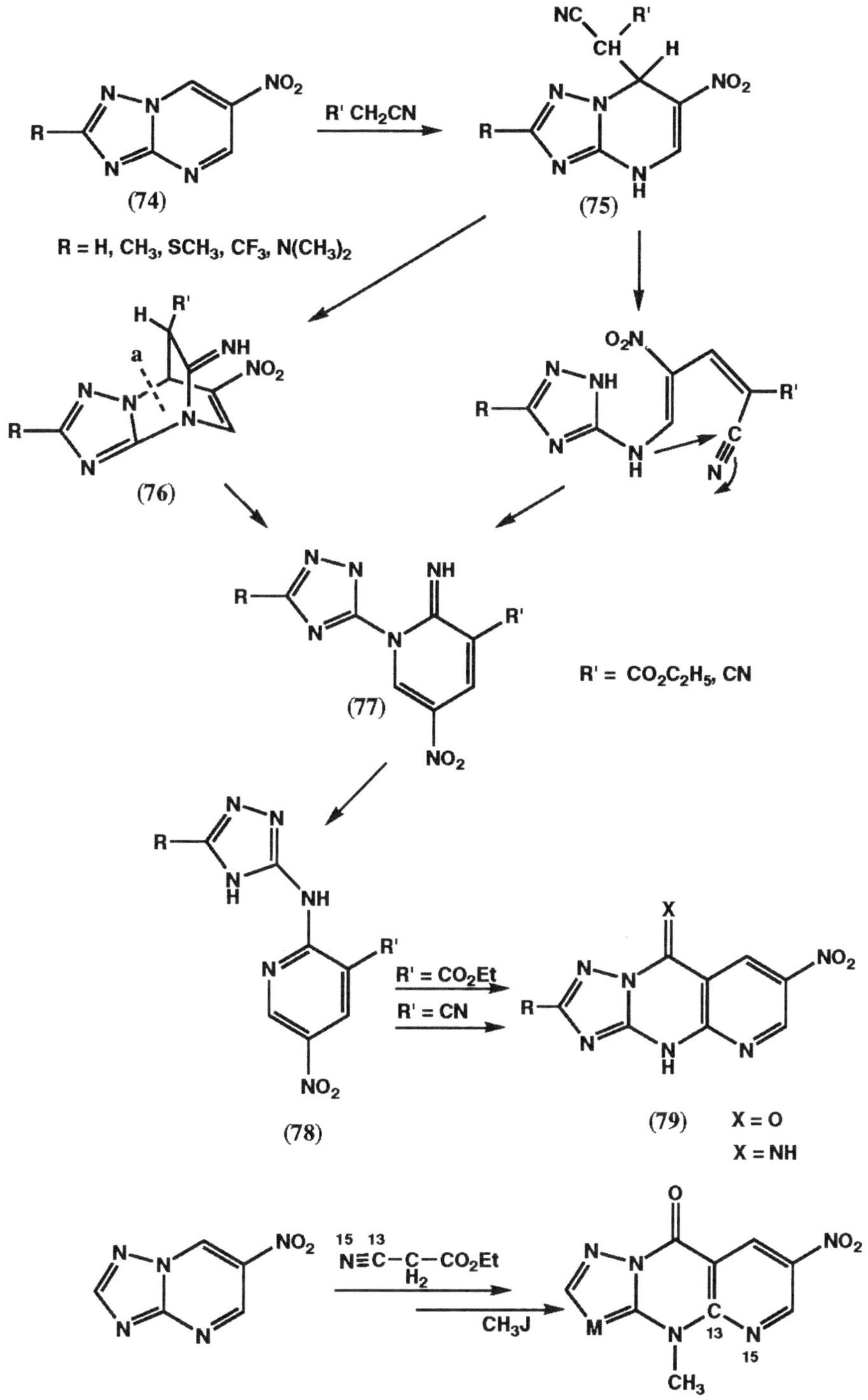

(74)
R = H, CH3, SCH3, CF3, N(CH3)2
R' CH2CN
(75)
(76)
(77)
R' = CO2C2H5, CN
(78)
R' = CO2Et
R' = CN
(79)
X = O
X = NH
CH3J

SCHEME IV.33

SCHEME IV.33A

1,3-Exoannular carbon–carbon exchange. As already discussed in the Introduction of this chapter (Section IV,A), rearrangements that involve the participation of one atom in the side chain can be categorized in two types: rearrangements between hetero atoms (see Scheme IV.1) and transformations that involve a 1,3-carbon–carbon exchange. Numerous examples of 1,3-exoannular heteroatom exchanges are found and have been discussed in Sections IV,B,2(A–D). The rearrangements involving an exchange between a ring carbon and side-chain carbon are less frequently encountered. They are observed in both five- and six-membered heterocycles, but mainly observed in pyrimidines, containing in position 5 a cyano functionality or a carbalkoxy group.

An early example of a degenerate ring transformation involving a base-induced interchange between a ring carbon and a carbonyl side-chain carbon is the conversion of 1-phenyl-2,3-dimethyl-4-formylpyrazolinone-5 (**85A**) into 1-phenyl-2-methyl-4-acetylpyrazolinone-5 (**85B**) (50LA84). The reaction can be explained by addition of the hydroxide ion to the α, β-unsaturated aldehyde function, ring opening between the 2,3-bond, and recyclization (Scheme IV.33b). Based on these results, the "antipyrinaldehyde" described in the literature has, not structure **85A**, (39G664; 40G410), but that of its rearranged product **85B**.

A very similar rearrangement has been discovered during base treatment of 4-hydroxymethylene-5-oxazolone to oxazole-4-carboxylic acid (60CB1033) (Scheme IV. 33b). Investigation of the mechanism of the reaction by isotopic ^{14}C-labeling (60JOC1819) revealed that when the carbonyl carbon of the oxazolone ring is labeled, in the rearranged oxazole-4-carboxylic acid the label is present on the carboxyl carbon. This degenerate ring transformation can also be described by the ANRORC mechanism, involving the initial attack of the base at the 2-position of the oxazolone ring (see Scheme IV.33b).

Another example of a 1,3-exoannular rearrangement, involving a 1,3-carbon exchange, is the base-catalyzed conversion of 2-amino-3-formylthiophenes into 3-cyanothiophenes [77JCR(S)294]. The reaction in-

SCHEME IV.33B

volves addition of the base and bond fission between sulfur and C-2, creating the possibility of a recyclization leading to an exchange of the carbon at C-2 with that of the carbon of the carboxaldehyde group (Scheme IV.33b).

A 1,3-exoannular carbon–carbon rearrangement in the pyrimidine series is the degenerate ring transformation observed when 1,3-dimethyl-5-cyanouracil (**86**) is treated with acetone in a basic solution. Two products were obtained: 1,3,7-trimethyl-pyrido[2,3-*d*]pyrimidine-2,4(1*H*,3*H*)-dione (**87**) (Scheme IV.34) and 1,3-dimethyluracil-5-carboxamide, formed by hydrolysis of the nitrile group in **81** (82JHC1261). The formation of **87** can be explained by an initial formation of an adduct at position C-6 in **86**, a position that is highly susceptible to nucleophilic attack because of the electron-withdrawing character of the cyano group. A subsequent retro-Michael reaction leads to ring opening of the C-6 adduct. Recyclization in the open-chain compound by attack of the terminal urea nitrogen on the cyano group affords the 6-aminouracil **88**. Intramolecular cyclization of the amino group with the keto function yields **87**. In this process it is the carbon of the cyano function that is incorporated into the original position 6 of the pyrimidine ring.

Since the pyrido[2,3-*d*]pyrimidine ring system is found to exist in a number of biologically active compounds, these rearrangements were extensively studied. Reaction of compound **86** with malononitrile leads to 7-amino-6-cyano-1,3-dimethylpyrido[2,3-*d*]pyrimidine-2,4(1*H*,3*H*)-dione (**90,** R = CN), and with ethyl cyanoacetate, 7-amino-6-ethoxycarbonyl-1,3-

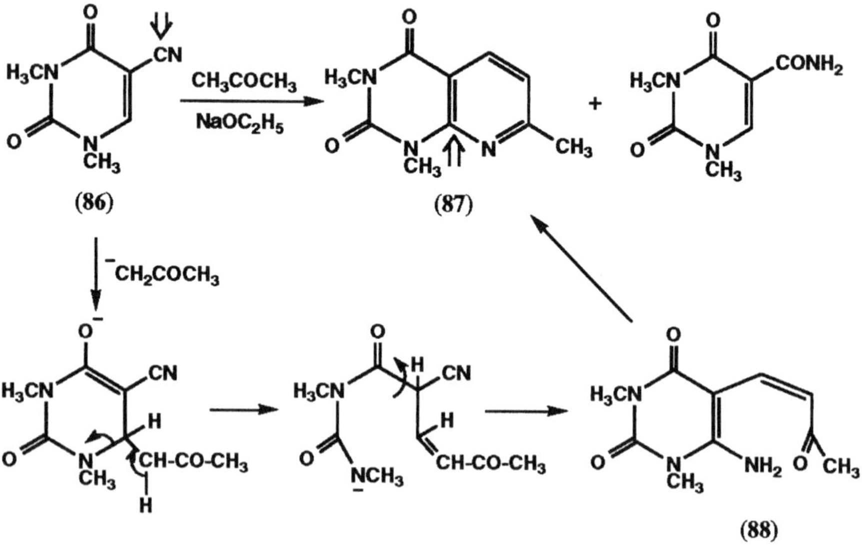

SCHEME IV.34

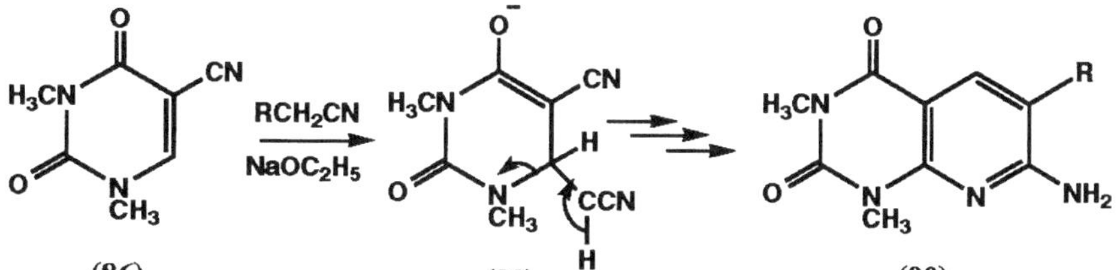

(86) (89) (90)

R = CN, $CO_2C_2H_5$

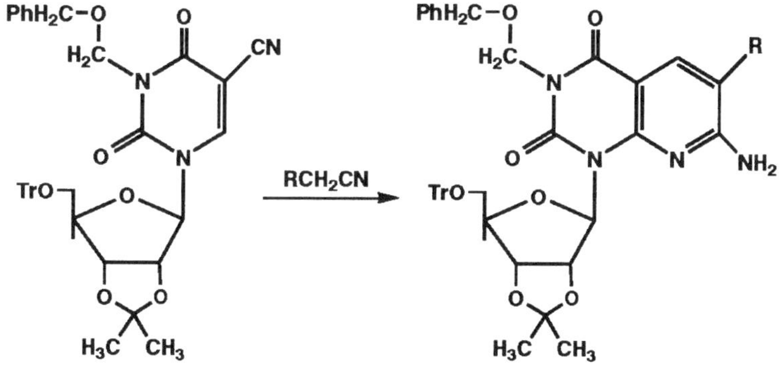

R = $CONH_2$, CN, $CO_2C_2H_5$

SCHEME IV.35

dimethylpyrido [2,3-*d*]pyrimidine-2,4(1*H*,3*H*)-dione (**90,** R = $CO_2C_2H_5$) is obtained (Scheme IV.35) (82JHC1261).

The formation of the adduct between **86** and the nitrile, i.e., **89,** occurs more readily than that between **86** and ketones, since an activated nitrile is a better nucleophile than a ketone. Since the α-proton in the adduct **89** is more acidic than the α-proton in the ketonic adduct, also the ring opening will occur more easily. The interchange of a nitrile carbon with the ring carbon of a pyrimidine ring was also observed with the 3-benzyloxymethyl-1-ribosyl-5-cyanouracil. With a series of activated nitriles, the protected bicyclic nucleosides are formed. After deprotection, the corresponding bicyclic nucleosides are obtained (Scheme IV.35).

Other examples of ring transformations involving the incorporation of the carbon of the cyano group into the heterocyclic ring have been observed in the chemistry of the nitrogen bridgehead compounds, the 1,3*a*,6*a*-triazaphenalenium salts (82TL2891). Reaction of the 1-dimethylamino-5-cyano-2,3*a*,6*a*-triazaphenalenium chloride (**91,** R' = H, i-C_3H_7, C_6H_5; R'' =

H, C_6H_5) with hydroxide yields a product whose structure was proved to be that of betaine **92**. It contains, besides an imino group at position 4 a formyl function at position 5 (Scheme IV.36) (83H1891).

Since the 2,3a,6a-triazaphenalenium salts can easily undergo Michael-type additions with a bisulfite or cyanide anion at position 4 (83H579), it is suggested that the degenerate ring transformation of **91** into **92** involves initial formation of the C-4 adduct. Ring opening, conversion into a rotational isomer, and an intramoleculair recyclization between the cyano group and the dihydropyridine ring nitrogen (highly nucleophilic because of the presence of the strong electron-donating character of the dimethylamino group) gives **92** (Scheme IV.36). Similar rearrangements were observed with the 5-ethoxycarbonyl compound, yielding betaine **93** (83H1891), and with hydrazine (82TL2891).

The addition at C-4 is also supported by quantum chemical calculations showing that the minimum of the nucleophilic moleculair potential map is located near the triple bridgehead C-9b, and that the interaction potential at C-4 is only slightly higher than at C-9b. However, addition of the hydroxide at C-9b is less favored, since it leads to considerable distortion (73JA6894; 82TL2891). An interesting extension of the ring transformations as discussed in Scheme IV.35 is the transformation of **91** into the

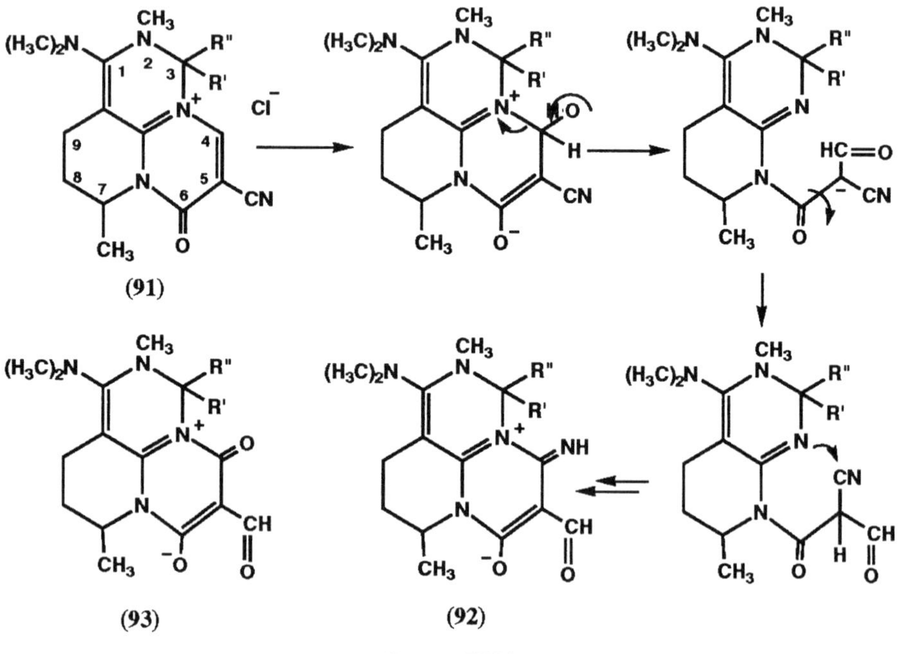

SCHEME IV.36

tetracyclic system **94** on treatment with malononitrile (Scheme IV.37) (85H2549). Again an addition of the nitrile to C-4 is involved, followed by ring opening and reclosure. Because of the presence of the amino group in a position adjacent to the dicyanoethylidene group, ring cyclization into **94** occurs. So, the formation of **94** is the result of three subsequent reactions: ring cleavage and two cyclizations, one caused by the original and one by the newly formed nitrile group.

A very similar reaction has been reported to occur when the 4-cyano-1-thia-2a,5a-diazaacenaphthene **95** reacts with malononitrile in a basic solu-

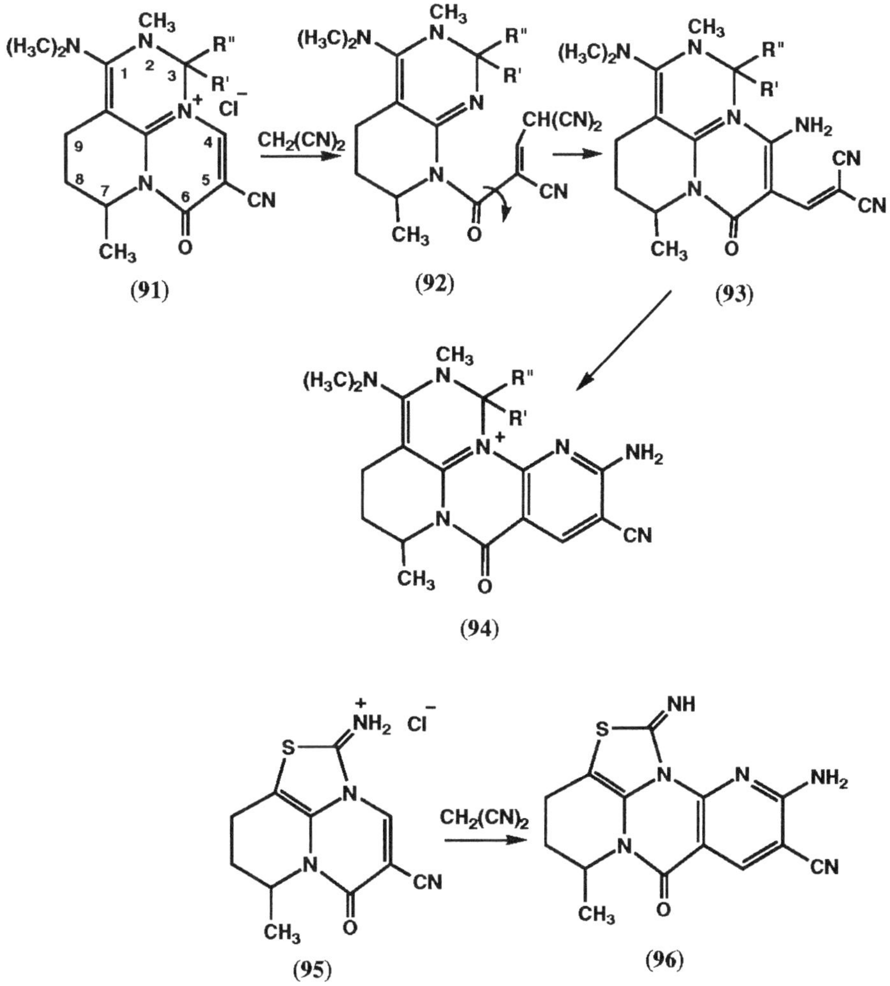

SCHEME IV.37

tion at room temperature (85H2289; 86H69). The tetracycle **96** is obtained (Scheme IV.36); its formation also involves an aminodicyanoethylidene derivative as intermediate.

Interesting chemistry concerning 1,3-carbon exchange rearrangements was reported on base treatment of 5-cyano-1,2-dihydro-2-imino-1-methylpyrimidine (**97**) with a base. With dilute ammonia, the expected Dimroth rearrangement product 5-cyano-2-(methylamino)pyrimidine is formed (Scheme IV.38). The intermediary acyclic compound **98** could be isolated and satisfactorily analyzed. However, when the 5-cyano compound **97** was treated with alkali (pH 14) fast formation of 6-amino-5-formyl-1,2-dihydro-2-imino-1-methylpyrimidine (**99**) took place [66JCS(C)164]. The interchange between the carbon of the nitrile and the ring carbon C-6 takes place via intermediate **98,** which reacts in the strong alkaline medium to **99** by addition of the methylamino group to the nitrile group. That in a strong basic solution the addition of the amino group to the formyl group in **98** is less favored than the addition of the aminogroup to the nitrile is probably due to the intrinsic property of the aldehyde group to form in strong alkali a hydrate deactivating the aldehyde group for cyclization by water elimination. Being an α-N-methylated imine, compound **99** undergoes a further rearrangement into 5-carbamoyl-2-methylaminopyrimidine in strong alkali. This rearrangement is slow because of the presence of the electron-donating amino group at position 6, retarding the addition of the hydroxide ion to C-6. It is of interest to mention that the 5-cyano-1,2-dihydro-2-imino-1,4,6-trimethylpyrimidine gives a normal Dimroth rearrangement into 5-

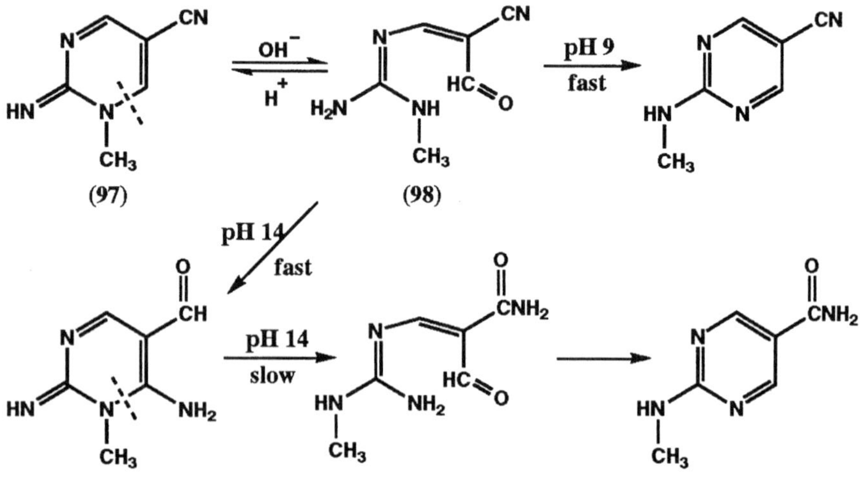

SCHEME IV.38

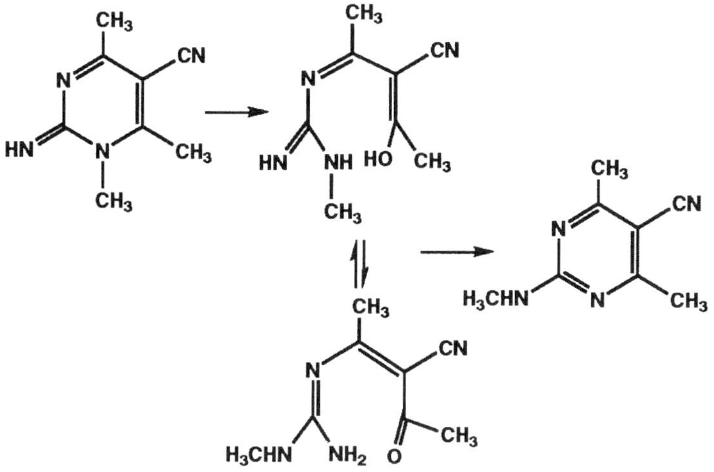

SCHEME IV.39

cyano-4,6-di-methyl-2-(methylamino)pyrimidine [67JCS(C)903] (Scheme IV.39). In the ring-opened intermediate **101,** no hydrate formation in the acetyl group takes place and recyclization by addition of the amino group to the carbonyl functionality is the sole process.

C. Degenerate Ring Transformations Involving the Participation of Two Atoms of a Side Chain

(1,2)–(4,5) Exoannular rearrangements, categorizing the interchange of two atoms in the side chain with two identical atoms of the heterocyclic ring, have been observed in both five- ($n = 2$) and six-membered heterocycles ($n = 3$). These degenerate rearrangements can be schematized as given in Scheme IV.2.

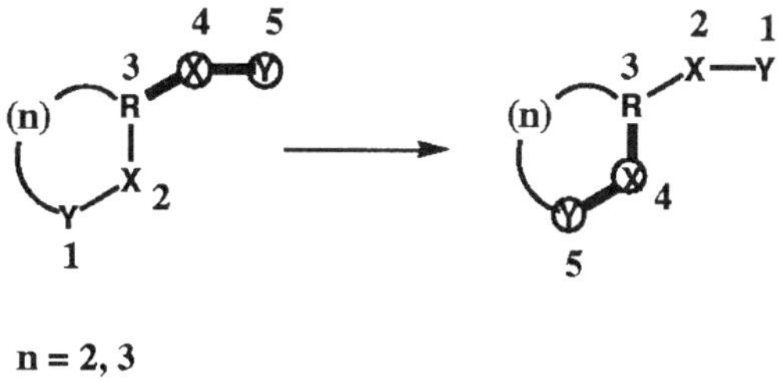

SCHEME IV.2

1. FIVE-MEMBERED HETEROCYCLES

a. *Oxazoles and Isoxazoles*

Among the five-membered heterocyclic compounds, the thermal interconversion of 4-carbonyl-substituted oxazoles **102** into 4-carbonyl-substituted oxazoles **104** is an interesting, well-documented example of a degenerate rearrangement involving "exchange" between the C–C–O side chain and the C–C–O fragment of the oxazole ring (Scheme IV.40) (75CRV389, 75JOC1521; 79CRV181). This interconversion is known as the Cornforth rearrangement, named for its discoverer (49MI1).

The thermally induced Cornforth rearrangement can be rationalized by postulating the dicarbonylnitrile ylid **103** as intermediate. It is especially the natures of the substituents R' and R'' that determine the direction of the rearrangement. For example, heating of 5-alkoxy-4-(aminocarbonyl)oxazole (**102**, R' = OAlkyl, R'' = NR_2) readily gives in a good yield rearrangement into the isomeric 5-(substituted amino)-4-(alkoxycarbonyl) oxazole (**104**, R' = OAlkyl, R'' = NR_2) (Scheme IV.40).

Further supporting evidence for this kind of degenerate interconversion was obtained by using deuterium-labeled oxazoles. It was observed that heating of 2-phenyl-5-methoxy-4-[(methoxy-d_3)carbonyl]oxazole leads to scrambling of the trideuteriomethyl group at position 5 and the methyl ester group at position 4 (Scheme IV.41) (75JOC1521).

A similar degenerate ring interconversion, but under photolytic conditions, has been observed with isoxazoles (Scheme IV.42) (75JA6484; 76HCA2074). Irradiation of 3-phenyl-5-methyl-4-(acetyl-d_3)isoxazole leads to a 1:1 equilibrium mixture, in which the trideuteriomethyl group is scrambled over the positions 4 and 5. The reaction involves photolytic cleavage of the N–O bond, leading to a resonance-stabilized diradical intermediate **105**, in which ring closure occurs (Scheme IV.42). In a similar way, irradiation of 3,5-diphenyl-4-acetylisoxazole leads to the formation of 3-phenyl-4-benzoyl-5-methylisoxazole.

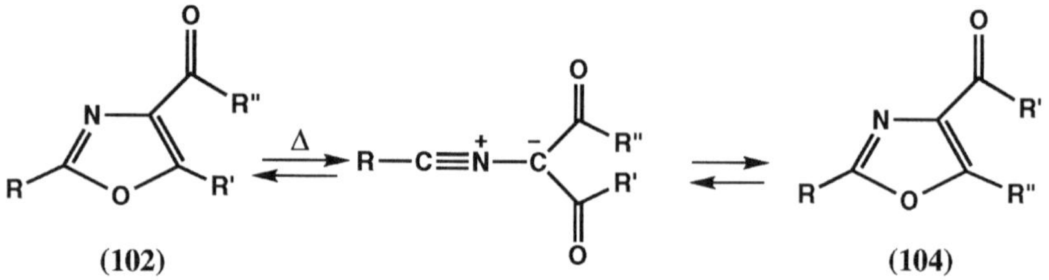

SCHEME IV.40

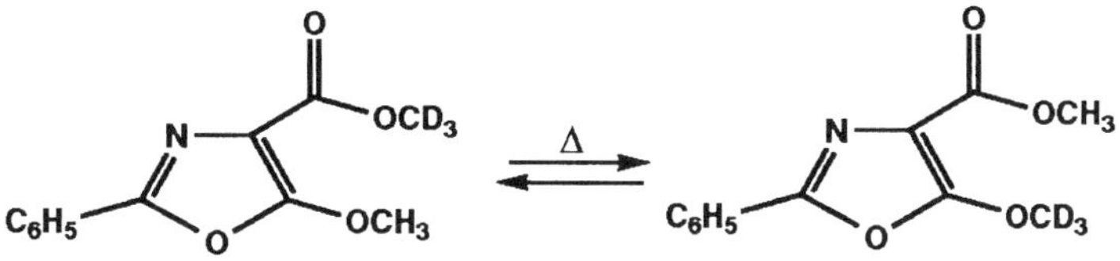

SCHEME IV.41

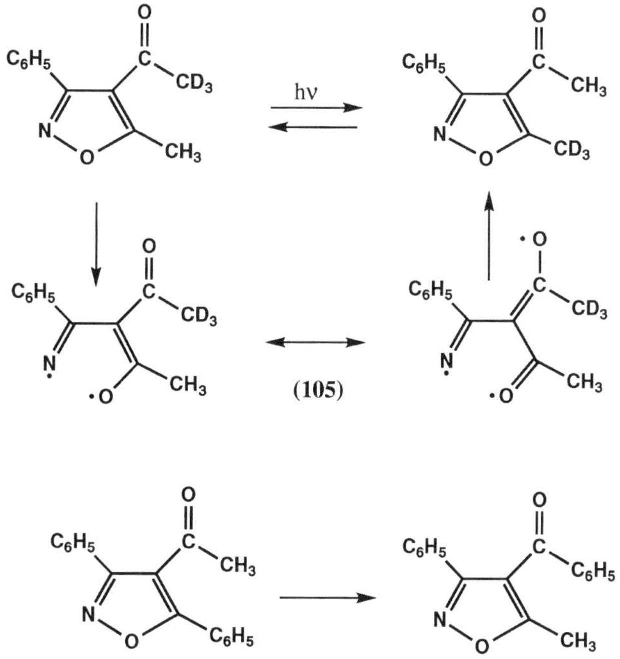

SCHEME IV.42

b. *1,2,3-Triazoles*

4-Iminomethyl-1-phenyl-1,2,3-triazoles (**106**, R = CH$_3$) and their structural isomer 4-(iminophenyl)-1-methyl-1,2,3-triazole (**108**, R = CH$_3$) are interconvertible when they are heated in DMSO at 80°C (Scheme IV.43) (84JHC627; 89JHC701; 90JHC2021). The equilibrium position is dependent on the electronic properties of the substituent R. When R is alkyl, a benzyl or anisyl compound **108** is favored; when R is *p*-chlorophenyl or *p*-nitrophenyl, **106** is the favored isomer. The hydrazone **106** (R = NH$_2$) or oxime **106** (R = OH) do not rearrange. A diazoimine **107** can be postulated as intermediate, a species whose structure is similar to the one proposed

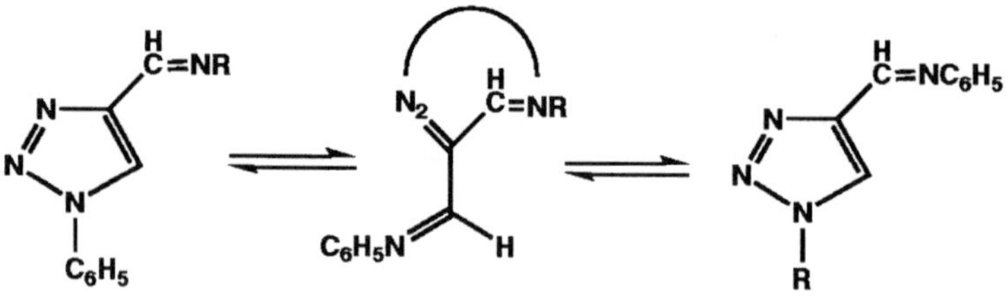

SCHEME IV.43

in the conversion of 2-amino-1-aryl-1,2,3-triazole into 2-arylamino-1,2,3-triazole (Section IV,B,1,a).

All the compounds that undergo this type of rearrangement are found to have the C=N bond in the *E*-configuration (as evidenced by the ^{13}CH=NH coupling constants) (90JHC2021). It has been argued that apparently a *cis* relationship between the triazole ring and the imino-nitrogen lone pair is a necessary structural requirement for rearrangement. In the light of this mechanism it is understandable that the hydrazone **106** (R = NH$_2$) and oxime **106** (R = OH) are reluctant to rearrange because of intramoleculair hydrogen bonding of the NH$_2$ or OH group with N-3 of the triazole ring. This is confirmed by ^{13}C NMR spectroscopy (90JHC2021).

Following the results of a mechanistic study of the azidomethine-1*H*-tetrazole isomerization (see Section IV,B,1,a), which are based on *ab initio* calculations (82JHC943), in the cyclization process of the intermediary diazoimine a bending of the diazofunction must take place because of the formation of the lone electron pair on the central nitrogen atom. This is accompanied by a π-electron flow toward the imine function and the formation of a σ-bond at the expense of the lone pair of the imino nitrogen (Figure IV.1) (90JHC2021).

Interesting synthetic applications of these degenerate rearrangements are the preparation of 1-alkyl-4-formyl-1,2,3-triazole **110** from 1-phenyl-4-

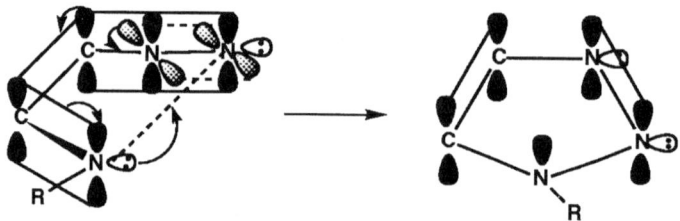

FIG. IV.1

formyl-1,2,3-triazole (**109**) via the intermediary formation of the 4-alkylimino derivative, thermal rearrangement, and acid hydrolysis (Scheme IV.44). In particular, this rearrangement reaction allows an easy entry to the synthesis of sterically hindered derivatives, such as (**110**, $R = t\text{-}C_4H_9, i\text{-}C_3H_7$). The thermal conversion of 5-hydroxy-4-iminomethyl-1,2,3-triazole into 1,2,3-triazole-4-carboxamide is another useful application of this type of rearrangement (Scheme IV.44).

Although a variety of degenerate ring transformations were reported involving 1H-1,2,3-triazoles, only in the early 1990s were similar reactions found with 4-amino-1-sulfonylated 1H-benzotriazoles (92JOC190). Apparently, the benzotriazole ring is less susceptible toward ring opening than the 1H-1,2,3-triazoles. When a solution of 4-amino-5-methyl-1-(p-toluenesulfonyl)benzotriazole (**111**) was refluxed in toluene for 3 hours, 7-methyl-4-(p-toluenesulfonylamino)benzotriazole (**112**) was isolated in 92% yield. The rearrangement is irreversible, indicating that the 7-methyl isomer is more stable than the 5-methyl isomer (Scheme IV.45).

Sulfonylation of 4-(p-toluenesulfonylamino)benzotriazole (**113**) with

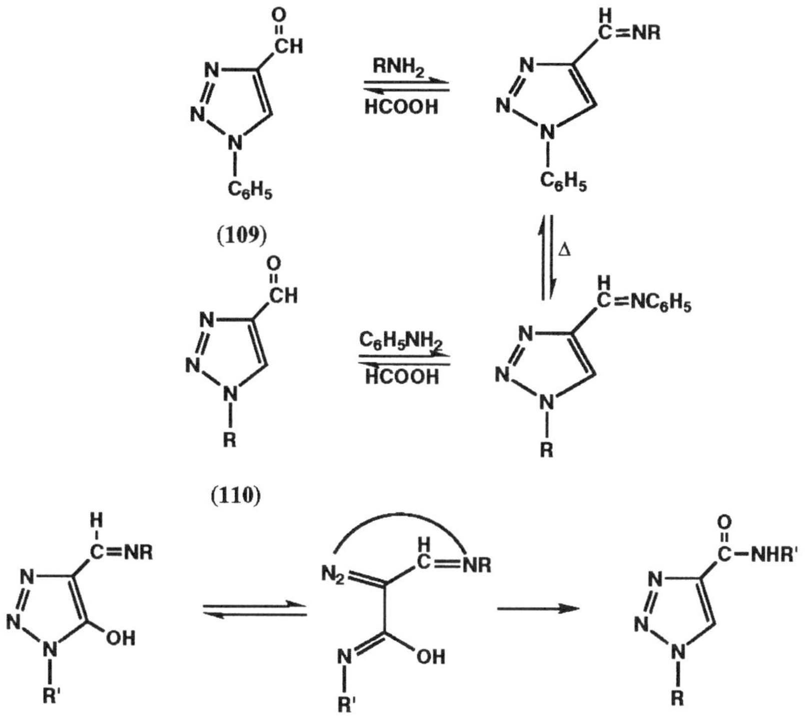

SCHEME IV.44

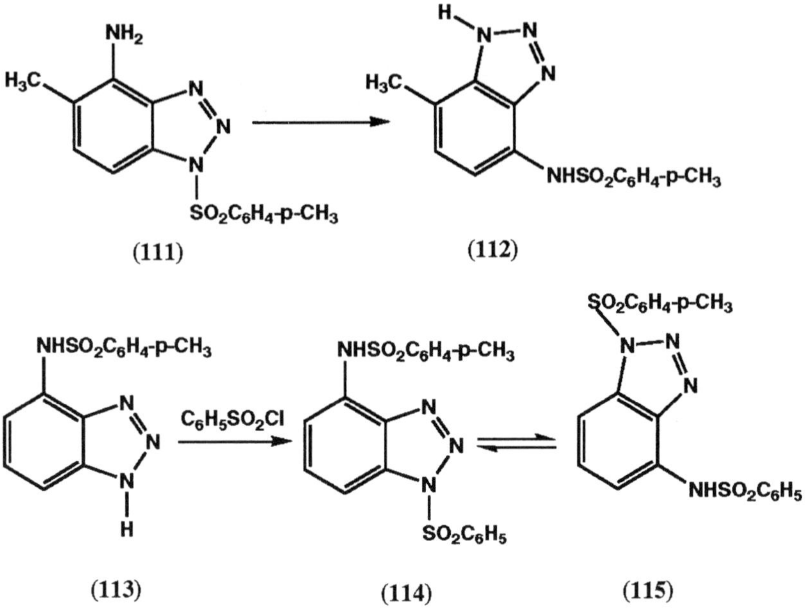

SCHEME IV.45

benzenesulfonyl chloride gave an inseparable mixture of the bissulfonylated compounds **114/115**.

Coalescence studies on the equilibrium **114/115** by recording the ^{1}H NMR spectra at different temperatures revealed that the coalescence temperature for the NH protons was found to be 65°C, and 70°C for the CH$_3$ protons. Based on these coalescence temperatures and the chemical shift differences of the two separate peaks in the slow exchange region (82MI2), the free energy of activation $\Delta G^{\neq}$ was calculated as 20.8 kcal/mol.

In order to exclude the possibility of an intermoleculair process in these degenerate ring transformations, crossover experiments with a mixture of two bissulfonylated benzotriazoles, i.e., 1-benzenesulfonyl-4-(benzenesulfonylamino)benzotriazole (**116**) and 4-(toluenesulfonylamino)-1-(p-toluenesulfonyl)benzotriazole (**117**) (Scheme IV.46), in DMSO-d_6 showed in the ^{1}H as well as in the ^{13}C NMR spectra the superposition of each of the isomers (92JOC190). No exchange of the arylsulfonyl groups was observed, justifying the conclusion that the rearrangement is not an intermolecular but an intramolecular process.

It was concluded that the rearrangement follows the same pattern as described in Scheme IV.5 for the Dimroth rearrangement of 5-amino-1,2,3-triazoles, i.e., the involvement of the intermediacy of a diazoimine (Scheme IV.46).

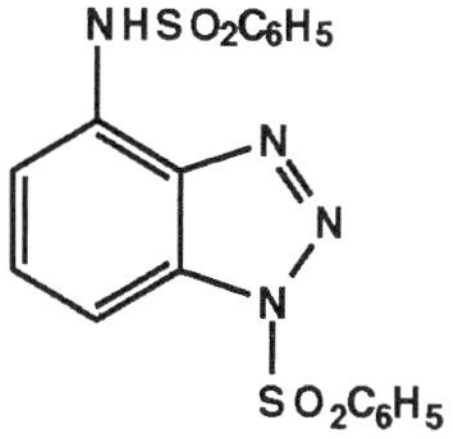

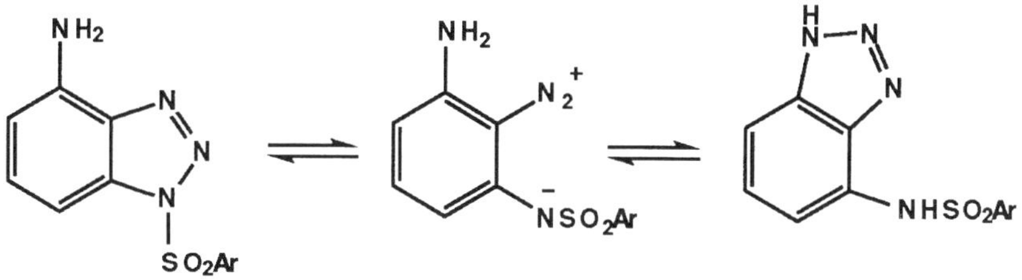

SCHEME IV.46

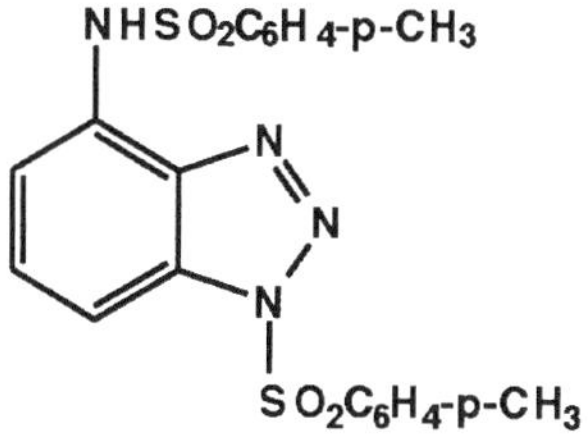

2. SIX-MEMBERED HETEROCYCLES

Degenerate ring transformations involving the interchange of two atoms in the side chain with two identical ring atoms have also frequently been encountered in azine chemistry. The six-membered heterocycles usually undergo these ring transformations in the presence of a base. The function of the base is twofold: It may form with the π-deficient six-membered substrate a 1:1 covalent σ-adduct, and it may induce the ring opening of the σ-adduct by proton abstraction.

Several degenerate ring transformation reactions have been reported involving the base-induced "exchange" of a C–N-containing sidechain, located at the 3-position of the ring nitrogen, with that of the C(2)–N(1) part of the azine ring (Scheme IV.2). Side-chain carbon–nitrogen functionality is present in the nitrile group, the carbonamide, and the amidine group, and all three groups have been found to take part in these interchange reactions. These degenerate ring transformations are schematically represented in Scheme IV.47.

a. *Pyridines*

An intriguing and informative example of a rearrangement involving C–N interchange has been reported to take place when the 1-methyl-

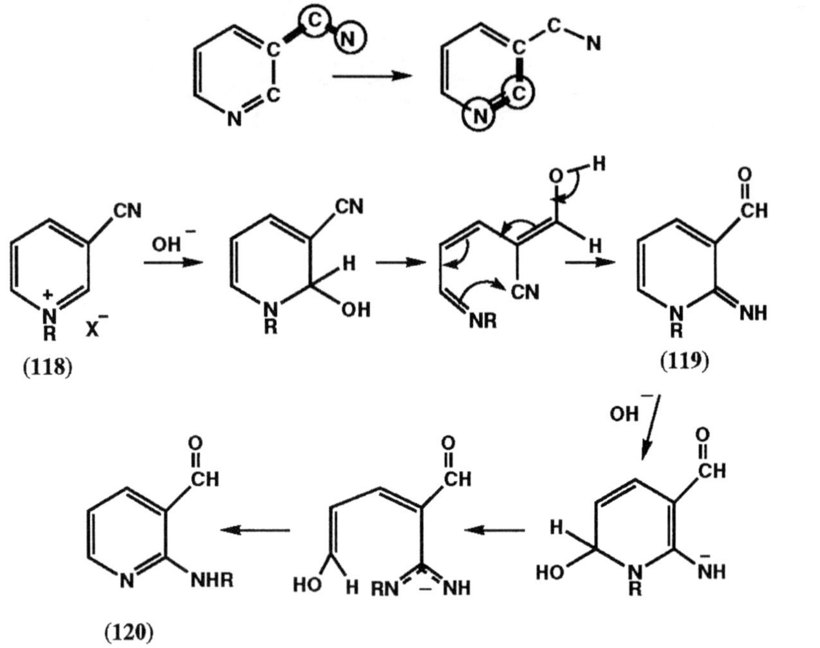

SCHEME IV.47

3-cyanopyridinium salt (**118**, $R = CH_3$) reacts with a base [72JCS(B)1892; 79KGS28, 79T809, 79T2591; 80T785]. The product is identified as 3-formyl-2-(methylamino)pyridine (**120**, $R = CH_3$). Its formation can only be understood when in this rearrangement two successive ANRORC processes are involved. First an process, which involves initial pseudobase formation at C-2, a subsequent ring opening and recyclization into 3-formyl-2-imino-1-methyl-1,2-dihydropyridine (**119**), and secondly a base-induced amidine (Dimroth) rearrangement of **119** into product **120** (Scheme IV.47). Compound **120** is also obtained when 1-ethyl-3-cyanopyridinium iodide (**118** $R = C_2H_5, X^- = I^-$) reacts with methylamine (79KGS28). It requires an ethylamine–methylamine exchange, probably in one of the open-chain intermediates. The 2-cyano- and 4-cyano-1-methylpyridinium salts do not undergo this remarkable rearrangement reaction; only hydrolysis into the corresponding carbonamides takes place.

A similar C–N interchange has also been reported with 3,5-dicyano-1,2,6-trimethylpyridinium salts (**121**), leading to the formation of 3-acetyl-2-(methylamino)pyridine derivative **122** (Scheme IV.48) (95T8599).

The same type of rearrangement could be observed with 3-cyano-5-nitropyridinium salt (**123**) when it reacts with methylamine, affording **124**

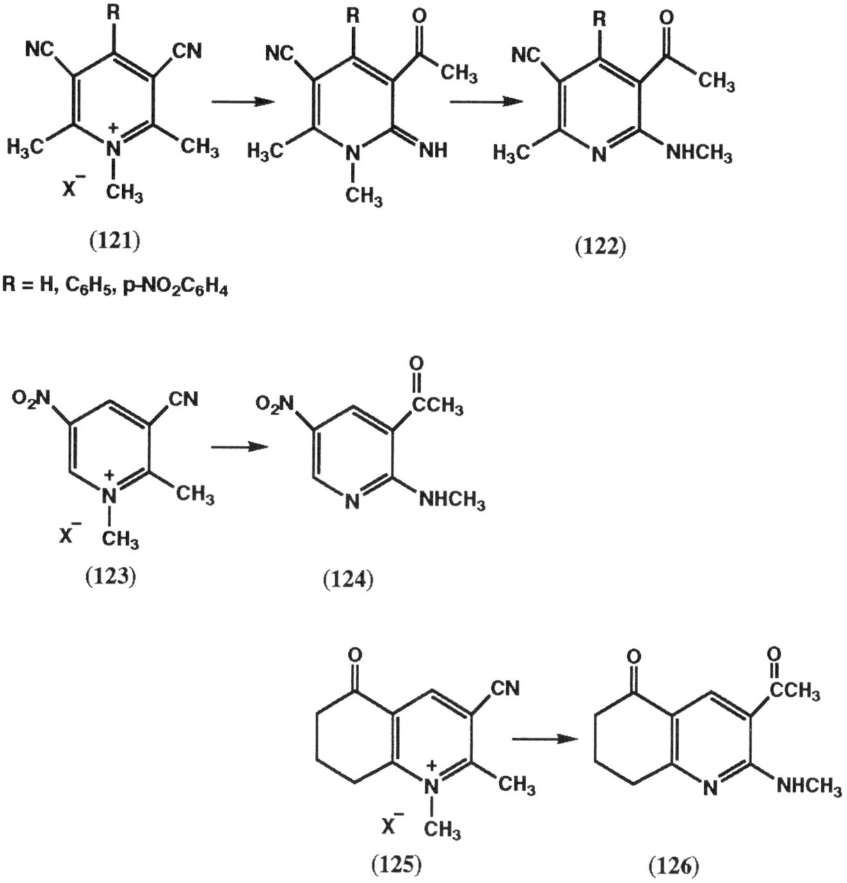

SCHEME IV.48

(95T8599), and with 3-cyano-1,2-dimethyl-tetrahydro-4-oxo-tetrahydroquinolinium salt (**125**), yielding 3-acetyl-2-(methylamino)-4-oxotetrahydroquinoline (**126**) on treatment with a base (Scheme IV.48) [93JCR(S)350].

The involvement of the aminocarbonyl group at position 3 of the pyridine ring to act as a C–N function for interchange with the C(2)–N part of the pyridine ring is demonstrated by the base-induced conversion of 1,3-di(aminocarbonyl)pyridinium salt **127** into 3-formylpyridone-2 (**130**) (71B2313) (Scheme IV.49). In principle this degenerate ring transformation can occur via two different covalent σ-adducts: (i) via the σ-adduct at C-2 **128**, formed between the highly electron-deficient C-2 position in the pyridinium salt and the hydroxide ion, and (ii) via the covalent C-6 σ-adduct **129**, formed by addition of the hydroxide ion at C-6. Both pseudobases can

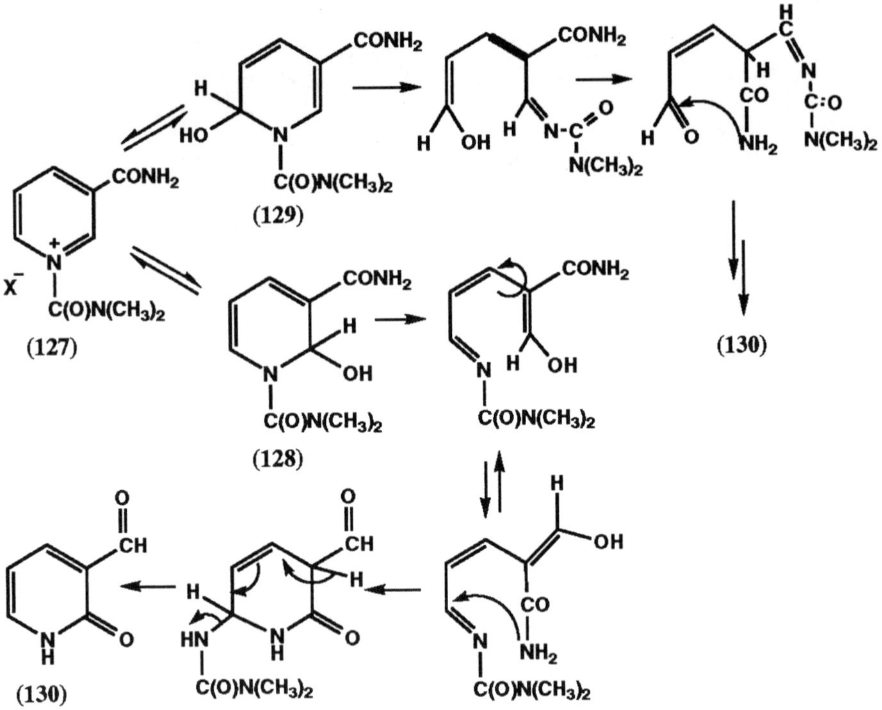

SCHEME IV.49

undergo the base-induced ring opening, formation of rotational isomers, ring closure, and aromatization (ANRORC process) (Scheme IV.49). No NMR measurements have been performed at various temperatures, which could give some information whether these adducts are formed in a kinetically or thermodynamically controlled process.

In the pyridine *N*-oxide series a C–N exchange rearrangement was also observed when 3-aminocarbonyl-1-methoxypyridinium salt (**131**) reacts with ammonia or alkali, 3-methoxyiminomethylpyridin-2(1*H*)-one (**132**) being obtained (74T4055). This conversion takes place via the ANRORC pathway; the ring opening occurs between N-1 and C-6 after initial addition of the hydroxide ion at C-6 (Scheme IV.50).

The involvement of an amidine group as C–N partner for interchange with the C(2)–N part of the pyridine ring has been observed during base treatment of the 3-formamidopyridinium salt; 2-amino-3-formylpyridine is formed in a reasonable yield (Scheme IV.50) [79ZN(B)1019]. The reaction follows the same pathway as described in Scheme IV.49.

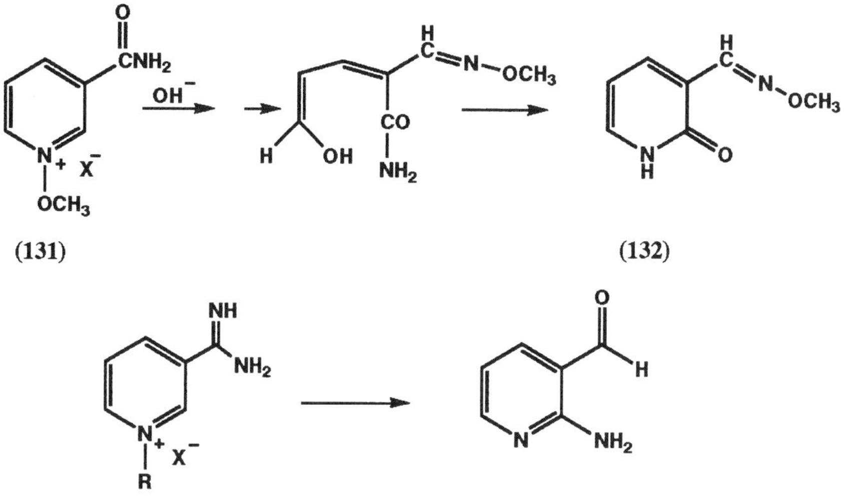

DEGENERATE RING TRANSFORMATIONS 199

(131) (132)

SCHEME IV.50

D. Degenerate Ring Transformations Involving Participation of Three Atoms of a Side Chain

Ring transformations involving the participation of three atoms present in a side chain can formally be considered as a (1,2,3)–(5,6,7) exoannular exchange. They are schematically represented in Scheme IV.3. In Scheme IV.3 the atoms *A, B, C, D,* and *R* may represent combinations of carbon, nitrogen, oxygen, and sulfur atoms. Since the first reports on this type of rearrangement [67JCS(C)2005], many other examples of these rearrangements have been frequently encountered, mostly in five-membered heterocycles with all kinds of different combinations of heteroatoms (74MI1; 81AHC(56)141; 92AHC(56)49). For a degenerate ring transformation, it is not only required that the atoms *DCB in* the ring and *outside* the ring be

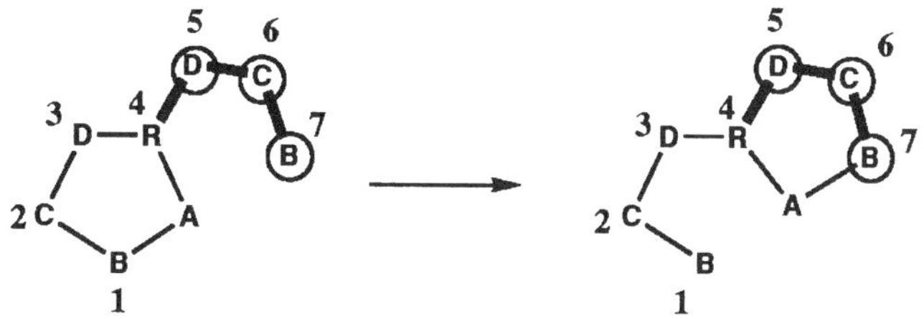

SCHEME IV.3

identical; the *sequence* of the atoms *DCB* in the side chain connected to R has to be the same as that for the atoms *DCB* in the heterocyclic ring.

Almost all of these reactions can formally be considered as proceeding by a nucleophilic attack of atom *B* on the pivotal center *A* and bond-breaking between the ring atoms *A* and *B* [S_Ni mechanism]. It has been argued [71BSF1925; 79CRV181; 80AG(E)947] that when the three-atom side chain has the character of an allyl-type side chain, being conjugated with the heterocyclic ring, the reaction can be viewed as a 6π-assisted hetero electrocyclization. The continuous π-electron system extended over the reaction partners in the transition state is decisive for such a process [76JCS(PI)315]. As intermediate a bicyclic species can be pictured with subsequent or concerted breaking of the *A–B* bond. When the pivotal atom is S, a thiapentalene-type intermediate or reversible bond switch mechanism, involving a hypervalent sulfur, has been suggested, the borderline of no bond resonance (80MI1) (see Section IV,D,2).

Rearrangements are also found with five-membered heterocycles, containing a *saturated* side chain of three atoms [79JCR(S)64; 82JCS(PI)759; 92JCS(PI)3069; 93CB1835; 94H(37)2051].

1. Degenerate Ring Transformations Involving Nitrogen as Pivotal Atom

a. *1,2,4-Oxadiazoles (RA = CN, DCB = NCO, Scheme IV.3)*

One of the earlier examples of a degenerate ring transformation involving the participation of a three-atom side chain is the interconversion of the isomers 3-benzoylamino-5-methyl-1,2,4-oxadiazole (**133**, X = H) and 3-acetylamino-5-phenyl-1,2,4-oxadiazole (**134**, X = H) (Scheme IV.51) (75JHC985; 89H(29)737). Heating each of these compounds at 181°C furnished an identical equilibrium mixture of isomers, in which mixture isomer **134** (X = H) is predominant ($K_{134/133}$ at 181°C is about 6.4). These data indicate that the conversion of isomer **133** (X = H) into **134** (X = H) is dri-

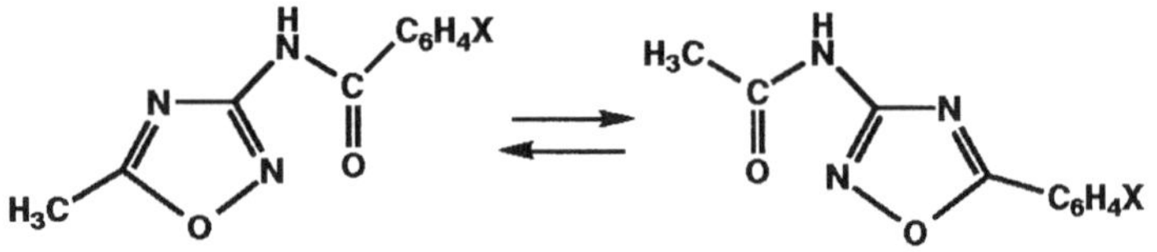

(133) ⇌ (134)

Scheme IV.51

ven in the direction of the thermodynamically more stable compound. The higher thermodynamic stability of isomer **134** ($X = H$) is certainly due to the resonance stabilization of the diaryloid system in isomer **134** ($X = H$), in which the phenyl group at C-5 is conjugated with the 1,2,4-oxadiazole ring (75JHC985; 89H(29)737).

This interconversion can also be performed in solvents, and the rate of the isomerization is dependent on the solvent used. In the dipolar aprotic solvent DMSO the rate of the reaction is fast, but in methanol, acetone, or dioxane the rate is low. However, the value of the equilibrium constant is scarcely influenced by the solvent ($K_{134/133} = 6\text{–}10$) (75JHC985). This is not too surprising, since the equilibrium position is controlled by the relative thermodynamic stability of the isomers, which is a function of their heats of formation and of solvation. Undoubtedly, the heat of formation is the more important factor to the thermodynamic stability (75JHC985).

Since in these interconversions the nucleophilicity of the carbamoyl oxygen pays a decisive role, it is not surprising that heating of 3 trichloroacetylamino-5-methyl-1,2,4-oxadiazole above its melting point (127°C) or in DMSO (60°C, with and without potassium *t*-butoxide) did not result in a rearranged product; the compound was recovered unchanged (Scheme IV.52) (75JHC985). Apparently the nucleophilicity of the carbamoyl oxygen is too low to be able to achieve N–O bond fission.

A study has been carried out on the influence of substituents X present in the aryl moiety of the isomers **133** and **134** on the equilibrium position between **133** and **134** in solutions of CD_3OD, and between anions **133**$^-$ and **134**$^-$ in solutions of CD_3OD, containing potassium *t*-butoxide (see Scheme IV.53 and Table IV.1) (95T5133).

From the results of these detailed measurements, the following interesting conclusions were drawn:

- The position of the equilibrium is quite different whether neutral or anionic forms are involved.
- Under neutral conditions (CD_3OD), the effect of the substituents on K_N is very limited, and for all substituents the isomer **134** is always favored ($K_N > 1$).

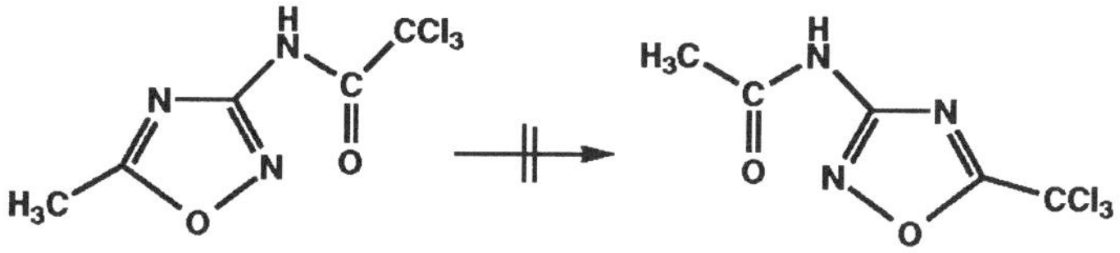

SCHEME IV.52

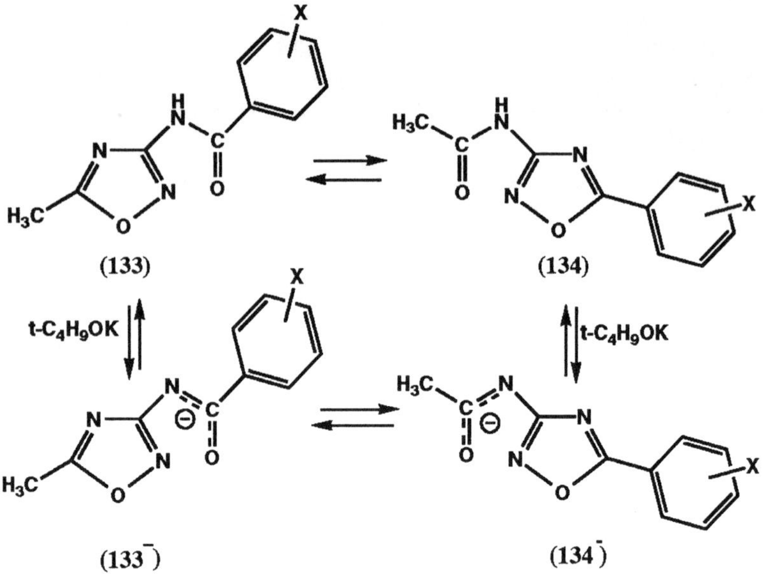

SCHEME IV.53

- In comparison with $X = H$ ($K_N = 10.1$), electron-withdrawing substituents in the *meta* or *para* position shift K_N values to lower than 10.1 (what means more formation of the isomer **133**, while electron-releasing substituents favor the formation of the isomer **134** ($K_N > 10.1$).

TABLE IV.1
Equilibrium Constants and Composition (%) of the Mixture **133** and **134** Obtained in the Rearrangement **133** ⇌ **134** at 313 K in Neutral Solutions (K_N) and in Basic Solution (K_B)

Substituent X	CD$_3$OD ratio 133/134	CD$_3$OD K_N 133/134	CD$_3$OD/ t-C$_4$H$_9$OK ratio 133$^-$/134$^-$	CD$_3$OD/ t-C$_4$H$_9$OK K_B 133$^-$/134$^-$
p-OCH$_3$	7/93	13.3	22/78	3.55
p-CH$_3$	8/92	11.5	30/70	2.33
H	9/91	10.1	38/62	1.63
p-Cl	11/89	8.09	67/33	0.493
p-CF$_3$	14/86	6.14		
p-CN	15/85	5.67	90/10	0.111
p-NO$_2$	16/84	5.25	92/8	0.0870
m-OCH$_3$	11/89	8.09	41/59	1.44
m-CH$_3$	9/91	10.1	32/68	2.13
m-Cl	16/84	5.25	78/22	0.282
m-CF$_3$	17/83	4.88		
m-NO$_2$	22/78	3.55	93/7	0.0753

- The K_B values for the equilibrium between the anionic isomers 133^- and 134^- show a wide range: from 0.0753 (X = m-NO_2) to 3.55 (X = p-OCH_3).

In basic solution the ratio of isomer 133^- to isomer 134^- is always more dominant present than in the neutral solution, because of the stronger negative charge stabilization in the aryl group. When an electron-withdrawing group is present in the aryl group, it is easily understood that 133^- strongly prevails in the equilibrium.

Advantage can be taken of such different K values in neutral and basic solutions. When the 5-aryl-1,2,4-oxadiazoles **134** containing a strong electron-withdrawing group in the phenyl ring are dissolved in a basic solution, a mixture is obtained that is highly enriched with isomer 133^-. When operating in neutral conditions isomer **134** is always enriched in the mixture obtained (95T5133).

A dynamic ^1H NMR study of the equilibrium of 3-acetylamino-5-methyl-1,2,4-oxadiazole **135α/135β** (Scheme IV.54) revealed that heating in DMSO at 190°C did not lead to coalescence of the two sharp signals for the ring methyl and the acetamido methyl, indicating a $\Delta G^{\neq}$ value higher than 25 kcal/mol (75JHC1327). However, when the compound was heated in DMSO, containing potassium t-butoxide, at around 110°C the methyl signals coalescence into a singlet; on cooling the original spectrum is reproduced. From the coalescence temperature and the frequency separation of the methyl singlets the free energy of activation of 19.6 ± 0.4 kcal/mol was calculated (75JHC1327; 81AHC251).

It has been suggested that a symmetrical intermediate or transition state as represented in Scheme IV.54 could be involved, in which the new and the old nitrogen–oxygen bonds are formed and broken to the same extent. A semiempirical approach for calculating the free energy of activation for the degenerate rearrangement of the anion of 3-acetylamino-5-methyl-1,2,4-

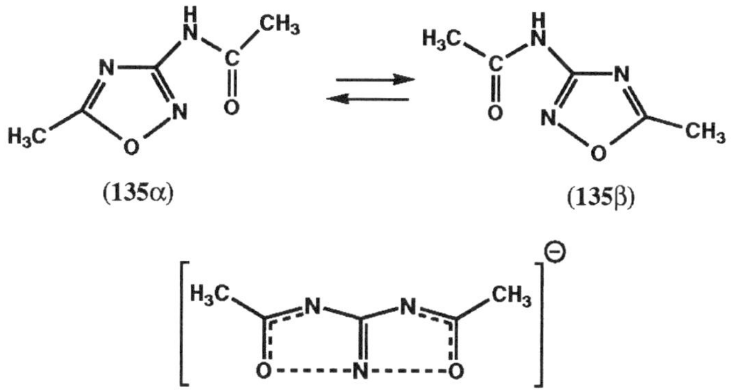

SCHEME IV.54

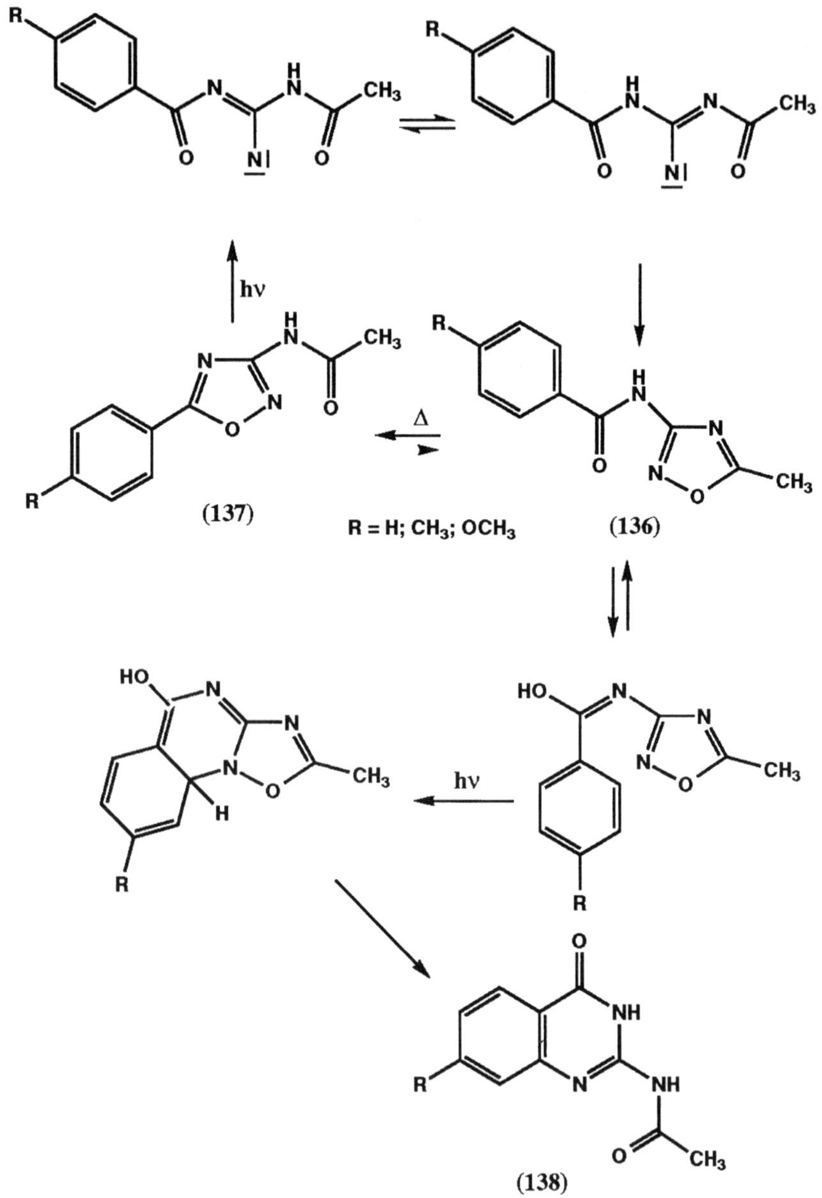

SCHEME IV.55

oxadiazole has shown a considerable difference between the experimental value (19.6 kcal/mol) and the calculated one (MNDO: 71.6 kcal/mol; AM1: 47.6 kcal/mol). This difference can be (partly) attributed to the solvent effect (91H(32)1547).

Interesting similarities but also differences in rearrangement pattern were observed on irradiation of the isomers 3-aroylamino-5-methyl-1,2,4-oxadiazoles (**136**) and 3-acetyl- amino-5-aryl-1,2,4-oxadiazoles (**137**) in anhydrous methanol at 254 nm. Prolonged irradiation of **136** (R = H, CH$_3$, OCH$_3$) and **137** (R = H, CH$_3$) gave in good yields ring transformation into the corresponding 2-acetylamino-7-R-quinazolin-4-ones **138**. However it was also observed that irradiation of compound **137** (R = OCH$_3$) furnished besides the quinazolin-4-one **138** (R = OCH$_3$) the rearranged isomer **136** (R = OCH$_3$) (Scheme IV.55). HPLC studies of the composition of the photoreaction mixture as a function of the irradiation time led to the conclusion that the 3-acetylamino compound **137** is converted into the quinazolin-4-one **138** via the 3-aroylamino compound **136**. The photoinduced rearrangement of **137** into **136** may be explained via the intermediacy of the tautomeric ring-opened nitrene. The formation of the quinazolin-4-one from the 3-aroylamino compound **136** was suggested to involve a tricyclic intermediate in which by N–O bond fission and prototropic shift the 2-acetylaminoquinazolone-4 **138** is obtained (Scheme IV.55) (89H(29)1301).

b. *1,2,5-Oxadiazoles (RA = CN, DCB = CNO, Scheme IV.3)*

It has been reported that the Z-isomer of the 3-aminofurazan, containing at position 4 an *N,N*-disubstituted amidoxime group (**139**, R' = R'' = CH$_3$, R'–R'' = (CH$_2$)$_4$, R'–R'' = (CH$_2$)$_5$) on heating with a base at 120°–140°C rearranges into furazan **141**, containing the *N,N*-disubstituted amino group at position 4 and the unsubstituted amino group in the amidoxime group at

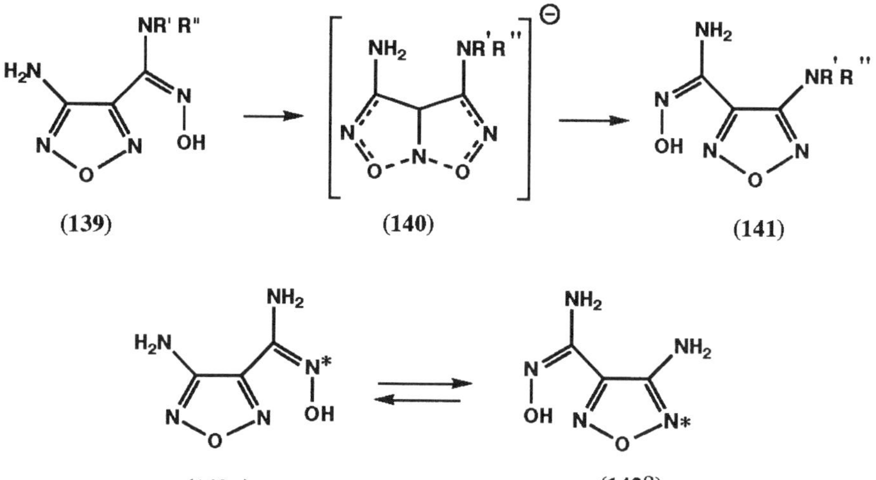

SCHEME IV.56

position 5 (Scheme IV.56) (90CHE1199; 91CHE102). The reversed reaction has not been observed.

This degenerate ring transformation can be considered to occur via the intermediacy of the bicyclic anionic species **140**. When the rearrangement was carried out with **142**, containing the ^{15}N-labeled N,N-unsubstituted amidoxime group, the product obtained showed (as determined by mass spectroscopy) the presence of equimolar amounts of the two isomers **142α** and **142β**, unequivocally proving the degenerate ring transformation concept (Scheme IV.56) (88CHE1410).

A similar base-induced rearrangement has been reported to occur at room temperature with the Z-isomer of 3-methyl-4-benzoylfuroxan oxime (**143**), yielding the 3-(α-nitroethyl)-4-phenylfurazan (**144**) (82G181), and with 3,4-diformaldoxime furoxan (**145**), which gives 3-(α-nitroacetaldoxime)furazan (**146**) (Scheme IV.57) (81AHC(2G)251).

The degenerate rearrangement does not occur with 3-methyl-4-acetyloximefuroxan (**147**) (37G388, 37G518), probably due to the unfavorable E-geometry, as found by X-ray crystallography [87JCS(P2)523].

Descriptor frequence values [ν'_{sr} (cm^{-1})] are claimed to be helpful in predicting the direction of these rearrangements (96ZOK1742). These values were developed for oxazoles, 1,2,4-oxadiazoles, and furazanes. The calculated descriptor values [ν'_{sr}(cm^{-1})] for the E- and Z-isomers of 4-aminofurazan 3-carboxamidoximes [E-I and Z-I, $R_2 = (CH_2)_4$, $(CH_3)_2$] and their rearranged 3-(substituted amino)furazan 4-carboxamidoximes [II,

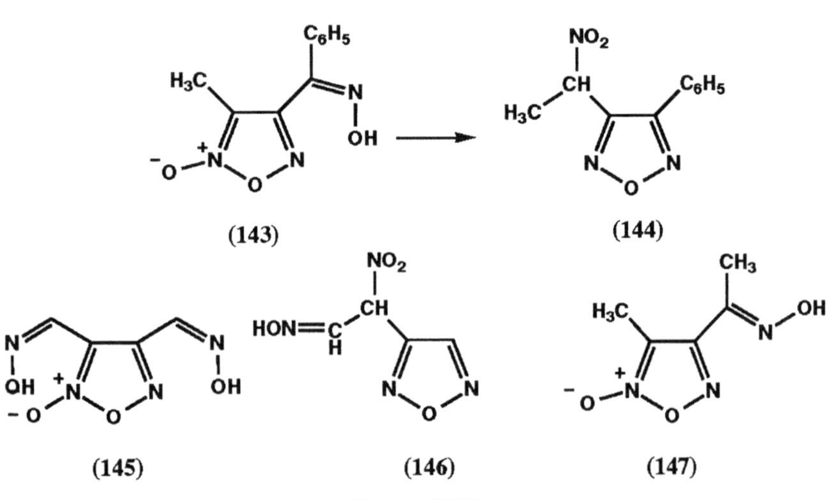

SCHEME IV.57

TABLE IV.2
v'_{sr} (cm^{-1}) FOR SOME AMINOFURAZAN CARBOXAMIDOXIMES

	E-1	Z-1	II
$R_2 = (CH_2)_4$	1100	1088	1039
$R_2 = (CH_3)_2$	1202	1188	1011

$R_2 = (CH_2)_4$, $(CH_3)_2$] are given in Table IV.2. These calculations are based on a fixed standard geometry of these molecules (and temperature of 20°C).

As one can see from Table IV.2, these calculated frequencies decrease in the sequence of the furazans E-I, Z-I, II. It is stated (96ZOK1742) that these data show that isomerization can take place from E-I to Z-I. The degenerate ring transformation of Z-I into II is also predicted by these frequency data, which are in agreement with the experiment. For a further discussion on this subject one is referred to the original literature and the references cited therein.

2. DEGENERATE RING TRANSFORMATIONS INVOLVING SULFUR AS PIVOTAL ATOM

a. *1,2,4-Thiadiazoles (RA = CS; DCB = NCN, Scheme IV.3)*

It has been reported that the aluminum chloride–catalyzed reaction of 5-amino-3-methyl-1,2,4-thiadiazole (**148**) with an aliphatic or aromatic nitrile RCN [$R = CH_2Cl, CH_2CH_3, CH(CH_3)_2, C(CH_3)_3, p\text{-}CH_3C_6H_4, C_6H_5, p\text{-}ClC_6H_4$] always yielded a reaction product that after purification and recrystallization showed in the ^{1}H NMR spectrum (CDCl$_3$, 34°C), besides the NMR signals of the substituent R, two pairs of methyl singlets (see Table IV.3) (79AG176, 79JA5857; 81H1155). This unexpected result was rationalized by assuming the occurrence of a ring transformation of the first reaction product **149** into **150** leading in fact to an equilibrium mixture of **149** and **150**. This equilibrium is facilitated by bond switch with participa-

TABLE IV.3

δ-Values of the Ring Methyl Singlet in Isomer **149** and the Amidine Methyl Singlet in Isomer **150**; the Ratio **150/149** at Equilibrium

Substituent R	CH$_3$ singlet isomer **149**	CH$_3$ singlet isomer **150**	Ratio **150/149**
CH$_2$Cl	2.55	2.22	2.14
CH$_2$CH$_3$	2.52	2.21	1.68
CH(CH$_3$)$_2$	2.52	2.21	2.53
C(CH$_3$)$_3$	2.50	2.21	8.57
p-CH$_3$C$_6$H$_4$	2.55	2.23	5.99
C$_6$H$_5$	2.57	2.23	14.5
p-ClC$_6$H$_4$	2.58	2.26	≈50

tion of π-hypervalent S^{IV}, involving the intermediacy of the 3aλ^4-1,3,4,6-tetraazathiapentalene **151** with a linear N–S^{IV}–N group (Scheme IV.58). It is questionable whether structures such as **151** are indeed real intermediates or whether the reaction proceeds through a transition state whose structure resembles structure **151**. Participation of hypervalent sulfur in this degenerate rearrangement is of crucial importance, since it is demonstrated that this rearrangement does not occur with the structurally related 5-amidino-1,2,4-oxadiazoles (79JA5857).

The equilibrium position is dependent or the substituent R. The ratio **150/149** increases when group R becomes larger and when in the aryl group

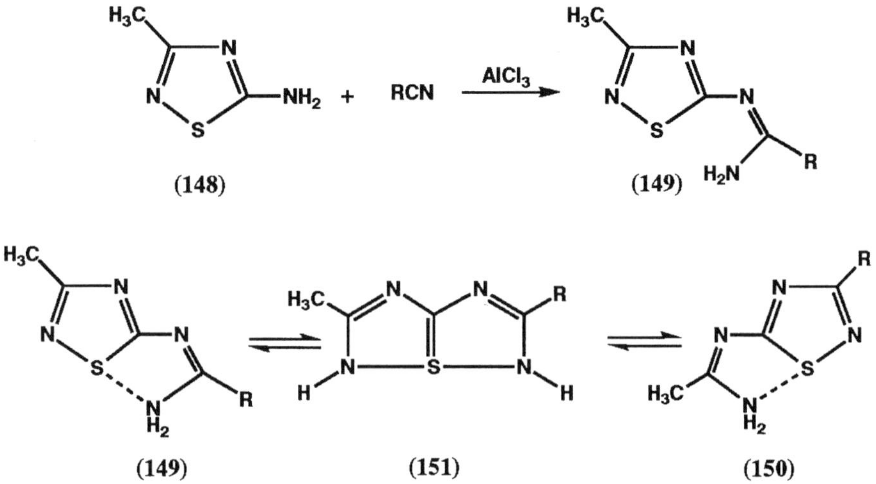

Scheme IV.58

a substituent is present with an electron-withdrawing character. Apparently the conjugation of the aryl group with the heterocycle (diaroyl effect) is more favored than in the arylamidino group. This effect has also been observed in the 1,2,4-oxadiazole series (Section IV,D,1,a). It has also been found that isomer **150** is more favored in solvents with lone-pair electrons; for $R = CH_2Cl$ the following ratios **150/149** were established: deuterochloroform 2.14; benzene 2.23; chlorobenzene 2.08; acetone-d_6 4.47; dimethylsulfoxide-d_6 5.74; methanol-d_4 6.53. The ratio **150/149** is also temperature-dependent. At higher temperatures isomer **149** is more favored, but the isomer **150** still remains predominant (ratio **150/149** > 1)(Scheme IV.3) (79JA5857).

Further support for the degenerate character of the bond-switch process is obtained from the experiment that from the reaction of **148** with benzonitrile in the presence of $AlCl_3$ the same compound is formed as from 5-amino-3-phenyl-1,2,4-thiadiazole and acetonitrile (89BCJ479). Unambiguous evidence for the bond switch in the rearrangement of the 5-amidino-1,2,4-thiadiazoles has been obtained by dynamic NMR spectroscopic measurements (87BSB827). When group R in **149** and **150** is methyl, both isomers are identical (**149** = **150**). The NMR-spectrum of **149** (R = CH_3)/**150** (R = CH_3) in DMSO shows two singlets, one being ascribed to the ring methyl and the other to the methyl of the amidino group. On heating up to 170°C these two methyl groups do not coalesce. However, when an equimolar amount of potassium t-butoxide is added to the solution, it was observed that already at room temperature the two methyl signals coalesce, suggesting a fast equilibrium between the anionic species **149** and **150,** for which a symmetrical species can be suggested (Scheme IV.59). In the presence of catalytic amounts of base the methyl signals coalesce at 90°C; the free energy of activation of this degenerate ring transformation was calculated as 17.7 kcal mol^{-1}.

The occurrence of the bond switch in the rearrangement has also been studied with ^{15}N-labeled compounds. When 5-amino-3-methyl 1,2,4-

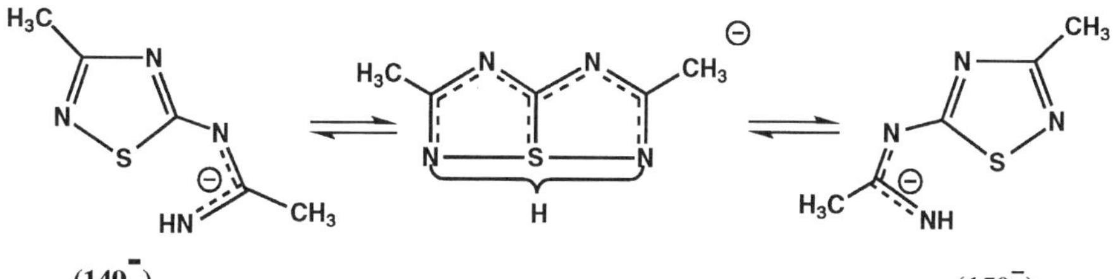

SCHEME IV.59

thiadiazole was converted into the corresponding imidate **153** and this compound was treated with ^{15}N-labeled ammonia, a mixture of the side-chain ^{15}N-labeled amidino-1,2,4-thiadiazole **154** and the ring ^{15}N-labeled 1,2,4-thiadiazole **155** was formed (Scheme IV.60) (89BCJ479). It was proved by proton decoupled ^{15}N NMR spectroscopy that both ^{15}N peaks have almost equal heights.

Extensive studies on bond switching revealed that this process is strongly accelerated by acid. ^{1}H NMR spectroscopy of a solution of isomers **154/155** in CDCl$_3$–DMSO-d_6, containing 1 equivalent of trifluoroacetic acid, showed the presence of only one methyl signal. From this result and those obtained by ^{15}N NMR spectroscopy of the ^{15}N-labeled compounds (89BCJ479) it was correctly concluded that the observed species with one methyl signal could not have the protonated thiapentalene structure **156** (R = CH$_3$) (Scheme IV.61), but that one deals with the monoprotonated species having structure **154-H$^+$**, being in rapid equilibrium with **155-H$^+$**.

It has also been shown (89BCJ479) that it is *not* **154-H$^+$** that undergoes the rearrangement into the equilibrated species **155-H$^+$**, but the monoprotonated species **154'-H$^+$**, which in a rate-determining step equilibrate with **155'-H$^+$** (Scheme IV.62). In **154'-H$^+$/155'-H$^+$** the N–S bond is weakened by protonation at N-2 of the ring and allows the sulfur atom to accept an electron from the unprotonated amidine nitrogen. It has been calculated that the sulfur atom moves back and forth in this bond switching process by 0.38 Å along the N–S–N bond (Scheme IV.62).

In light of the preceding discussions, it is not too surprising that in the reaction of 3,4-dimethyl-5-imino-Δ^2-1,2,4-thiadiazoline (**157**) with an imino ester instead of 3,4-dimethyl-Δ^2-1,2,4-thiadiazoline (**158**), its rearranged, more stable aromatic product *N*-methyl-*N*-[5-(3-methyl-1,2,4-thiadiazol-5-yl)]acetamidine (**159**) is obtained (79AG176). Interestingly **159** converts in acidic medium into the 1,2,4-thiadiazolium salt **160** (Scheme IV.63). This conversion has been explained by an initial protonation at N-2 of the 1,2,4-thiadiazole ring in **159**, from which the 3aλ^4-thia-1,3,4,6-tetra-

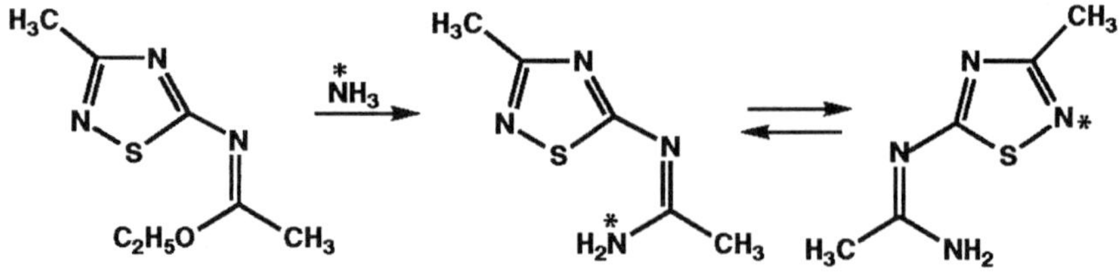

(153)   (154)   (155)

SCHEME IV.60

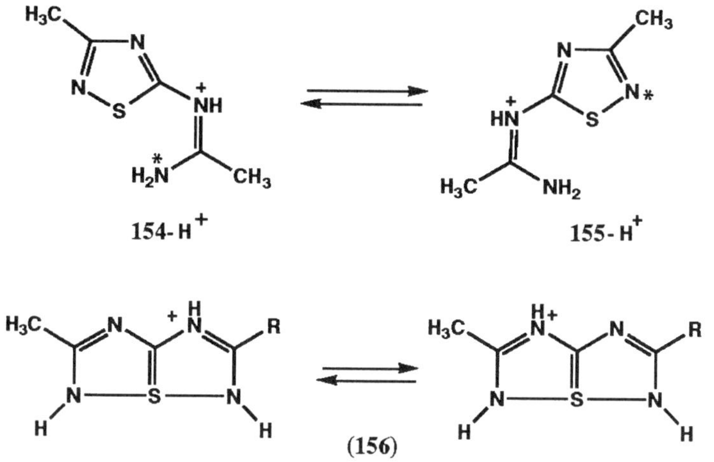

SCHEME IV.61

azapentalenium salt **161** can be formed. A fast prototropic rearrangement leads to the thermodynamically more stable salt **160**.

Similar degenerate ring transformations have been observed when 1,2,4-thiadiazolo[4,5-*a*]pyrimidine **162** reacts with a series of nitriles (benzoni-

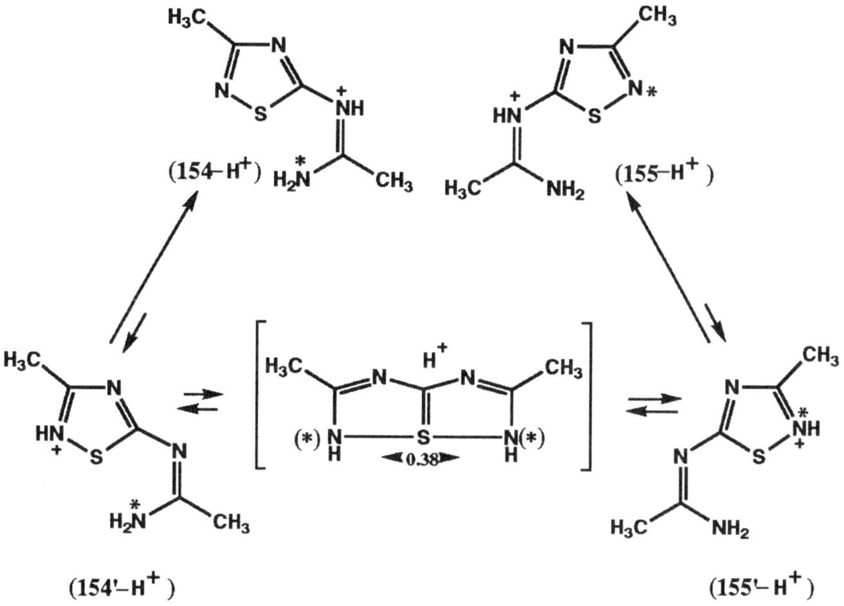

SCHEME IV.62

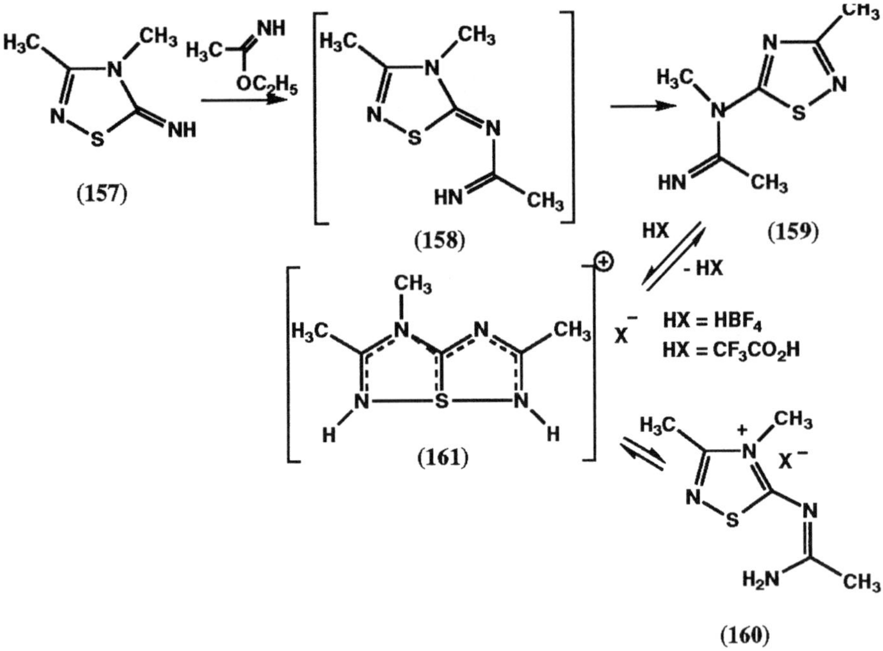

SCHEME IV.63

trile, 4-nitrobenzonitrile, 4-cyanopyridine, methoxyacetonitrile). It leads to incorporation of the reacting nitrile yielding as reaction product the 1,2,4-thiadiazolo[4,5-*a*]pyrimidine **163** with concomitant elimination of acetonitrile [93JCS(PI)1753]. When **162** is refluxed with trideuterioacetonitrile, the trideuteriomethyl analogue **163** ($R = CD_3$) is obtained, which can be reconverted into **162** on heating with acetonitrile. This result demonstrates the reversibility of the process (Scheme IV.64).

As suggested in previous similar investigations, as intermediate have been proposed the non-stable, high-energy species **164**, formed in low concentration in a reversible cycloaddition-elimination process of **161** with nitriles. This process either occurs in a concerted manner (pathway A) or involves a four-step sequence with the intermediacy of a bipolar species (pathway B). It has been argued that the reaction does not proceed through a thiatetra-azapentalene intermediate, but instead involves a transition state that resembles structure **164**.

When **162** reacts with isoselenocyanate RNCSe in boiling toluene, elimination of acetonitrile takes place with concomitant addition of 2 mol of isoselenocyanates to give the diselenones **163B** (94HAC149). This degenerate ring transformation reaction involves intermediate **163A**

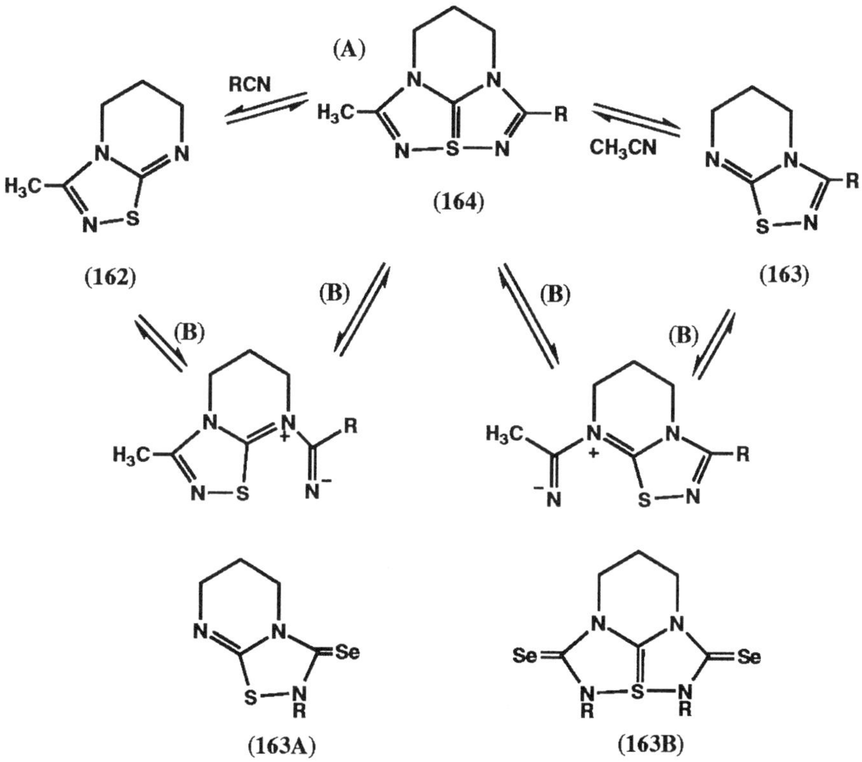

SCHEME IV.64

(Scheme IV.64). The same rearrangement was reported for the reaction with isocyanates (RNCO) and isothiocyanates (RCNS) (96HAC97).

b. *1,2,3-Thiadiazoles (RA = CS, DCB = CNN, Scheme IV.3)*

An interesting degenerate rearrangement was observed when in attempts to prepare the 5-diazomethyl-4-alkoxycarbonyl-1,2,3-thiadiazole (**167**) from its precursor, the tosylhydrazone **165** or the oxime **166**, not **167** but the rearranged 5-(α-alkoxycarbonyldiazomethyl)-1,2,3-thiadiazole (**168**) was obtained (Scheme IV.65) [82TL1103; 83JCS(CC)588].

It has been argued that this rearrangement may occur via a bond-switch process in **167**. In this process the sulfur acts as a nucleophilic center, which is opposite to the electrophilic behavior of the pivotal sulfur in, for example, the conversion of **149** into **150**. The alternative intermediacy of a bipolar sulfur tetra-azapentalene structure or a ring-opened intermediate in

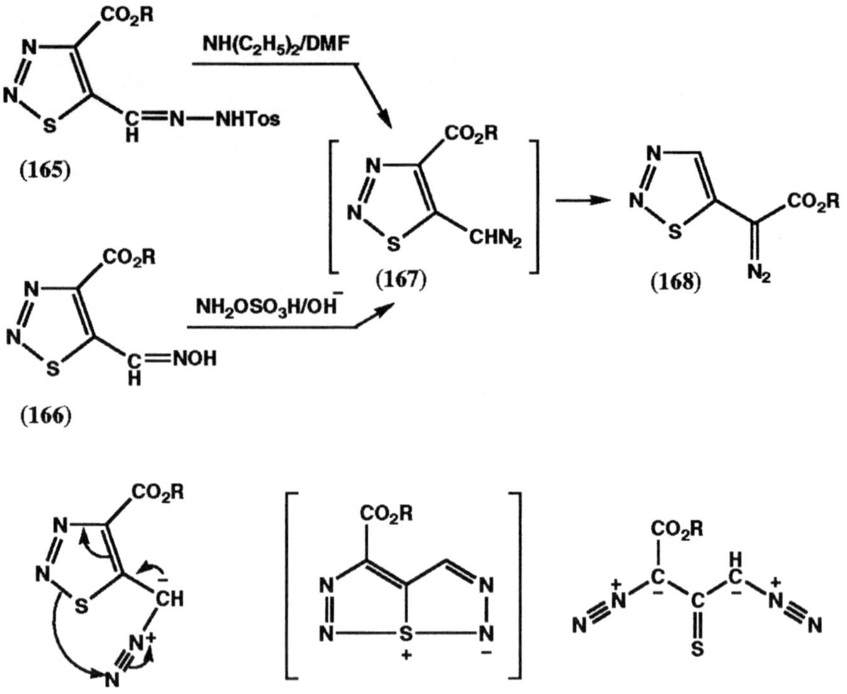

SCHEME IV.65

which the thioketone function is flanked by two 1,3-dipoles is also advanced. Recyclization occurs in the direction of the most stable isomer (Scheme IV.65).

c. *Isothiazoles (RA = CS; DCB = CCN, Scheme IV.3) and Thiazoles (RA = CS; DCB = CNC, Scheme IV.3)*

Ring degenerate equilibria involving participation of π-bonded S^{IV} have been found with 5-(β-amino-β-arylvinyl)isothiazoles. Treatment of 5-methyl-3-arylisothiazole with lithium diisopropyl amide and subsequently with an aromatic nitrile leads to products whose ^{1}H NMR spectra show, when the aryl groups are different, the presence of *two* singlets for the vinyl hydrogens (the protons of the heterocyclic ring are buried in the aromatic region). This NMR result was considered as a strong indication that the product 5-(β-amino-β-arylvinyl)-3-arylisothiazole is in fact a mixture of isomers **169** and **170** (Scheme IV.66) (84JA2713; 85JA2721). In agreement with this result, the ^{1}H NMR-spectra of the compounds in which both aryl groups are the same ($Ar^1 = Ar = C_6H_5$, $p\text{-}ClC_6H_4$) show only one singlet

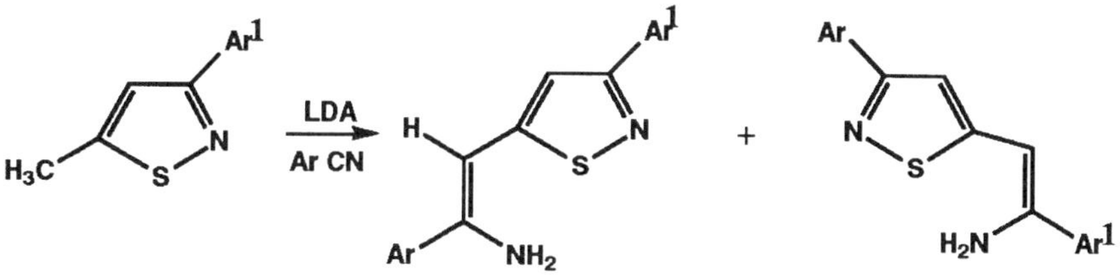

SCHEME IV.66

Ar¹ = p-ClC₆H₄   Ar = p-CH₃C₆H₄
Ar¹ = p-ClC₆H₄   Ar = p-FC₆H₄
Ar¹ = p-ClC₆H₄   Ar = Cl₂C₆H₃

Ar¹ = C₆H₅   Ar = C₆H₅
Ar¹ = p-ClC₆H₄   Ar = p-ClC₆H₄

for the vinyl hydrogen. This result excludes the alternative possibility that the product is just a mixture of geometrical Z- and E-isomers around the C=C double bond. The ratio **169/170** is only slightly dependent on the substituent and the solvent (C_6H_6, $CDCl_3$, DMSO-d_6). It varies between 58/42 and 47/53 (85JA2721).

Further evidence for the occurrence of the degenerate rearrangement in these compounds was obtained by ^{15}N NMR spectroscopy. The Z-form of 3-p-chlorophenyl-5-(β-p-chlorophenyl-β-[^{15}N-amino])vinylisothiazole (**171**) showed in the non-proton-decoupled ^{15}N spectrum a double triplet for the amino nitrogen and a singlet for the heterocyclic ring nitrogen. It confirms that **171** is in fact an equilibrium mixture of **171** and **172** (Scheme IV.67).

From studies of the kinetics of the bond switching, the equilibration, the

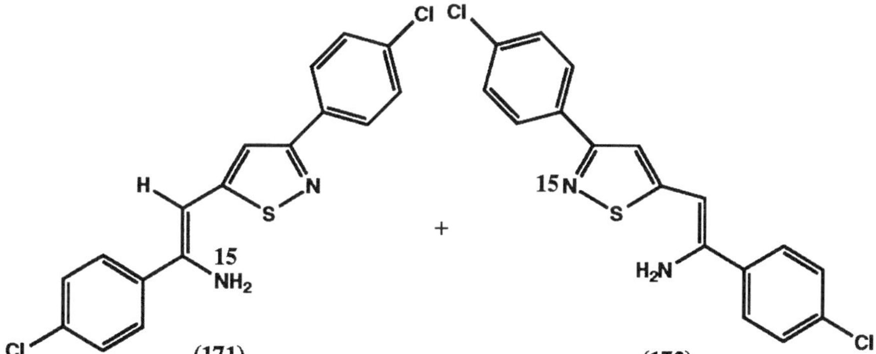

SCHEME IV.67

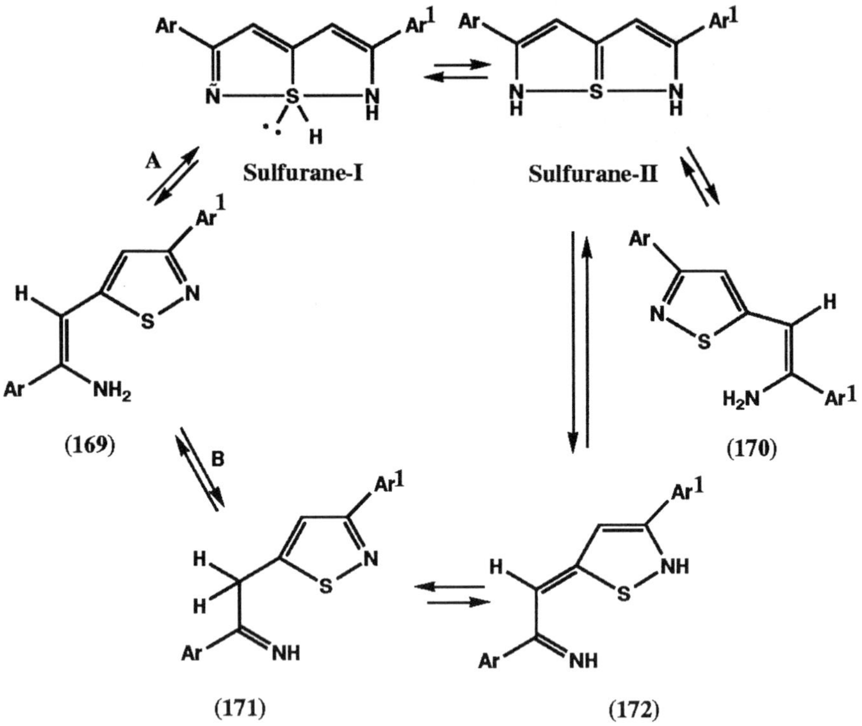

SCHEME IV.68

influence of the concentration of substrates, acids, and bases, and the influence of solvents and substituents, it was concluded that in the reversible rearrangement of **169** into **170**, the hypervalent sulfurane-II acts as intermediate (path A), which can be obtained via the transition state sulfurane-I. Alternatively, it might be formed via the imine **171/172** (path B) (84JA2713; 85JA2721), in which by a 1,5-sigmatropic shift of hydrogen sulfurane-II is formed (Scheme IV.68).

A remarkable base-induced degenerate rearrangement was observed during treatment of 4-amino-5-ethoxycarbonyl-2-cyanimino-3-R-thiazoline **173** ($R = C_6H_5$, $CH_2CH = CH_3$). It yields 4-amino-5-cyano-2-(R-amino)thiazole **176** (81M1393). This conversion was described to involve the ring-opened cyano intermediate **174A/174B**, being formed by ring opening between N-3 and C-4 in the thiazole ring and simultaneous dehydrogenation of the amino (imino) substituent into the cyano group. In intermediate **174**, deprotonation and a Thorpe carbon–carbon cyclization between the carbon of the N–CN group and the carbon of the cyano functionality of the cyanoimino group yields the 1,3-thiazoline **175**. Aromatization occurs by

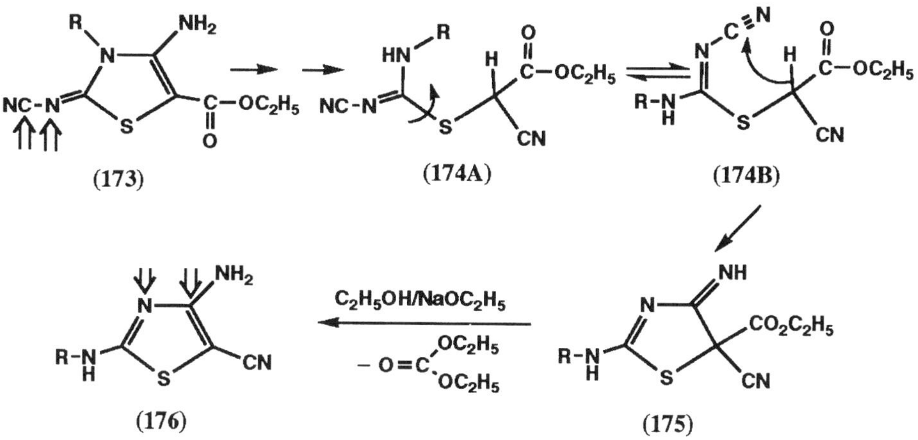

SCHEME IV.69

loss of the carbethoxy group as the diester of carbonic acid (Scheme IV.69). Similar ring cyclizations have been reported before (71TL1075).

d. *1,2 Dithioles (RA = CS, DCB = CCS, Scheme IV.3)*

A nice example of a degenerate rearrangement has been reported for the complex pentacarbonyltungsten(0)-thioaldehyde-1,2-dithiol **177** (R^1 = H) or the thioketone–1,2-dithiol complex **177** (R^1 = CH_3). These compounds (obtained by the reaction of silver nitrate with tetraethylammonium pentacarbonyliodotungstate(0) in the presence of 1,6,6aλ^4-trithiapentalenes) appeared to exist as an equilibrium mixture of the two isomers **177/178** (Scheme IV.70) [83JCS(CC)289]. The symmetrically substituted 1,2-dithiol (R^1 = CH_3, R^2 = H) displays fluxional behavior as appears by NMR spectroscopy

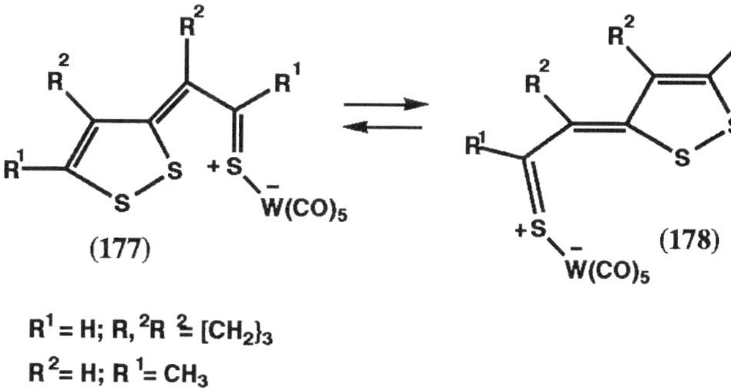

R^1 = H; R, ^{2}R ≗ [CH_2}$_3$
R^2 = H; R^1 = CH_3

SCHEME IV.70

in chloroform solution. At −20°C two sharp signals were observed for the ring hydrogens and two sharp signals for the methyl groups. The signals of the ring hydrogen coalesce at +1°C and appear as a sharp singlet at +33°C; both methyl signals coalesce at +4°C and give rise to one sharp singlet at 33°C. It has been concluded that the fluxional process is intramolecular.

e. *Thiophenes (RA = CS, DCB = CCC, Scheme IV.3)*

A degenerate ring transformation, involving the replacement of three carbon atoms of the thiophene ring by three carbon atoms of the side chain, has been reported to occur when methyl α-cyano-β-(2-thienyl) acrylate (**179A**) was allowed to stand overnight in morpholine or piperidine at room temperature (71JOC2196). The product obtained was assigned the structure methyl 2-cyano-5-(4-methoxycarbonyl-5-amino-2-thienyl)-2,4-pentadienoate **179B**. The role of the base was twofold: first to liberate cyanoacetic ester from the acrylate side-chain, and second to achieve ring opening by attack at position 5 of the thiophene ring (Scheme IV. 70A). The ring opening can formally be described as indicated in the scheme, although different intermediates can be envisaged.

A similar base-catalyzed ring opening–ring closure sequence has been observed with α-cyano-β-furylacrylic esters, which also led to the formation of a 5-aminofuran derivative (70JOC1234).

SCHEME IV.70A

3. DEGENERATE RING TRANSFORMATIONS INVOLVING CARBON AS PIVOTAL ATOM

a. *Pyrroles (RA = CN, DCB = CNN, Scheme IV.3)*

A remarkable, intriguing rearrangement that possibly could be framed in the context of this chapter is the one observed in the biosynthesis of the chlorinated fungicide pyrrolnitrin. This compound is produced from the amino acid tryptophan **179** by the bacterial cells of *Pseudomonas pyrrocinia* [85BBA181; 91MI2; 92MI1; 94MI3]. A proposal for the biosynthesis

(Scheme IV.71) shows the involvement of two chlorinating steps: first, the chlorination of compound **179** at position 7 into **180**, and second, the chlorination of 4-(2-amino-3-chlorophenyl)pyrrole (**183**) at position 3, forming **184**. Two chlorinating enzymes responsible for these regiospecific chlorinations have been identified (85BBA181).

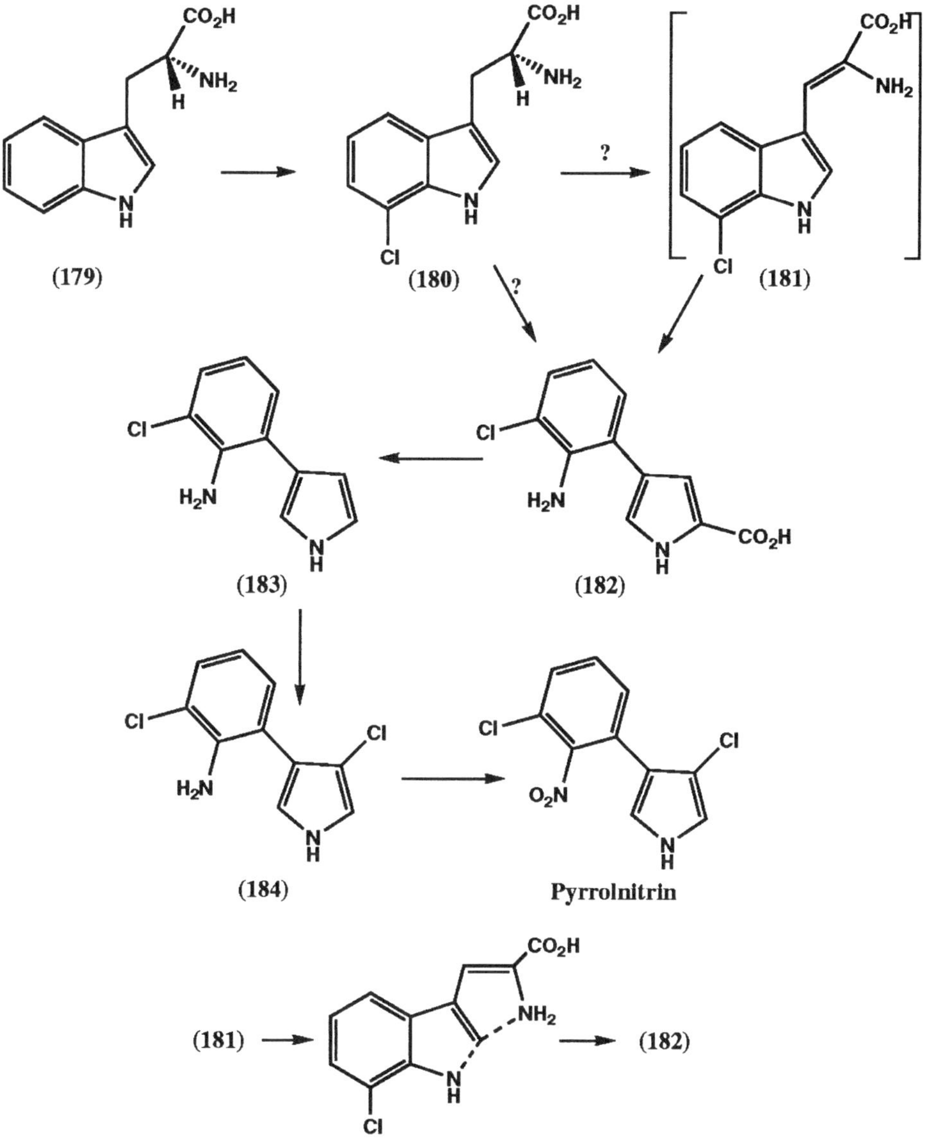

SCHEME IV.71

The decarboxylation of **182** into **183** and the oxidative conversion of the amino group in **184** to pyrrolnitrin are well-established biochemical processes. The least understandable part in the overall scheme is the formation of 2-carboxy-4-(2-amino-3-chlorophenyl)pyrrole (**182**). The presence of the amino group in the phenyl ring seems to justify the conclusion that an opening of the pyrrole ring has taken place. That this ring opening should occur spontaneously under physiological conditions seems highly unlikely, and this is very probably an enzyme-mediated reaction. It is tempting to describe the rearrangement via the oxidative formation of the dehydro intermediate **181**, which can convert by a 6-π assisted heterocyclization into **182**. Further investigations need to be carried out before firm conclusions can be drawn.

b. *1,2,4-Triazoles (RA = CN, DCB = CNN, Scheme IV.3)*

In the 1,2,4-triazole series, base-induced degenerate rearrangements have been reported with the aroylarylazo-1,2,4-triazolium salts **185** (87MI3) (Scheme IV.72). The reaction product is 3-aroyl-1-aryl-1,2,4-triazole **187**, and its formation can be described as involving an ANRORC mechanism, initiated by an addition of the nucleophile at C-5, ring opening into **186**, and subsequent heterocyclization with the former CNN side chain of the rearranging 1,2,4-triazole.

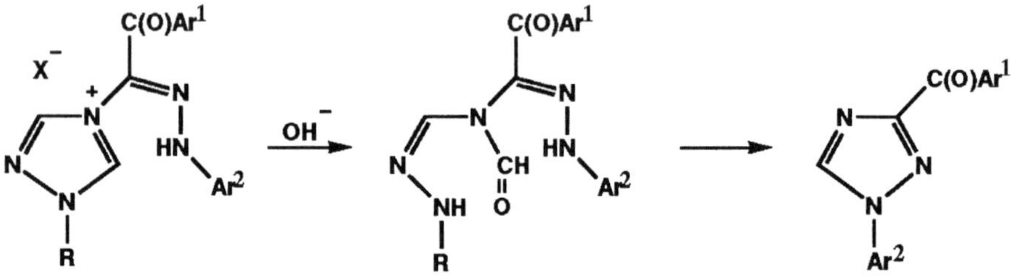

SCHEME IV.72

c. *1,2,3-Triazoles (RA = CN, DCB = CNN, Scheme IV.3)*

The involvement of a three-atom side chain in the rearrangement has been reported with 5-diazomethyl-4-methoxycarbonyl-1,2,3-triazoles **188**, containing a strong electron-withdrawing aryl group (4-nitrophenyl or 2,4-dinitrophenyl) on position N-1 (88T461). The rearrangement leads to incorporation of the CNN side chain, yielding 1-aryl-5-diazomethoxycarbonyl-1,2,3-triazole **190** (Scheme IV.73). The electron-withdrawing character of the group at N-1 weakens the N–N bond and promotes the ring

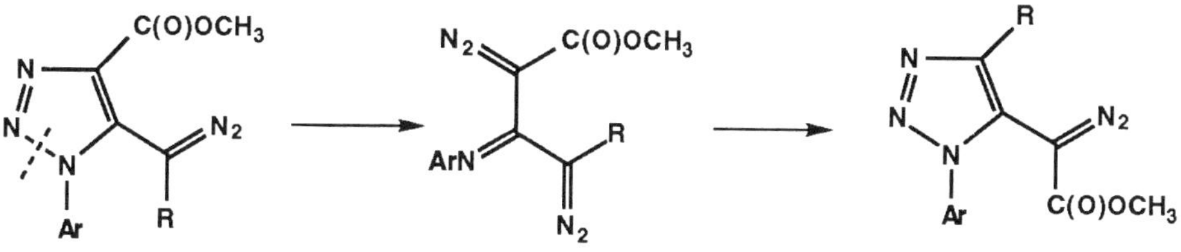

(188) (189) (190)

SCHEME IV.73

opening, as has also been observed in the Dimroth rearrangement of 5-aminotetrazoles (see Section IV,B,1,a).

This degenerate rearrangement has not been observed with 5-diazomethyl-1,4-diphenyl-1,2,3-triazole. The diazo function decomposes and the carbene formed reacts, when benzene is used as solvent, into a cycloheptatriene. The requirement to have the ester function present at C-4 apparently has to do with stabilization of the diazo function.

It has been reported that reaction of 1-aryl-5-chlorotetrazole with sodium azide gives the corresponding 5-azidotetrazole **191** (67JOC3580). This azido compound is under certain conditions in equilibrium with the imino-bisazido compound (**192a/192b**) (Scheme IV.74). It has been argued (84JHC627) that if the energy barrier for rotation around the C = N bond in **192a** is not too high during this process, it would lead to **193,** being identical to **191:** an interesting and typical example of a degenerate ring transformation. ^{15}N-labeling experiments can prove whether this equilibrium between **191** and **193** indeed exists.

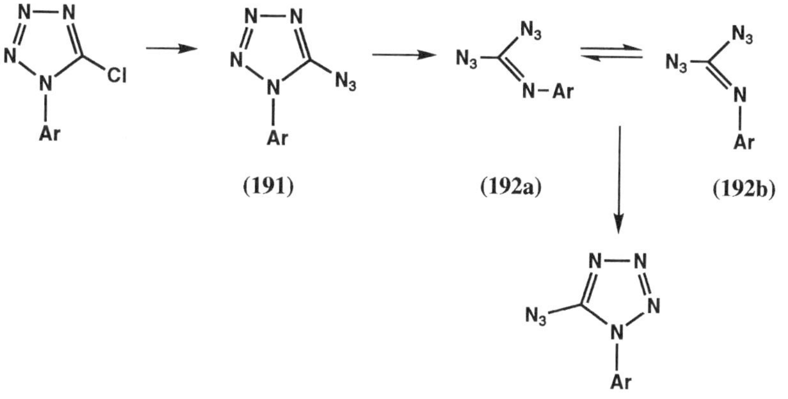

(191) (192a) (192b)

(193)

SCHEME IV.74

References

1888CB867	B. Rathke, *Chem. Ber.* **21,** 867 (1888).
1898CB542	E. Fisher, *Chem. Ber.* **31,** 542 (1898).
03LA361	Th. Zincke, *Justus Liebigs Ann. Chem.* **330,** 361 (1903).
04LA296	Th. Zincke, G. Heuser, and W. Möller, *Justus Liebigs Ann. Chem.* **333,** 296 (1904).
05LA365	Th. Zincke and W. Würker, *Justus Liebigs Ann. Chem.* **341,** 365 (1905).
06CB863	K. Dost, *Chem. Ber.* **39,** 863 (1906).
08CB1346	H. Th. Bucherer and J. Schenkel, *Chem. Ber.* **41,** 1346 (1908).
09LA183	O. Dimroth, *Justus Liebigs Ann. Chem.* **364,** 183 (1909).
10CB2597	J. Schenkel, *Chem. Ber.* **43,** 2597 (1910).
10CB2939	F. Reitzenstein and W. Breuning, *Chem. Ber.* **43,** 2939 (1910).
14MI1	A. E. Chichibabin and O. A. Zeide, *Zh. Russ. Fis.-Khim. O-va.* **46,** 1216 (1914) [*CA.* **9,** 1901 (1915)].
23CB758	W. König, G. Ebert, and K. Centher, *Chem. Ber.* **56,** 758 (1923).
27LA39	O. Dimroth and W. Michaelis, *Justus Liebigs Ann. Chem.* **459,** 39 (1927).
28CB1223	A. E. Chichibabin and A. W. Kirssanow, *Chem. Ber.* **61,** 1223 (1928).
28JA3311	E. F. Cornell, *J. Am. Chem. Soc.* **56,** 3311 (1928).
36BSF1600	A. Kirssanoff and I. Poliakowa, *Bull. Soc. Chim. Fr.* **2,** 1600 (1936).
37G388	G. Tappi, *Gazz. Chim. Ital.* **67,** 388 (1937).
37G518	G. Ponzio and G. Tappi, *Gazz. Chim. Ital.* **67,** 518 (1937).
37JOC411	F. W. Bergstrom, *J. Org. Chem.* **2,** 411 (1937).
39G664	G. Losco and M. Passerin, *Gazz. Chim. Ital.* **69,** 664 (1939).
39JPJ18	E. Otiai and M. Karri, *J. Pharm. Soc. Jpn.* **59,** 18 (1939).
40G410	G. Losco and M. Passerin, *Gazz. Chim. Ital.* **70,** 410 (1940).
41JPC19	W. König, *J. Prakt. Chem.* **70,** 19 (1941).
43RTC207	G. Tappi, *Recl. Trav. Chim. Pays-Bas* **62,** 207 (1943).
47JCS738	G. Newberry and W. Webster, *J. Chem. Soc.,* 738 (1947).
49MI1	J. W. Cornforth, *in* "The Chemistry of Penicillin." p. 688. Princeton University Press, Princeton, NJ, 1949.
50LA84	K. Bodendorf, and A. Popelak, *Justus Liebigs Ann. Chem.* **566,** 84 (1950).
51MI1	L. F. Audrieth and B. A. Ogg, "The Chemistry of Hydrazine." Wiley, New York, 1951.
53JA3290	J. D. Roberts, H. E. Simmons, Jr., L. A. Carlsmith, and C. W. Vaughan, *J. Am. Chem. Soc.* **75,** 3290 (1953).
53JOC779	W. G. Finnegan, R. A. Henry, and E. Lieber, *J. Org. Chem.* **18,** 779 (1953).
53JOC1283	R. M. Herbst and W. L. Garbrecht, *J. Org. Chem.* **18,** 1283 (1953).
53LA123	H. Lettré, W. Haede, and E. Ruhbaum, *Justus Liebigs Ann. Chem.* **579,** 123 (1953).
54CB68	J. Goerdeler, A. Huppertz, and K. Wember, *Chem. Ber.* **87,** 68 (1954).

54JA88	R. A. Henry, W. G. Finnegan, and E. Lieber, *J. Am. Chem. Soc.* **76**, 88 (1954).
54JCS1017	H. C. Carrington, A. F. Crowther, and G. J. Stacey, *J. Chem. Soc.*, 1017 (1954).
54JOC1570	C. C. Cheng and R. K. Robins, *J. Org. Chem.* **24**, 1570 (1954).
55JA3752	H. C. Brown and D. H. McDaniel, *J. Am. Chem. Soc.* **77**, 3752 (1955).
55JA4540	G. E. Hall, R. Piccolini, and J. D. Roberts, *J. Am. Chem. Soc.* **77**, 4540 (1955).
56JA601	J. D. Roberts, D. A. Semenov, H. E. Simmons, Jr., and L. A. Carlsmith, *J. Am. Chem. Soc.* **78**, 601 (1956).
56JOC654	E. Lieber, T. S. Chao, and C. N. R. Rao, *J. Org. Chem.* **22**, 654 (1956).
57JA5962	E. Lieber, C. N. R. Rao, and T. S. Chao, *J. Am. Chem. Soc.* **79**, 5962 (1957).
58JOC1912	R. M. Herbst and J. M. Klingbeil, *J. Org. Chem.* **23**, 1912 (1958).
58CLY1131	R. Lukes and J. Jizba, *Chem. Listy* **52**, 1131 (1958).
59JA5650	R. B. Angier and W. V. Curran, *J. Am. Chem. Soc.* **81**, 5650 (1959).
59E412	K. Lempert and M. Lempert-Sréter, *Experientia* **15**, 412 (1959).
59GEP958197	S. Skraup, Ger. Pat. 958, 197 (1959) [*CA*. **53**, 8177 (1959)].
60AG91	R. Huisgen and J. Sauer, *Angew. Chem.* **72**, 91 (1960).
60CB1033	F. Korte and K. Storiko, *Chem. Ber.* **93**, 1033 (1960).
60E107	J. Breuer and K. Lempert, *Experientia* **16**, 107 (1960).
60JA1609	H. K. Nagy, A. J. Tomson, and J. P. Horwitz, *J. Am. Chem. Soc.* **82**, 1609 (1960).
60JA3147	E. C. Taylor and P. K. Loeffler, *J. Am. Chem. Soc.* **82**, 3147 (1960).
60JA3629	M. Panar and J. D. Roberts, *J. Am. Chem. Soc.* **82**, 3629 (1960).
60JCS539	P. Brookes and P. D. Lawley, *J. Chem. Soc.*, 539 (1960).
60JCS3540	R. J. Grout and M. W. Partridge, *J. Chem. Soc.*, 3540 (1960).
60JOC1043	R. Kitawaki and K. Sugino, *J. Org. Chem.* **25**, 1043 (1960).
60JOC1819	C. G. Stuckwisch and D. D. Powers, *J. Org. Chem.* **25**, 1819 (1960).
60MI1	R. Huisgen, in "Organometalic Chemistry" (H. Zeiss, ed.). pp. 36–87, Reinhold, New York, 1960.
60T29	R. Huisgen, W. Mack, and L. Möbius, *Tetrahedron* **9**, 29 (1960)
61ACS991	K. E. Jensen and C. Pedersen, *Acta Chem. Scand.* **15**, 991 (1961).
61AG65	T. Kaufmann and F. P. Boettcher, *Angew. Chem.* **73**, 65 (1961).
61JCS1298	D. J. Brown and J. S. Harper, *J. Chem. Soc.*, 1298 (1961).
61LA66	S. Hünig and W. Lampe, *Justus Liebigs Ann. Chem.* **647**, 66 (1961).
61N828	D. J. Brown, *Nature (London)* **189**, 828 (1961).
61RTC1376	M. J. Pieterse and H. J. den Hertog, *Recl. Trav. Chim. Pays-Bas*, **80**, 1376 (1961).
62CB1528	T. Kaufmann and F. P. Boettcher, *Chem. Ber.* **95**, 1528 (1962).
62JOC883	E. Shaw, *J. Org. Chem.* **27**, 883 (1962).
62JOC2478	G. B. Elion, *J. Org. Chem.* **27**, 2478 (1962).
62JOC2622	E. C. Taylor and R. V. Ravindranathan, *J. Org. Chem.* **27**, 2622 (1962).
62TL387	R. Huisgen, R. Grashey, and R. Krischke, *Tetrahedron Lett.*, 387 (1962).
63AG604	R. Huisgen, *Angew. Chem.* **75**, 604 (1963).
63CB534	J. Goerdeler and W. Roth, *Chem. Ber.* **96**, 534 (1963).
63JCS1276	D. J. Brown and J. S. Harper, *J. Chem. Soc.*, 1276 (1963).

REFERENCES

63JCS1284	D. D. Perrin, *J. Chem. Soc.*, 1284 (1963).
63MI1	K. Lempert, M. Lempert-Sréter, and J. Breuer, *Period. Polytech.* **7,** 7 (1963) [*CA.* **59,** 1209a (1963)].
64AG206	Th. Kaufmann, *Angew. Chem.* **11,** 206 (1964).
64JCS3591	A. Fisher, W. J. Galloway, and J. Vaughan, *J. Chem. Soc.*, 3591 (1964).
64JOC1762	T. Ueda and J. J. Fox, *J. Org. Chem.* **29,** 1762 (1964).
64JOC1770	T. Ueda and J. J. Fox, *J. Org. Chem.* **29,** 1770 (1964).
64MI1	W. McNutt, *in* "Pteridine Chemistry" (W. Pleiderer, ed.), p. 342. Pergamon, Oxford, 1964.
64MI2	D. J. Brown and J. S. Harper, *in* "Pteridine Chemistry" (W. Pleiderer and E. C. Taylor, eds.). Pergamon, Oxford, 1964.
64TL1577	H. L. Jones and D. L. Beveridge, *Tetrahedron Lett.*, 1577 (1964).
64TL2093	H. C. van der Plas and G. Geurtsen, *Tetrahedron Lett.*, 2093 (1964).
64ZOK1745	N. N. Vereshehagina and I. Ya. Postovskii, *Zh. Obshch. Khim.* **34,** 1745 (1964).
65AJC471	D. D. Perrin and I. H. Pitman, *Aust. J. Chem.* **18,** 471 (1965).
65AHC121	H. J. den Hertog and H. C. van der Plas, *Adv. Heterocycl. Chem.* **4,** 121 (1965).
65AHC145	R. C. Shepherd and J. L. Frederick, *Adv. Heterocycl. Chem.* **4,** 145 (1965).
65AHC180	R. C. Shepherd and J. L. Frederick, *Adv. Heterocycl. Chem.* **4,** 180 (1965).
65B54	I. Wempen, G. B. Brown, T. Ueda, and J. J. Fox, *Biochemistry* **4,** 54 (1965).
65JA395	B. Woodward and R. Hoffmann, *J. Am. Chem. Soc.* **87,** 395 (1965).
65JCS5542	D. J. Brown and J. S. Harper, *J. Chem. Soc.*, 5542 (1965).
65JCS6659	M. D. Metha, D. Miller, and E. F. Mooney, *J. Chem. Soc.*, 6659 (1965).
65JCS7071	D. D. Perrin and I. H. Pitman, *J. Chem. Soc.*, 7071 (1965).
65JOC2858	E. C. Taylor and A. McKillopp, *J. Org. Chem.* **30,** 2858 (1965).
65RTC1569	H. J. den Hertog, H. C. van der Plas, M. J. Pieterse, and J. W. Streef, *Recl. Trav. Chim. Pays-Bas* **84,** 1569 (1965).
65T2205	R. Eisenthal and A. R. Katritzky, *Tetrahedron* **21,** 2205 (1965).
65TL555	H. C. van der Plas, *Tetrahedron Lett.*, 555 (1965).
66AG548	R. Neidlein and E. Henkelbach, *Angew. Chem.* **78,** 548 (1966).
66JA4766	J. A. Zoltewicz and C. L. Smith, *J. Am. Chem. Soc.* **88,** 4766 (1966).
66JCS(C)164	D. J. Brown and M. N. Paddon-Row, *J. Chem. Soc. C*, 164 (1966).
66JCS(C)1163	D. J. Brown, B. T. England, and J. S. Harper, *J. Chem. Soc. C*, 1163 (1966).
66JPR293	H. Beyer and E. Thieme, *J. Prakt. Chem.* **31,** 293 (1966)
66MI1	S. C. Wait, Jr. and J. W. Wesley, *J. Mol. Spectrosc.* **10,** 25 (1966).
66RTC1101	H. C. van der Plas, B. Haase, B. Zuurdeeg, and M. C. Vollering, *Recl. Trav. Chim. Pays-Bas* **85,** 1101 (1966).
66TL4517	H. W. van Meeteren and H. C. van der Plas, *Tetrahedron Lett.*, 4517 (1966).
66ZC181	M. Wahren, *Z. Chem.* **6,** 181 (1966).
67CB3671	E. Winterfeld and J. M. Nelke, *Chem. Ber.* **100,** 3671 (1967).
67JA6911	B. Singh and E. J. Ullman, *J. Am. Chem. Soc.* **89,** 6911 (1967).
67JCS(C)903	D. J. Brown and M. N. Paddon-Row, *J. Chem. Soc. C*, 903 (1967).

67JCS(C)1922	D. J. Brown and B. T. England, *J. Chem. Soc. C*, 1922 (1967).
67JCS(C)1928	D. J. Brown and M. N. Paddon-Row, *J. Chem. Soc. C*, 1928 (1967).
67JCS(C)2005	A. J. Boulton, A. R. Katritzky, and A. M. Hamid, *J. Chem. Soc. C*, 2005 (1967).
67JCS(CC)55	R. A. Abramovitch, G. M. Singer, and A. R. Vinutha, *J. Chem. Soc., Chem. Commun.*, 55 (1967).
67JOC1151	J. C. Parham, J. Fissekis, and G. B. Brown, *J. Org. Chem.* **32**, 1151 (1967).
67JOC3580	J. C. Kauer and W. A. Sheppard, *J. Org. Chem.* **32**, 3580 (1967).
67MI1	R. W. Hoffmann, "Dehydrobenzene and Cycloalkenes." Academic Press, New York, 1967.
67RTC15	H. W. van Meeteren and H. C. van der Plas, *Recl. Trav. Chim. Pays-Bas* **86**, 15 (1967).
67RTC187	H. J. den Hertog and D. J. Buurman, *Recl. Trav. Chim. Pays-Bas* **86**, 187 (1967).
67T2775	R. Eisenthal and A. R. Katritzky, *Tetrahedron* **23**, 2775 (1967).
67TL337	J. A. Zoltewicz and G. M. Kaufmann, *Tetrahedron Lett.*, 337 (1967).
68AJC2813	D. J. Brown and B. T. England, *Aust. J. Chem.* **21**, 2813 (1968).
68B3453	J. B. Macon and R. Wolfenden, *Biochemistry* **7**, 3453 (1968).
68JA4319	E. H. Cordes and W. P. Jencks, *J. Am. Chem. Soc.* **90**, 4319 (1968).
68JA4328	C. G. Swain and E. C. Lupton, Jr., *J. Am. Chem. Soc.* **90**, 4328 (1968).
68MI1	D. J. Brown, in "Mechanism of Molecular Migrations" (B. S. Thyugarajan, ed.) Vol. 1, p. 209. Wiley-Interscience, New York, 1968.
68T441	M. Wahren, *Tetrahedron* **24**, 441 (1968).
68TL9	H. C. van der Plas, P. Smit, and A. Koudijs, *Tetrahedron Lett.*, 9 (1968).
69IZV655	E. A. Arutyunyan, V. I. Gunar, E. P. Grachera, and S. I. Zavyalov, *Izv. Akad. Nauk SSSR, Ser. Khim.*, 655 (1969).
69JA2590	W. Adam, A. Grimison, and R. Hoffmann, *J. Am. Chem. Soc.* **91**, 2590 (1969).
69JCS(B)330	M. R. Crampton and M. El-Ghariani, *J. Chem. Soc. B*, 330 (1969).
69JCS(CC)1387	W. D. Crow and C. Wentrup, *J. Chem. Soc., Chem. Commun.*, 1387 (1969).
69JOC1405	J. A. Zoltewicz and G. M. Kaufmann, *J. Org. Chem.* **34**, 1405 (1969).
69MI1	H. J. den Hertog and H. C. van der Plas, in "Chemistry of Acetylenes" (H. G. Viehe, ed.), Chapter 17. Dekker, New York, 1969.
69MI2	H. G. Viehe, in "Chemistry of Acetylenes" (H. G. Viehe, ed.), Chapter 12. Dekker, New York, 1969.
69RTC426	H. C. van der Plas and B. Zuurdeeg, *Recl. Trav. Chim. Pays-Bas* **88**, 426 (1969).
69RTC1156	H. C. van der Plas, B. Zuurdeeg, and H. W. van Meeteren, *Recl. Trav. Chim. Pays-Bas* **88**, 1156 (1969).
69RTC1391	J. W. Streef and H. J. den Hertog, *Recl. Trav. Chim. Pays-Bas* **88**, 1391 (1969).
69T4291	A. R. Katritzky and E. Lunt, *Tetrahedron* **25**, 4291 (1969).
69ZC241	M. Wahren, *Z. Chem.* **7**, 241 (1969).
70AJC51	E. N. Cain and R. N. Warrener, *Aust. J. Chem.* **23**, 51 (1970).
70CI(L)926	Y. Tamura and N. Tsjusimoto, *Chem. Ind. (London)*, 926 (1970).

70CRV667	M. J. Strauss, *Chem. Rev.* **70,** 667 (1970).
70IJC1055	C. M. Gupta, A. P. Bhaduri, and N. M. Khanna, *Indian J. Chem.* **8,** 1055 (1970)
70IZV904	E. A. Arutyunyan, V. I. Gunar, and S. I. Zavyalov, *Izv. Akad. Nauk SSSR, Ser. Khim.*, 904 (1970).
70JA7463	K. K. Kim and J. F. Bunnett, *J. Am. Chem. Soc.* **92,** 7463 (1970).
70JA7465	K. K. Kim and J. F. Bunnett, *J. Am. Chem. Soc.* **92,** 7465 (1970).
70JOC383	M. J. Strauss, T. C. Jensen, H. Schram, and K. Connor, *J. Org. Chem.* **35,** 383 (1970).
70JOC1234	H. Yasuda, T. Hayashi, and H. Midorikawa, *J. Org. Chem.* **35,** 1234 (1970).
70RTC129	H. C. van der Plas and A. Koudijs, *Recl. Trav. Chim. Pays-Bas* **89,** 129 (1970).
70RTC680	H. C. van der Plas and H. Jongejan, *Recl. Trav. Chim. Pays-Bas* **89,** 680 (1970).
71B2313	S. C. Johnson and C. C. Guilbert, *Biochemistry* **10,** 2313 (1971).
71BSF1925	J. Elguero, *Bull. Soc. Chim. Fr.*, 1935 (1971).
71JCS(C)2357	A. Albert and K. Ohta, *J. Chem. Soc. C*, 2357 (1971).
71JCS(CC)674	G. Tennant and R. J. S. Vevers, *J. Chem. Soc., Chem. Commun.*, 674 (1971).
71JOC1705	T. J. van Bergen and R. Kellogg, *J. Org. Chem.* **36,** 1705 (1971).
71JOC2196	H. Yasuda and H. Mirodikawa, *J. Org. Chem.* **36,** 2196 (1971).
71KGS571	V. A. Ershov and I. Ya. Postovskii, *Khim. Geterotsikl. Soedin.* **4,** 571 (1971).
71MI1	A. F. Pozharskii and A. M. Simonov, "Chichibabin Amination of Heterocycles." Izd. Rost. Univ., Rostov-on-Don, USSR, 1971.
71MI2	J. H. Lister, *in* "The Chemistry of Heterocyclic Compounds. Fused Pyrimidines" (D. J. Brown, ed.), Part II. Wiley-Interscience, New York, 1971.
71MI3	Y. Tamura, N. Tsujimoto, and M. Ma-no, *Chem. Pharm. Bull. Jpn.* **19,** 130 (1971).
71RTC105	H. W. van Meeteren and H. C. van der Plas, *Recl. Trav. Chim. Pays-Bas* **90,** 105 (1971).
71RTC1239	J. de Valk and H. C. van der Plas, *Recl. Trav. Chim. Pays-Bas* **90,** 1239 (1971).
71TL1075	D. E. L. Corrington, K. Clarke, and R. M. Scrowston, *Tetrahedron Lett.*, 1075 (1971).
71TL2349	M. J. Strauss and H. Schram, *Tetrahedron Lett.*, 2349 (1971).
72ACR139	J. F. Bunnett, *Acc. Chem. Res.* **5,** 139 (1972).
72CB2963	G. Himbert and M. Regitz, *Chem. Ber.* **105,** 2963 (1972).
72CJC917	J. W. Bunting and W. G. Meathrel, *Can. J. Chem.* **50,** 917 (1972).
72JA682	J. A. Zoltewicz and L. S. Helmick, *J. Am. Chem. Soc.* **94,** 682 (1972).
72JCS(B)1892	I. H. Blanch and K. Fretheim, *J. Chem. Soc. B*, 1892 (1972).
72JHC865	Y. Tamura, Y. Mike, T. Honda, and M. Ikeda, *J. Heterocycl. Chem.* **9,** 865 (1972).
72JHC1235	A. Rosowsky and N. Papathanasopoulos, *J. Heterocycl. Chem.* **9,** 1235 (1972).
72RTC841	H. J. den Hertog and D. J. Buurman, *Recl. Trav. Chim. Pays-Bas* **91,** 841 (1972).
72RTC850	P. J. Lont and H. C. van der Plas, *Recl. Trav. Chim. Pays-Bas* **91,** 850 (1972).

72RTC449	P. J. Lont, H. C. van der Plas, and A. J. Verbeek, *Recl. Trav. Chim. Pays-Bas* **91**, 449 (1972).
72RTC1414	J. de Valk and H. C. van der Plas, *Recl. Trav. Chim. Pays Bas* **91**, 1414 (1972).
72S571	G. Himbert and M. Regitz, *Synthesis*, 571 (1972).
72T3299	M. R. Crampton, M. A. El Ghariani, and H. A. Khan, *Tetrahedron* **11**, 3299 (1972).
72UK1788	O. P. Swaika and V. N. Artyomov, *Usp. Khim.* **41**, 1788 (1972).
73CJC1965	J. W. Bunting and W. G. Meathrel, *Can. J. Chem.* **51**, 1965 (1973).
73JA6894	H. Nakatsuji, T. Kuwata, and A. Yoshida, *J. Am. Chem. Soc.* **95**, 6894 (1973).
73JCS(PI)2659	A. Albert, *J. Chem. Soc., Perkin Trans. I*, 2659 (1973)
73JCS(PI)2758	W. Pendergast, *J. Chem. Soc., Perkin Trans. I*, 2758 (1973).
73JOC658	J. A. Zoltewicz and L. S. Helmick, *J. Org. Chem.* **38**, 658 (1973).
73JOC1947	J. A. Zoltewicz, L. S. Helmick, T. M. Oestrich, R. W. King, E. Kandetzki, *J. Org. Chem.* **38**, 1947 (1973).
73JOC1949	J. A. Zoltewicz, T. Oestreich, J. K. O'Halloran, and L. S. Helmick, *J. Org. Chem.* **38**, 1949 (1973).
73JOC2247	M. H. Wilson and J. A. McCloskey, *J. Org. Chem.* **38**, 2247 (1973).
73MI1	H. C. van der Plas, "Ring Transformations of Heterocycles," Vols. 1 and 2. Academic Press, London and New York, 1973.
73RTC145	J. de Valk and H. C. van der Plas, *Recl. Trav. Chim. Pays-Bas* **92**, 145 (1973).
73RTC311	P. J. Lont and H. C. van der Plas, *Recl. Trav. Chim. Pays-Bas* **92**, 311 (1973).
73RTC442	J. de Valk, H. C. van der Plas, and J. W. A. de Bode, *Recl. Trav. Chim. Pays-Bas* **92**, 442 (1973).
73RTC449	P. J. Lont and H. C. van der Plas, *Recl. Trav. Chim. Pays-Bas* **92**, 449 (1973).
73RTC460	J. de Valk, H. C. van der Plas, F. Jansen, and A. Koudijs, *Recl. Trav. Chim. Pays-Bas* **92**, 460 (1973).
73RTC471	J. de Valk and H. C. van der Plas, *Recl. Trav. Chim. Pays-Bas* **92**, 471 (1973).
73RTC708	P. J. Lont, H. C. van der Plas, and A. van Veldhuizen, *Recl. Trav. Chim. Pays-Bas* **92**, 708 (1973).
73RTC711	H. C. van der Plas and A. Koudijs, *Recl. Trav. Chim. Pays-Bas* **92**, 711 (1973).
73RTC970	J. Pomorski, H. J. den Hertog, D. J. Buurman, and N. H. Bakker, *Recl. Trav. Chim. Pays-Bas* **92**, 970 (1973).
73RTC1020	A. P. Kroon and H. C. van der Plas, *Recl. Trav. Chim. Pays-Bas* **92**, 1020 (1973).
73RTC1232	J. P. Geerts, H. C. van der Plas, and A. van Veldhuizen, *Recl. Trav. Chim. Pays-Bas* **92**, 1232 (1973).
73SC99	F. D. Ho, *Synth. Commun.* **3**, 99 (1973).
74AHC33	T. L. Cilchrist and G. E. Gymer, *Adv. Heterocycl. Chem.* **16**, 33 (1974).
74CB2513	G. Himbert and M. Regitz, *Chem. Ber.* **107**, 2513 (1974).
74CB3408	J. Schnekenburger and D. Heber, *Chem. Ber.* **107**, 3408 (1974)
74CJC303	J. W. Bunting and W. G. Meathrel, *Can. J. Chem.* **52**, 303 (1974).
74CJC951	J. W. Bunting and W. G. Meathrel, *Can. J. Chem.* **52**, 951(1974).

REFERENCES 229

74CJC962	J. W. Bunting and W. G. Meathrel, *Can. J. Chem.* **52,** 962(1974).
74CJC975	J. W. Bunting and W. G. Meathrel, *Can. J. Chem.* **52,** 975 (1974).
74CJC981	J. W. Bunting and W. G. Meathrel, *Can. J. Chem.* **52,** 981 (1974).
74JCS(PI)627	A. R. McCarthy, W. D. Ollis, and C. A. Ramsden, *J. Chem. Soc., Perkin Trans. I,* 627 (1974).
74JCS(PI)638	W. D. Ollis and C. A. Ramsden, *J. Chem. Soc., Perkin Trans. I,* 638 (1974).
74MI1	A. J. Boulton, in "Lectures in Heterocyclic Chemistry" (R. N. Castle and L. B. Townsend, eds.), Vol. 2, p. S-45. Heterocorporation: Orem, UT, 1974.
74MI2	H. C. van der Plas, in "Lectures in Heterocyclic Chemistry" (R. N. Castle and L. B. Townsend, eds.), Vol. 2, p. S-83. Heterocorporation: Orem, UT, 1974.
74RC1233	W. Czuba and H. Paradowska, *Rocz. Chem.* **48,** 1233 (1974).
74RTC58	R. Peereboom, H. C. van der Plas, and A. Koudijs, *Recl. Trav. Chim. Pays-Bas* **93,** 58 (1974).
74RTC111	A. P. Kroon and H. C. van der Plas, *Recl. Trav. Chim. Pays-Bas* **93,** 111 (1974).
74RTC114	E. A. Oostveen, H. C. van der Plas, and H. Jongejan, *Recl. Trav. Chim. Pays-Bas* **93,** 114 (1974).
74RTC166	W. J. van Zoest and H. J den Hertog, *Recl. Trav. Chim. Pays-Bas* **93,** 166 (1974)
74RTC195	H. J. den Hertog, H. Boer, J. W. Streef, F. C. A. Vekemans, and W. J. van Zoest, *Recl. Trav. Chim. Pays-Bas* **93,** 195 (1974).
74RTC198	G. M. Sanders, M. van Dijk, and H. J. den Hertog, *Recl. Trav. Chim. Pays-Bas* **93,** 198 (1974).
74RTC225	H. C. van der Plas, M. C. Vollering, H. Jongejan, and B. Zuurdeeg, *Recl. Trav. Chim. Pays-Bas* **93,** 225 (1974).
74RTC227	A. P. Kroon and H. C. van der Plas, *Recl. Trav. Chim. Pays-Bas* **93,** 227 (1974).
74RTC231	J. P. Geerts, C. A. H. Rasmussen, H. C. van der Plas, and A. van Veldhuizen, *Recl. Trav. Chim. Pays-Bas* **93,** 231 (1974).
74RTC277	R. Peereboom, and H. C. van der Plas, *Recl. Trav. Chim. Pays-Bas* **93,** 277 (1974).
74RTC281	R. Peereboom and H. J. den Hertog, *Recl. Trav. Chim. Pays-Bas* **93,** 281 (1974).
74RTC325	A. P. Kroon, H. C. van der Plas, and G van Garderen, *Recl. Trav. Chim. Pays-Bas* **93,** 325 (1974).
74T4055	J. Schnekenburger and D. Heber, *Tetrahedron* **30,** 4055 (1974).
74TL3201	A. P. Kroon and H. C. van der Plas, *Tetrahedron Lett.,* 3201 (1974).
75AP594	J. Schnekenburger and D. Heber, *Arch. Pharm. (Weinheim, Ger.)* **308,** 594 (1975).
75CPB844	M. Maeda and Y. Kawazoe, *Chem. Pharm. Bull.* **23,** 844 (1975).
75CRV389	I. J. Turchi and M. J. S. Dewar, *Chem. Rev.* **75,** 389 (1975).
75JA6484	A. Padwa, E. Chen, and A. Ku, *J. Am. Chem. Soc.* **97,** 6484 (1975).
75JOC1521	M. J. S. Dewar and I. J. Turchi, *J. Org. Chem.* **40,** 1521 (1975).
75JHC985	N. Vivona, G. Cusmano, M. Ruccia, and D. Spinelli, *J. Heterocycl. Chem.* **12,** 985 (1975).
75JHC1327	N. Vivona, M. Ruccia, G. Cusmano, M. L. Marino, and D. Spinelli, *J. Heterocycl. Chem.* **12,** 1327 (1975).
75KGS1400	A. M. Kost, R. S. Sagitullin, and G. G. Damagulyan, *Khim. Geterotsikl. Soedin,* 1400 (1975).

75MI1	A. Nagel and H. C. van der Plas, *J. Pharm. Bull. Jpn.* **23,** 2678 (1975).
75MI2	R. M. J. Ings, J. A. Macfadzean, and W. E. Ormerod, *Xenobiotica* **5,** 223 (1975).
75MI3	G. Haflinger, *in* "Chemistry of Amidines and Imidates" (S. Patai, ed.), Chapter 1. Wiley, New York, 1975.
75OMR86	J. P. Geerts, H. C. van der Plas, and A. van Veldhuizen, *Org. Magn. Reson.* **7,** 86 (1975).
75RTC45	A. Nagel, H. C. van der Plas, and A. van Veldhuizen, *Recl. Trav. Chim. Pays-Bas* **94,** 45 (1975).
75RTC204	A. Rykowski and H. C. van der Plas, *Recl. Trav. Chim. Pays-Bas* **94,** 204 (1975).
75SC119	J. P. Kutney and R. Greenhouse, *Synth. Commun.* **5,** 119 (1975).
75UP1	J. P. Geerts and H. C. van der Plas, unpublished results (1975).
76CB370	G. Himbert, D. Frank, and M. Regitz, *Chem. Ber.* **109,** 370 (1976).
76HCA2074	K. Dietliker, P. Gilgen, H. Heimgartner, and H. Schmid, *Helv. Chim. Acta* **59,** 2074 (1976).
76JA1259	R. Harder and C. Wentrup, *J. Am. Chem. Soc.* **98,** 1259 (1976).
76JCS(PI)315	A. Sultan Afridi, A. R. Katritzky, and C. A. Ramsden, *J. Chem. Soc., Perkin Trans. I*, 315 (1976).
76JOC160	H. Sliwa and A. Tartar, *J. Org. Chem.* **41,** 160 (1976).
76JOC1303	J. A. Zoltewicz, L. S. Helmick, and J. K. O'Halloran, *J. Org. Chem.* **41,** 1303 (1976).
76JOC2621	U. Berg, R. Gallo, and J. Metzger, *J. Org. Chem.* **41,** 2621 (1976).
76MI1	W. L. F. Armarego, *in* "The Fused Pyrimidines" (D. J. Brown, ed.), Part I. Wiley-Interscience, New York, 1976.
76MI2	J. Fleming, "Frontier Orbitals and Organic Chemical Reactions," Wiley, Chichester, 1976.
76OMR607	J. P. Geerts, J. Nagel, and H. C. van der Plas, *Org. Magn. Reson.* **8,** 607 (1976).
76RTC113	G. Simmig, H. C. van der Plas, and C. A. Landheer, *Recl. Trav. Chim. Pays-Bas* **95,** 113 (1976).
76RTC125	G. Simmig and H. C. van der Plas, *Recl. Trav. Chim. Pays-Bas* **95,** 125 (1976).
76RTC209	E. A. Oostveen, H. C. van der Plas, and H. Jongejan, *Recl. Trav. Chim. Pays-Bas* **95,** 209 (1976).
76RTC282	F. Roeterdink and H. C. van der Plas, *Recl. Trav. Chim. Pays-Bas* **95,** 282 (1976).
76TL3337	F. Roeterdink and H. C. van der Plas, *Tetrahedron Lett.*, 3337 (1976).
76TL4717	H. Sliwa and A. Tartar, *Tetrahedron Lett.*, 4717 (1976).
77AG(E)572	R. Huisgen, *Angew. Chem., Int. Ed. Engl.* **16,** 572 (1977).
77CB373	G. Grenner and H. L. Schmidt, *Chem. Ber.* **110,** 373 (1977).
77CHC210	V. N. Novikov, A. F. Pozharskii, and V. N. Doron'kin, *Chem. Heterocycl. Compd.* **12,** 210 (1977).
77H205	A. Nagel and H. C. van der Plas, *Heterocycles* **7,** 205 (1977).
77JA4899	M. J. S. Dewar and W. Thiel, *J. Am. Chem. Soc.* **99,** 4899 (1977).
77JCR(S)294	O. Meth-Con and B. Narine, *J. Chem. Res., Synop.*, 294 (1977).
77JHC537	K. Hirota, K. A. Watanabe, and J. J. Fox, *J. Heterocycl. Chem.* **14,** 537 (1977).

REFERENCES

77KGS821	O. P. Shkurko and V. P. Mamaev, *Khim. Geterotsikl. Soedin.*, 821 (1977).
77RTC68	E. A. Oostveen and H. C. van der Plas, *Recl. Trav. Chim. Pays-Bas* **96**, 68 (1977).
77RTC101	C. A. H. Rasmussen and H. C. van der Plas *Recl. Trav. Chim. Pays-Bas* **96**, 101 (1977).
77RTC183	E. A. Oostveen and H. C. van der Plas, *Recl. Trav. Chim. Pays-Bas* **96**, 183 (1977).
77UP1	R. J. Platenkamp, J. P. Geerts, and H. C. van der Plas, unpublished results (1977).
77UP2	H. Jongejan and H. C. van der Plas, unpublished results (1977).
78ACR462	H. C. van der Plas, *Acc. Chem. Res.* **11**, 462 (1978).
78H33	H. C. van der Plas, *Heterocycles* **9**, 33 (1978).
78H108	E. Marsuura, M. Ariga, and Y. Tondor, *Heterocycles* **9**, 108 (1978).
78JCS(CC)652	A. R. Butler, C. Glidewell, and D. C. Liles, *J. Chem. Soc., Chem. Commun.*, 652 (1978).
78JHC445	A. D. Counotte-Potman and H. C. van der Plas, *J. Heterocycl. Chem.* **15**, 445 (1978).
78JHC1121	C. A. H. Rasmussen, H. C. van der Plas, P. Grotenhuis, and A. Koudijs, *J. Heterocycl. Chem.* **15**, 1121 (1978).
78JOC1193	K. Hirota, K. A. Watanabe, and J. J. Fox *J. Org. Chem.* **43**, 1193 (1978).
78JOC1673	H. C. van der Plas, A. van Veldhuizen, M. Wozniak, and P. Smit, *J. Org. Chem.* **43**, 1673 (1978).
78JOC2682	J. P. Geerts and H. C. van der Plas, *J. Org. Chem.* **43**, 2682 (1978).
78JOC4951	G. L'abbé, A. Timmerman, C. Martens, and S. Toppet, *J. Org. Chem.* **43**, 4951 (1978).
78KGS867	H. C. van der Plas, *Khim. Geterotsikl. Soedin.*, 867 (1978).
78RCR1042	A. F. Pozharski, A. M. Simonov, and V. N. Doron'kin, *Russ. Chem. Rev. (Engl. Transl.)* **47**, 1042 (1978).
78RTC273	A. Rykowski, H. C. van der Plas, and A. van Veldhuizen, *Recl. Trav. Chim. Pays-Bas* **97**, 273 (1978).
78RTC288	C. A. H. Rasmussen and H. C. van der Plas, *Recl. Trav. Chim. Pays-Bas* **97**, 288 (1978).
78TL2021	A. Nagel and H. C. van der Plas, *Tetrahedron Lett.*, 2021 (1978).
78TL3841	C. A. H. Rasmussen and H. C. van der Plas, *Tetrahedron Lett.*, 3841 (1978).
79AG176	K. Akiba, S. Arai, T. Tsuchiya, Y. Yamamoto, and F. Iwasaki, *Angew. Chem.* **91**, 176 (1979).
79BCJ1225	T. Kinoshita, S. Sato, and C. Tamura, *Bull. Chem. Soc. Jpn.* **52**, 1225 (1979).
79CRV181	E. C. Taylor and I. J. Turchi, *Chem. Rev.* **79**, 181 (1979).
79JA3982	F. C. Schaefer and G. A. Peters, *J. Am. Chem. Soc.*, **101**, 3982 (1979).
79JA5857	K. Akiba, T. Kobayashi, and S. Arai, *J. Am. Chem. Soc.* **101**, 5857 (1979).
79JCR(S)64	D. Korbonitz, E. M.-Bako, and K. Horvath, *J. Chem. Res., Synop.*, 64 (1979).
79JCR(S)316	A. F. Cuthbertson, C. Glidewell, H. D. Holden, and D. C. Liles, *J. Chem. Res., Synop.*, 316 (1979).

79JOC3140	N. J. Kos, H. C. van der Plas, and A. van Veldhuizen, *J. Org. Chem.* **44**, 3140 (1979).
79JOC3982	W. K. Chung, C. K. Chu, K. A. Watanabe, and J. Fox, *J. Org. Chem.* **44**, 3982 (1979).
79JOC4677	J. Breuker and H. C. van der Plas, *J. Org. Chem.* **44**, 4677 (1979).
79KGS28	A. N. Kost, R. S. Sagitullin, and S. P. Gromov, *Khim. Geterotsikl. Soedin.*, 28 (1979).
79RTC5	C. A. H. Rasmussen and H. C. van der Plas, *Recl. Trav. Chim. Pays-Bas* **98**, 5 (1979).
79T809	F. M. Moracci, F. Liberatore, S. Tortorella, and B. Di Rienzo, *Tetrahedron* **35**, 809 (1979).
79T2591	F. M. Moracci, S. Tortorella, B. Di Rienzo, and F. Liberatore, *Tetrahedron* **35**, 2591 (1979).
79TL1337	I. Hermecz, J. Engler, Z. Mészáros, and G. Tóth, *Tetrahedron Lett.*, 1337 (1979).
79ZN(B)1019	W. H. Gundel, *Z. Naturforsch. B* **43**, 1019 (1979).
80AG(E)947	R. Huisgen, *Angew. Chem., Int. Ed. Engl.* **19**, 947 (1980).
80JA2451	F. Pittner, T. Mizon, G. Pittner, and M. Wilchek, *J. Am. Chem. Soc.* **102**, 2451 (1980).
80JA6159	C. Wentrup and H.-W. Winter, *J. Am. Chem. Soc.* **102**, 6159 (1980).
80JCR(S)114	A. R Butler, C. Glidewell, I. Hussain, and P. R. Maw, *J. Chem. Res., Synop.*, 114 (1980).
80JHC1733	Y. A. Ibrahim, M. M. Eid, and S. A. L. Abdel-Hady, *J. Heterocycl. Chem.* **17**, 1733 (1980).
80JOC881	A. Rykowski and H. C. van der Plas, *J. Org. Chem.* **45**, 881 (1980).
80JOC2942	N. J. Kos and H. C. van der Plas, *J. Org. Chem.* **45**, 2942 (1980).
80JOC3097	J. D. Reinheimer, L. L. Mayle, G. G. Dolnikowski, and J. T. Gerig, *J. Org. Chem.* **45**, 3097 (1980).
80KGS98	A. N. Kost, R. S. Sagitullin, and S. P. Gromov, *Khim. Geterotsikl. Soedin.*, 98 (1980).
80MI1	C. T. Pedersen, *Sulfur Rep.* **1**, 1 (1980).
80S589	J. Becher, *Synthesis*, 589 (1980).
80T785	F. M. Moracci, B. Di Rienzo, S. Tortorella, and F. Liberatore, *Tetrahedron* **36**, 785 (1980).
80UP1	R. J. Platenkamp, E. A. Oostveen, F. Roeterdink, and H. C. van der Plas, unpublished results (1980).
80WCH491	H. C. van der Plas, *Wiad. Chem.* **34**, 491 (1980).
81AHC1	R. K. Smalley, *Adv. Heterocycl. Chem.* **29**, 1 (1981).
81AHC141	M. Ruccia, N. Vivona, and D. Spinelli, *Adv. Heterocycl. Chem.* **29**, 141 (1981).
81AHC251	G. Gasco and A. J. Boulton, *Adv. Heterocycl. Chem.* **29**, 251 (1981).
81BSB615	M. Regitz, B. Arnold, D. Danion, H. Schubert, and G. Fusser, *Bull. Soc. Chim. Belg.* **90**, 615 (1981).
81CPB2516	C. Kashima, A. Katoh, Y. Yokota, and I. Omoto, *Chem. Pharm. Bull.* **29**, 2516 (1981).
81H1041	N. J. Kos, K. Breuker, H. C. van der Plas, and A. van Veldhuizen, *Heterocycles* **15**, 1041 (1981).
81H1155	K. Akiba, A. Noda, K. Ohkata, T. Akiyama, and Y. Yamamoto, *Heterocycles* **15**, 1155 (1981).
81JA1598	I. Saito, H. Sugiyama, S. Ito, N. Furakawa, and T. Matsuura, *J. Am. Chem. Soc.* **103**, 1598 (1981).

REFERENCES

81JHC953	Y. A. Ibrahim, M. M. Eid, M. A. Badawy, and S. A. L. Abdel-Hady, *J. Heterocycl. Chem.* **18**, 953 (1981).
81JOC2138	A. D. Counotte-Potman, H. C. van der Plas, and A. van Veldhuizen, *J. Org. Chem.* **46**, 2138 (1981).
81JOC3509	K. Breuker, N. J. Kos, H. C. van der Plas, and A. van Veldhuizen, *J. Org. Chem.* **46**, 3509 (1981).
81JOC3805	A. Counotte-Potman, H. C. van der Plas, and A. van Velhuizen, *J. Org. Chem.* **46**, 3805 (1981).
81JOC5102	A. D. Counotte-Potman, H. C. van der Plas, A. van Veldhuizen, and C. A. Landheer, *J. Org. Chem.* **46**, 5102 (1981).
81M1393	U. Hain and P. Hartung, *Monatsh. Chem.* **112**, 1393 (1981).
81MI1	I. Saito, H. Sugiyama, N. Furakawa, and T. Matsuura, *Nucleic Acids Res., Symp. Ser.* **10**, 61 (1981).
81T3423	A. N. Kost, S. P. Gromov, and R. S. Sagatullin, *Tetrahedron* **37**, 3423 (1981).
81TH1	N. J. Kos, Ph.D. Thesis, Agricultural University, Wageningen (1981).
81TL2365	I. Saito, H. Sugiyama, N. Furakawa, and T. Matsuura, *Tetrahedron. Lett.*, 2365 (1981).
81TL2409	K. Hirota, Y. Kitade, and S. Senda, *Tetrahedron Lett.*, 2409 (1981).
82CPB1942	C. Kashima, A. Katoh, Y. Yokota, and I. Omoto, *Chem. Pharm. Bull.* **30**, 1942 (1982).
82CRV223	F. Terrier, *Chem. Rev.* **82**, 223 (1982).
82G181	A. J. Boulton, F. J. Frank, and M. R. Huckstep, *Gazz. Chim. Ital.* **112**, 181 (1982).
82JCS(PI)759	D. Korbonitz, I. Kanzel-Szvoboda, and K. Horvath, *J. Chem. Soc., Perkin Trans. I*, 759 (1982).
82JHC41	H. A. Daboun and Y. A. Ibrahim, *J. Heterocycl. Chem.* **19**, 41 (1982).
82JHC673	A. Rykowski and H. C. van der Plas, *J. Heterocycl. Chem.* **19**, 673 (1982).
82JHC943	M. T. Nguyen, G. Leroy, M. Sana, and J. Elguero, *J. Heterocycl. Chem.* **19**, 943 (1982).
82JHC1261	T.-L. Su and K. A. Watanabe, *J. Heterocycl. Chem.* **19**, 1261 (1982).
82JHC1285	H. Hara and H. C. van der Plas, *J. Heterocycl. Chem.* **19**, 1285 (1982).
82JHC1527	H. Hara and H. C. van der Plas, *J. Heterocycl. Chem.* **19**, 1527 (1982).
82JHC3739	A. Dlugozs, H. C. van der Plas, and A. van Veldhuizen, *J. Heterocycl. Chem.* **19**, 3739 (1982).
82JOC498	A. R. Katritzky, R. Awartan, and R. C. Patel, *J. Org. Chem.* **47**, 498 (1982).
82JOC963	K. Breuker, N. J. Kos, H. C. van der Plas, and A. van Veldhuizen, *J. Org. Chem.* **47**, 963 (1982).
82JOC1081	J. J. Fox, T.-L. Su, L. M. Stempel, and K. A. Watanabe, *J. Org. Chem.* **47**, 1081 (1982).
82JOC2856	C. H. Stam, A. D. Counotte-Potman, and H. C. van der Plas, *J. Org. Chem.* **47**, 2856 (1982).
82JOC2858	D. H. Oskin, G. P. Wooden, and R. A. Olofson, *J. Org. Chem.* **47**, 2858 (1982).
82MI1	A. T. Balaban *et al.*, eds., "Pyrylium Salts. Syntheses, Reactions and Physical Properties." Academic Press, New York, 1982.

82MI2	J. Sandström, "Dynamic NMR Spectroscopy," p. 96. Academic Press, New York, 1982.
82RTC342	S. A. G. F. Angelino, D. J. Buurman, H. C. van der Plas, and F. Müller, *Recl. Trav. Chim. Pays-Bas* **101**, 342 (1982).
82RTC367	J. Breuker and H. C. van der Plas, *Recl. Trav. Chim. Pays-Bas* **101**, 367 (1982).
82T3537	G. L'abbé, *Tetrahedron* **38**, 3537 (1982).
82T1405	T.-L. Su, K. A. Watanabe, and J. J. Fox, *Tetrahedron* **38**, 1405 (1982).
82TH1	J. Breuker, Ph.D. Thesis, Agricultural University, Wageningen (1982).
82TL1103	G. L'abbé, M. Deketele, and J.-P. Dekerk, *Tetrahedron Lett.*, 1103 (1982).
82TL2891	I. Bitter, B. Pete, I. Hermecz, G. Tóth, K. Simon, M. Czugler, and Z. Mészáros, *Tetrahedron Lett.*, 2891 (1982).
82TL3965	V. N. Charushin and H. C. van der Plas, *Tetrahedron Lett.* **23**, 3965 (1982).
83AHC(33)95	H. C. van der Plas, M. Wozniak, and H. J. W. van den Haak, *Adv. Heterocycl. Chem.* **33**, 95 (1983).
83AHC(34)305	G. Illuminati and F. Stegel, *Adv. Heterocycl. Chem.* **34**, 305 (1983).
83CL715	Y. Tohda, M. Ariga, and E. Matsumura, *Chem. Lett.*, 715 (1983).
83H51	S. A. El-Bahaie, M. A. Badawy, S. A. Abdel-Hady, and Y. A. Ibrahim, *Heterocycles* **20**, 51 (1983).
83H579	I. Bitter, G. Tóth, B. Pete, I. Hermecz, K. Simon, and Z. Mészáros, *Heterocycles* **20**, 579 (1983).
83H1891	I. Bitter, G. Tóth, B. Pete, I. Hermecz, K. Simon, G. Náray-Szabó, and Z. Mészáros, *Heterocycles* **20**, 1891 (1983).
83JCS(CC)289	P. J. Pogorzelec and D. H. Reid, *J. Chem. Soc., Chem. Commun.*, 289 (1983).
83JCS(CC)588	G. L'abbé, M. Deketele, and J.-P. Dekerk, *J. Chem. Soc., Chem. Commun.*, 588 (1983).
83JCS(D)261	M. N. Hughes, J. R. Lusty, and H. L. Wallis, *J. Chem. Soc., Dalton Trans.*, 261 (1983).
83JHC9	M. Wozniak, H. C. van der Plas, and A. van Veldhuizen, *J. Heterocycl. Chem.* **20**, 9 (1983).
83JHC415	H. C. van der Plas, S. Baloniak, and H. Jongejan, *J. Heterocycl. Chem.* **20**, 415 (1983).
83JHC1259	A. Counotte-Potman and H. C. van der Plas, *J. Heterocycl. Chem.* **20**, 1259 (1983).
83JOC1354	H. C. van der Plas, V. N. Charushin, and A. van Veldhuizen, *J. Org. Chem.* **48**, 1354 (1983).
83MI1	M. M. Eid, M. A. H. Ghazala, and Y. A. Ibrahim, *Egypt. J. Chem.* **24**, 281 (1983).
83RTC359	M. Wozniak, H. C. van der Plas, M. Tomula, and A. van Veldhuizen, *Recl. Trav. Chim. Pays-Bas* **102**, 359 (1983).
83RTC373	V. N. Charushin and H. C. van der Plas, *Recl. Trav. Chim. Pays-Bas* **102**, 373 (1983).
83TL5763	D. Korbonits, K. Simon, and P. Kolonits, *Tetrahedron Lett.*, 5763 (1983).
84H289	K. A. Watanabe, T. L. Su, K. W. Pankiewicz, and K. Harada, *Heterocycles* **21**, 289 (1984).

84JA2713	Y. Yamamoto and K. Akiba, *J. Am. Chem. Soc.* **106,** 2713 (1984).
84JCS(PI)1859	K. Hirota, Y. Kitade, and S. Senda, *J. Chem. Soc., Perkin Trans. I,* 1859 (1984).
84JHC433	A. Rykowski and H. C. van der Plas, *J. Heterocycl. Chem.* **21,** 433 (1984).
84JHC627	G. L'abbé, *J. Heterocycl. Chem.* **21,** 627 (1984).
84JHC1403	M. A. Badawy, Y. A. Ibrahim, and A. M. Kadry, *J. Heterocycl. Chem.* **21,** 1403 (1984).
84MI1	A. R. Katritzky and C. W. Rees, eds., "Comprehensive Heterocyclic Chemistry," Vols. 1–8. Pergamon, Oxford, 1984.
84MI2	A. R. Katritzky and J. M. Lagowski, *in* "Comprehensive Heterocyclic Chemistry" (A. R. Katritzky and C. W. Rees, eds.), Vol. 5, p. 94. Pergamon, Oxford, 1984.
84MI3	F. Terrier, "Electron Deficient Aromatic and Heteroaromatic Base Interactions. The Chemistry of Anionic Sigma Complexes," pp. 78–85. Elsevier, Amsterdam, 1984.
84T433	S. A. G. F. Angelino, A. van Veldhuizen, D. J. Buurman, and H. C. van der Plas, *Tetrahedron* **40,** 433 (1984).
85BBA181	W. Wiesner, K.-H. van Pée, and E. de Boer, *Biochim. Biophys. Acta* **830,** 181 (1985).
85H2289	I. Bitter, B. Pete, G. Tóth, A. Almasy, I. Hermecz, K. Simon, and Z. Mészáros, *Heterocycles* **23,** 2289 (1985).
85H2549	I. Bitter, B. Pete, G. Tóth, I. Hermecz, and Z. Mészáros, *Heterocycles* **23,** 2549 (1985).
85JA2721	K. Akiba, K. Kashiwagi, Y. Ohyama, Y. Yamamoto, and K. Ohkata, *J. Am. Chem. Soc.* **107,** 2721 (1985).
85JHC155	A. M. Kadry and S. A. Mansour, *J. Heterocycl. Chem.* **22,** 155 (1985).
85JHC353	H. Tondys, H. C. van der Plas, and M. Wozniak, *J. Heterocycl. Chem.* **22,** 353 (1985).
85JHC1535	M. A. Badawy, S. A. L. Abdel-Hady, Y. A. Ibrahim, and A. M. Kadry, *J. Heterocycl. Chem.* **22,** 1535 (1985).
85JOC484	D. de Bie, B. Geurtsen, and H. C. van der Plas, *J. Org. Chem.* **50,** 484 (1985).
85MI1	H. C. van der Plas, *Janssen Chim. Acta* **3,** 23 (1985).
85S884	A. Rykowski and H. C. van der Plas, *Synthesis,* 884 (1985).
85T237	H. C. van der Plas, *Tetrahedron* **41,** 237 (1985).
86CCA89	M. Hamana, *Croat. Chem. Acta* **59,** 89 (1986).
86H69	I. Bitter, B. Pete, G. Tóth, I. Hermecz, K. Simon, and Z. Mészáros, *Heterocycles* **24,** 69 (1986).
86IJ67	K. Breuker, H. C. van der Plas, and A. van Veldhuizen, *Isr. J. Chem.* **27,** 67 (1986).
86JCS(PI)2163	D. Korbonits and P. Kolonits, *J. Chem. Soc., Perkin Trans. I,* 2163 (1986).
86JHC477	H. Sladowska, J. W. G. de Meester, and H. C. van der Plas, *J. Heterocycl. Chem.* **23,** 477 (1986).
86JOC71	D. A. de Bie, B. Geurtsen, and H. C. van der Plas, *J. Org. Chem.* **51,** 71 (1986).
86JOC4070	H. C. van der Plas, A. T. M. Marcelis, D. M. W. van den Ham, and J. W. Verhoeven, *J. Org. Chem.* **51,** 4070 (1986).

REFERENCES

86JST215	K. Horvath, D. Korbonitz, G. Náray-Szabó, and K. Simon, *J. Mol. Struct.* (*Theochem.*) **136**, 215 (1986).
86MI1	S. Riva, G. Carrea, F. M. Veronese, and A. F. Bückmann, *Enzyme Microb. Technol.* **9**, 556 (1986).
87BSB827	G. L'abbé, L. Beenaerts, F. Godts, and S. Toppet, *Bull. Soc. Chim. Belg.* **96**, 827 (1987).
87JCS(P2)523	R. Fruttero, R. Calvino, B. Ferrarotti, A. Gasco, S. Aime, R. Gobetto, C. Chiari, and G. Calestani, *J. Chem. Soc., Perkin Trans. 2*, 523 (1987).
87JHC1657	A. C. Brouwer and H. C. van der Plas, *J. Heterocycl. Chem.* **24**, 1657 (1987).
87JOC5643	M. Wozniak, A. Baranski, K. Novak, and H. C. van der Plas, *J. Org. Chem.* **52**, 5643 (1987).
87KGS488	S. P. Gromov, M. Bkhaumik, and Yu. G. Bundel, *Khim. Geterotsikl. Soedin.*, 488 (1987).
87KGS857	V. L. Rusinov, T. L. Pilicheva, A. A. Tumashov, and O. N. Chupakhin, *Khim. Geterotsikl. Soedin.*, 857 (1987).
87KGS1011	H. C. van der Plas, *Khim. Geterotsikl. Soedin.*, 1011 (1987).
87MI1	A. F. Bückmann, *Biocatalysis* **1**, 173 (1987).
87MI2	A. F. Bückmann, M. Morr, and M.-R Kula, *Biotechnol. Appl. Biochem.* **9**, 258 (1987).
87MI3	H. J. Timpe, V. Schikowiski, and K. D. Remmler, *Chem. Pap.* **41**, 259 (1987).
88AHC2	C. K. McGill and A. Rappa, *Adv. Heterocycl. Chem.* **44**, 2 (1988).
88CHE1410	V. G. Andrianov, V. G. Semenikhina, A. V. Eremeev, and A. P. Gaukhman, *Chem. Heterocycl. Compd.* (*Engl. Transl.*) **24**, 1410 (1988).
88H1623	A. F. Bückmann, *Heterocycles* **27**, 1623 (1988).
88JHC831	A. T. M. Marcelis, H. Tondys, and H. C. van der Plas, *J. Heterocycl. Chem.* **25**, 831 (1988).
88JOC382	R. S. Hosmane, B. B. Lim, and F. N. Burnett, *J. Org. Chem.* **53**, 382 (1988).
88KGS1213	M. A. Yurovskaya, A. Z. Afanas'ev, V. A. Chertkov, and Yu. G. Bundel, *Khim. Geterotsikl. Soedin.*, 1213 (1988).
88KG1570	E. V. Babaev, S. I. Bobrovskii, and Yu. G. Bundel, *Khim. Geterotsikl. Soedin.*, 1570 (1988).
88MI1	O. N. Chupakhin, V. N. Charushin, and A. I. Chernyshev, *Prog. NMR Spectrosc.* **20**, 95 (1988).
88T1	O. N. Chupakhin, V. N. Charushin, and H. C. van der Plas, *Tetrahedron* **44**, 1 (1988).
88T461	G. L'abbé and W. Dehaen, *Tetrahedron* **44**, 461 (1988).
89BCJ479	Y. Yamamoto and K. Akiba, *Bull. Chem. Soc. Jpn.* **62**, 479 (1989).
89BSB343	G. L'abbé, M. Bruynseels, L. Beenaerts, A. Vandendriessche, P. Delkeke, and S. Toppet, *Bull. Soc. Chim. Belg.* **98**, 343 (1989).
89CRV1841	R. W. Hoffmann, *Chem. Rev.* **89**, 1841 (1989).
89H737	S. Buscemi and N. Vivona, *Heterocycles* **29**, 737 (1989).
89H1301	S. Buscemi, G. Macaluso, and N. Vivona, *Heterocycles* **29**, 1301 (1989).
89JHC701	G. L'abbé and A. J. Vandendriessche, *J. Heterocycl. Chem.* **26**, 701 (1989).

89MI1	A. F. Bückmann and G. Carrea, *Adv. Biochem. Eng. Biotechnol.* **39**, 97 (1989).
89T2693	A. T. M. Marcelis and H. C. van der Plas, *Tetrahedron* **45**, 2693 (1989).
90AP949	M. S. Montavia, E. B. Pedersen, J. Suwinski, and C. M. Nielsen, *Arch. Pharm. (Weinheim, Ger.)* **323**, 949 (1990).
90BCJ2830	Y. Tohda, M. Eiraku, T. Nakagawa, Y. Usami, M. Ariga, T. Kawashima, K. Tani, H. Watanabe, and Y. Mori, *Bull. Chem. Soc. Jpn.* **63**, 2830 (1990).
90CHE1199	V. G. Andrianov and A. V. Eremeev, *Chem. Heterocycl. Compd. (Engl. Transl.)* **26**, 1199 (1990).
90DOK1127	A. V. Belik and D. V. Belousov, *Dokl. Akad. Nauk SSSR* **313**, 1127 (1990).
90JHC1441	J. Liebscher, A. Hassoun, H. C. van der Plas, and C. Stam, *J. Heterocycl. Chem.* **27**, 1441 (1990).
90JHC2021	L'abbé, M. Bruynseels, P. Delbeke, and S. Toppet, *J. Heterocycl. Chem.* **27**, 2021 (1990).
90KGS256	O. N. Chupakhin, V. L. Rusinov, T. L. Pilicheva, A. A. Tumashov, G. G. Alexandrov, E. O. Sidorov, and L. V. Karpin, *Khim. Geterotsikl. Soedin.*, 256 (1990).
90KGS1632	V. L. Rusinov, T. L. Pilicheva, A. A. Tumashov, G. G. Alexandrov, E. O. Sidorov, L. V. Karpin, and O. N. Chupakhin, *Khim. Geterotsikl. Soedin.*, 1632 (1990).
90LA653	M. Wozniak, A. Baranski, K. Novak, and H. Paradowska, *Liebigs Ann. Chem.*, 653 (1990).
90MI1	G. Ottolina, G. Carrea, S. Riva, and A. F. Bückmann, *Enzyme Microb. Technol.* **12**, 596 (1990).
90PJC813	E. Salwinska and J. Suwinski, *Pol. J. Chem.* **64**, 813 (1990).
90S713	O. N. Chupakhin, V. L. Rusinov, T. L. Pilicheva, and A. A. Tumashov, *Synthesis,* 713 (1990).
91CHE102	V. G. Andrianov, V. G. Semenikhina, and A. V. Eremeev, *Chem. Heterocycl. Compd. (Engl. Transl.)* **27**, 102 (1991).
91H1455	I. Hermecz, A. Horváth, T. Erös-Takácsy, and B. Podányi, *Heterocycles*, **32**, 1455 (1991).
91H1547	G. La Manna, S. Buscemi, V. Frenna, N. Vivona, and D. Spinelli, *Heterocycles* **32**, 1547 (1991).
91JHC783	I. Hermecz, T. Breining, J. Sessi, and B. Podanyi, *J. Heterocycl. Chem.* **28**, 783 (1991).
91JOC7038	P. Romea, M. Aragones, J. Garcia, and J. Vilarrasa, *J. Org. Chem.* **56**, 7038 (1991).
91KGS101	V. L. Rusinov, N. A. Klyuev, V. G. Baklykov, T. L. Pilicheva, and A. A. Tumashov, *Khim. Geterotsikl. Soedin.*, 101 (1991).
91LA875	M. Wozniak, A. Baranski, and B. Szpakiewicz, *Liebigs Ann. Chem.*, 875 (1991).
91MI1	F. Terrier, *in* "Nucleophilic Aromatic Displacement: The Influence of the Nitro Group" (H. Feuer, ed.), p. 257. Org. Nitro Chem. Ser., VCH Publishers, New York, 1991.
91MI2	K.-H. van Pée, *Biotechnol. Adv.* **8**, 1 (1991).
91MI3	J. Suwinski and W. Szczepankiewicz, *J. Labelled Compd. Radiopharm.* **31**, 159 (1991).
91MI4	A. T. M. Marcelis and H. C. van der Plas, *Trends Heterocycl. Chem.* **2**, 111 (1991).

91PJC323	M. Wozniak, K. Novak, and H. Paradowska, *Pol. J. Chem.* **63**, 323 (1991).
91PJC515	J. Suwinski and W. Szczepankiewicz, *Pol. J. Chem.* **65**, 515 (1991).
91PJC1071	E. Salwinska, J. Suwinski, and M. Bialecki, *Pol. J. Chem.* **65**, 1071 (1991).
91TA941	J. Suwinski and W. Szczepankiewicz, *Tetrahedron: Assymetry* **2**, 941 (1991).
92AHC49	N. Vivona, S. Buscemi, V. Frenna, and G. Cusmano, *Adv. Heterocycl. Chem.* **56**, 49 (1992).
92AP317	J. Suwinski and W. Szczepankiewicz, *Arch. Pharm.* (*Weinheim, Ger.*) **325**, 317 (1992).
92BSB67	E. V. Babaev and N. S. Zefirov, *Bull. Soc. Chim. Belg.* **101**, 67 (1992).
92JCS(PI)3069	D. Korbonits, E. Tóbias-Héjá, K. Simon, Gy. Kramer, and P. Kolonits, *J. Chem. Soc., Perkin Trans. I*, 3069 (1992).
92JHC1133	T. Kuroda, K. Hisamura, I. Matsukuma, H. Nishikawa, M. Morimoto, T. Ashizawa, N. Nakamizo, and Y. Otsuji, *J. Heterocycl. Chem.* **29**, 1133 (1992).
92JOC190	A. R. Katritzky, F-B. Ji, W-Q. Fan, J. K. Gallos, J. V. Greenhill, R. W. King, and P. Steel, *J. Org. Chem.* **57**, 190 (1992).
92KGS116	N. A. Klyuev, V. G. Baklykov, and. V. L. Rusinov, *Khim. Geterotsikl. Soedin.*, 116 (1992).
92KGS792	V. I. Terenin, E. V. Babaev, M. A. Yurovskaya, and Yu. G. Bundel, *Khim. Geterotsikl. Soedin.*, 792 (1992).
92KGS808	E. V. Babaev and N. S. Zefirov, *Khim. Geterotsikl. Soedin.*, 808 (1992)
92LA899	M. Wozniak, A. Baranski, K. Novak, and H. Paradowska, *Liebigs Ann. Chem.*, 899 (1992).
92MI1	M. C. R. Fransen and H. C. van der Plas, *Adv. Appl. Microbiol.* **37**, 41 (1992).
92MI2	G. G. Dolnikowski, *J. Am. Soc. Mass Spectrom.* **3**, 467 (1992).
92MI3	E. V. Babaev and N. S. Zefirov, *J. Math. Chem.* **11**, 65 (1992).
92PJC943	H. Llempen, E. Salwinska, J. Suwinski, and W. Szczepankiewicz, *Pol. J. Chem.* **66**, 943 (1992).
92PJC1623	E. Salwinska and J. Suwinski, *Pol. J. Chem.* **66**, 1623 (1992).
92TL3695	O. N. Chupakhin, V. L. Rusinov, A. A. Tumashov, E. O. Sidorov, and L. V. Karpin, *Tetrahedron Lett.*, 3695 (1992).
93ACS95	M. Wozniak and H. C. van der Plas, *Acta Chem. Scand.* **47**, 95 (1993).
93CB1835	I. Bata, D. Korbonits, P. Kolonits, B. Podányi, T. Takácsy-Erös, and K. Simon, *Chem. Ber.* **126**, 1835 (1993).
93JA2416	E. V. Babaev, D. E. Lushnikov, and N. S. Zefirov, *J. Am. Chem. Soc.* **115**, 2416 (1993).
93JCS(PI)1753	L.-L. Lai, T.-Ho Ngoi, D. H. Reid, R. H. Nicol, and J. B. Rhodes, *J. Chem. Soc., Perkin Trans. I*, 1753 (1993).
93JCR(S)350	A. Zandersons, G. P. Shkil, V. Lusis, D. Muceniece, R. Sagitullin, and G. Duburs, *J. Chem. Res., Synop.*, 350 (1993).
93LA7	M. Wozniak, A. Baranski, and B. Szpakiewicz, *Liebigs Ann. Chem.*, 7 (1993).
93LA823	M. Wozniak and M. Grzegozek, *Liebigs Ann. Chem.*, 823 (1993).

93ZOK789	V. L. Rusinov, A. A. Tumashov, E. O. Sidorov, L. V. Karpin, and O. N. Chupakhin, *Zh. Org. Khim.* **29**, 789 (1993).
94AHC49	N. Vivona, S. Buscemi, V. Frenna, and G. Cusmano, *Adv. Heterocycl. Chem.* **56**, 49 (1994).
94H(37)1511	J. Suwinski, W. Pawlus, E. Salwinska, and K. Swierczek, *Heterocycles* **37**, 1511 (1994).
94H(37)2051	D. Korbonits and K. Horváth, *Heterocycles* **37**, 2051 (1994).
94H(38)249	N. Nishiwaki, T. Matsunaga, Y. Tohda, and M. Ariga, *Heterocycles* **38**, 249 (1994).
94HAC97	Y. A. Ibrahim and A. H. M. Elwahy, *Heteroatom Chem.* **5**, 97 (1994).
94HAC149	L.-L. Lai, D. H. Reid, R. H. Nicol, and J. B. Rhodes, *Heteroatom Chem.* **5**, 149 (1994).
94IJC(B)881	Y. A. Ibrahim and A. H. M. Elway, *Indian J. Chem., Sect. B* **33B**, 881 (1994).
94KGS1649	H. C. van der Plas, *Khim. Geterotsikl. Soedin.*, 1649 (1994).
94LA19	D. Korbonits, E. Tóbiás-Héja, K. Simon, and P. Kolonits, *Liebigs Ann. Chem.*, 19 (1994).
94MI1	O. N. Chupakhin, V. N. Charushin, and H. C. van der Plas, "Nucleophilic Aromatic Substitution of Hydrogen," Chapter 3. Academic Press, San Diego, CA, 1994.
94MI2	E. Schlimme, F. G. Ott, and C. Kiesner, *Int. Dairy J.* **4**, 617 (1994).
94MI3	M. C. R. Franssen, *Biocatalysis* **10**, 87 (1994).
94PJC635	M. Grzegozek, M. Wozniak, and H. C. van der Plas, *Pol. J. Chem.* **68**, 635 (1994).
94PJC1115	Y. A. Ibrahim, A. H. M. Elwahy, and S. A. A. Gad, *Pol. J. Chem.* **68**, 1115 (1994).
94T5741	J. Suwinski and E. Salwinska, *Tetrahedron* **50**, 5741 (1994).
95H(40)441	V. L. Rusinov, O. N. Chupakhin, and H. C. van der Plas, *Heterocycles* **40**, 441 (1995).
95H(41)1399	A. F. Bückmann, V. Wray, and H. C. van der Plas, *Heterocycles* **41**, 1399 (1995).
95JA3665	A. Ariza, V. Bau, and J. Vilarrasa, *J. Am. Chem. Soc.* **117**, 3665 (1995).
95MI1	K. Walczak and J. Suwinski, *Pol. J. Appl. Chem.* **39**, 87 (1995).
95RTC13	M. Wozniak, M. Grzegozek, W. Roszkiewicz, and B. Szpakiewcz, *Recl. Trav. Chim. Pays-Bas* **114**, 13 (1995).
95T5133	S. Buscemi, V Frenna, N. Vivona, G. Patrillo, and D. Spinelli, *Tetrahedron* **51**, 5133 (1995).
95T8599	G. P. Shkil, V. Lusis, D. Muceniece, and R. Sagitullin, *Tetrahedron* **51**, 8599 (1995).
95UP1	M. Grzegorek, P. Kowalski, and H. C. van der Plas, unpublished results (1995).
96CPB967	T. Itaya, K. Ogawa, Y. Takada, and T. Fujii, *Chem. Pharm. Bull.* **44**, 967 (1996).
96H2607	W. Arnold, B. Büttelmann, M.-P. Heitz, and R. Wyler, *Heterocycles* **43**, 2607 (1996).
96HAC97	L.-L. Lai and D. H. Reid, *Heteroatom Chem.* **7**, 97 (1996).
96JA4009	R. A. Evans, M. W. Wong, and C. Wentrup, *J. Am. Chem. Soc.* **118**, 4009 (1996).

96JHC1847	P. R. Barilli, *J. Heterocycl. Chem.* **33,** 1847 (1996).
96LA641	M. Grzegozek, M. Wozniak, and H. C. van der Plas, *Liebigs Ann. Chem.,* 641 (1996).
96MI1	J. Suwinski and W. Szczepankiewicz, *J. Labelled Compd. Radiopharm.* **38,** 395 (1996).
96T14905	J. Suwinski, W. Szczepankiewicz, and E. M. Holt, *Tetrahedron* **52,** 14905 (1996).
96ZOK1742	A. V. Belik and E. V. Igoshina, *Zh. Org. Khim.* **32,** 1742 (1996).
97CPB832	T. Itaya, N. Ito, T. Kanai, and T. Fujii, *Chem. Pharm. Bull.* **45,** 832 (1997).
97JA7817	M. P. Groziak, L. Chen, L. Li, and P. D. Robinson, *J. Am. Chem. Soc.* **119,** 7817 (1997).
97JCS(PI)2261	N. Nishiwaki, H.-P. Wang, K. Matsuo, Y. Tohda, and M. Ariga, *J. Chem. Soc., Perkin Trans. I,* 2261 (1997).
97MI1	E. G. Atavin, V. O. Tikhonenko, and R. S. Sagitullin, *Abstr. Int. Mem. Postovsky Conf. Org. Chem.,* Abstr. 72, p. 115, Ekaterinenburg, Russia, 1997.
97MI2	E. V. Babaev, *Targets Heterocycl. Chem.* **1,** 107 (1997).
97M13	J. Kökösi, J. Almási, B. Podániy, M. Fehér, Zs. Böcskei, K. Simon, and I. Hermecz, *unpublished results.*
97S1277	N. Nishiwaki, Y. Tohda, and M. Ariga, *Synthesis,* 1277 (1997).
98JOC5801	W. M. F. Fabian, V. A. Bakulev, and C. O Kappe, *J. Org. Chem.* **63,** 5801 (1998).
98KGS967	M. Wozniak and M. Grzegozek, *Khim. Geterotsikl. Soedin.,* 967 (1998).

Index

A

Acetamidine, 139, 141
Acetophenone, 146
3-Acetylamino-5-aryl-1,2,4-oxadiazoles,
3-Acetylamino-5-methyl-1,2,4-oxadiazole, 203
3-Acetylamino-5-phenyl-1,2,4-oxadiazole, 4, 200
2-Acetylamino-7-R-quinazolin-4-ones, 204
1-Acetyl-5-anilino-1,2,3,4-tetrazole, 158
Acidic hydrogens, 24, 44
1-Alkoxy-7-alkyladenines, 178
5-Alkoxy-4-(aminocarbonyl) oxazole, 190
5-(α-Alkoxycarbonyldiazomethyl)-1,2,3-thiadiazole, 213
1-Alkyl-4-formyl-1,2,3-triazole, 192
1-Alkyl-2-iminopyridines, 3
2-Alkyl-5-alkylimino-4-aryl-3-arylimino-1,2,4-thiadiazolidines, 160
2-Alkyl(aryl)amino-3,4-dihydro-4-oxoquinazoline, 180
2-Alkylamino-1,2-dihydropyridine, 91
6-Alkylamino-5,6-dihydrothmine, 117
2-Alkylaminopyridines, 3, 5
5-Alkylamino-1,2,4-thiadiazole, 159
1,3-Allylic strain, 112
1,3 Ambident nucleophiles, 131, 134, 142
Amidinium salts, 141
2-Amino-3-alkyl-3,4-dihydro-4-oxoquinazoline, 180
4-Amino-3-arylamino-4,5-dihydro-5-oxo-1,2,4-triazine, 120
1-Amino-2-arylaminopyrimidinium salt, 109
2-Amino-1-aryl-1,2,3-triazole, 192
5-(β-Amino-β-arylvinyl)-3-arylisothiazole, 214

3-Amino-1,2,4-benzotriazine, 76
1-Amino-9-benzyl-6-iminopurine, 176
2-Amino-5-bromo-4-t-butylpyrimidine, 42
1-Amino-3-bromoisoquinolinide, 19
4-Amino-2-bromo-1,5-naphthyridine, 61
6-Amino-4-t-butylpyrimidine, 41, 52
2-Amino-4-t-butylpyrimidine, 52
1-(4-Aminobutyl)thymine, 119
Aminocaprolactam, 119
3-Aminocarbonyl-1-methoxypyridinium salt, 198
3-Aminocarbonyl-1-methylpyridinium salt, 92
5-Amino-4-carboxamido-1-(2-nitrophenyl)-1H-1,2,3-triazole, 15
7-Aminocholesterol tryptamine, 88
6-Amino-2-chloro-7-methylpurine, 58
4-Amino-6-chloro-1-methylpyrazolo[3,4-d]pyrimidine, 58
6-Amino-2-chloro-5-nitropyridine, 16
1-Amino-4-cyano-2-azabutadiene, 26
7-Amino-6-cyano-1,3-dimethylpyrido[2,3-d]pyrimidine-2,4(1H,3H)-dione, 184
1-Amino-4-cyano-2-nitrobuta-1,3-diene, 16
4-Amino-1-cyano-2-phenyl-1-aza-1,3-butadiene, 34
4-Amino-1-cyano-4-phenyl-1-aza-1,3-butadiene, 38
6-Amino-5-cyano-4-phenylpyrimidine, 26
2-Amino-4,6-dicyanopyrimidine, 38
2-Amino-3,4-dihydro-3-methyl-4-oxopteridine, 174
4-Amino-6,7-dihydro-6-oxo-1-methylpyrazolo[3,4-d]pyrimidine, 58
2-Amino-1,2-dihydro-4-phenyl-pyrimidine, 49
2-Amino-3,5-dinitropyridine, 18

4-Amino-3,5-dioxo-2,3,4,5-tetrahydro-1,2,4-triazine, 120
2-Amino-3,6-diphenylpyrazine, 66
4-Amino-5-ethoxycarbonyl-2-cyanimino-3-R-thiazoline, 216
2-Amino-3-formylthiophenes, 183
3-Aminofurazan, 204
4-Aminofurazan 3-carboxamidoximes, 206
1-Amino-6-hydroxylamino-1,3-diazahexa-1,3,5-triene, 108
1-Aminoinosine, 116
1-Aminoisoquinolinide, 19
2-Amino-4-methylpyrimidine, 52
5-Amino-3-methyl-1,2,4-thiadiazole, 207
4-Amino-5-methyl-1-(p-toluenesulfonyl)benzotriazole, 193
4-Amino-2-methyl-1,3,5-triazanaphthalene, 61
4-Amino-2-methylquinazoline, 61
4-Amino-5-nitro-6-imino-1-methylpyrimidine, 166
2-Amino-5-nitropyridine, 15, 16, 142
3-Aminophenanthro[9,10-e]1,2,4-triazine, 76
3-Amino-3-phenylacrylonitrile, 37
2-Amino-3-phenyl-5-nitropyridine, 142
2-Amino-4-phenylpyrimidine, 34, 37, 38, 47–51, 131
6-Amino-4-phenylpyrimidine, 21, 23, 47, 48, 49, 50
6-Amino-5-phenylpyrimidine, 51
6-Amino-2-phenylpyrimidine, 59
2-Amino-4(5)-phenylpyrimidines, 51, 131
5-Amino-2-phenylpyrazine, 67
2-Amino-6-phenylpurine, 61
3-Amino-5-phenyl-1,2,4-triazine, 74
4-Amino-2-phenyl-1,3,5-triazine, 76
6-Amino-4-piperidinopyrimidine, 43
2-Amino-4-piperidinopyrimidine, 42, 43
2-Aminopurine, 60
1-Aminopyridinium salts, 88
2-Aminopyrazine, 65
4-Aminopyridine, 9
2-Amino[^{15}N]pyridine, 14, 163
4-Aminopyrimidine 3-oxides, 170
4-Aminoquinazoline, 53, 56, 58
4-Amino-1-sulfonylated 1H-benzotriazoles, 193
5-Amino-1,2,3,4-tetrazole, 157, 158
1-Amino-4-(1-thymyl)propane, 118
5-Amino-1,2,3-triazoles, 194

2-Amino-4,6,7-triphenylpteridine, 63
Aminodechlorination, 17, 18, 22, 24, 32, 53, 61, 65, 72
Aminodeethoxylation, 101, 102
6-Amino-4,5-dihydro-4-oxo-1-methylpyrazolo-[3,4-d] pyrimidine, 58
6-Amino-5-formyl-1,2-dihydro-2-imino-1-methylpyrimidine, 188
Aminodefluorination, 63, 72
Aminodehalogenation, 21, 28, 33, 34, 37,58, 62, 63, 71, 75
Aminodehydrogenation, 41, 45, 46, 47, 50, 52, 62, 67, 76, 77, 79
Aminodehydroxylation, 15
Aminohydroxy replacement, 92
Aminodemethylthiolation, 62, 63, 69
Aminodenitration, 125
Aminodeoxogenation, 56, 75
α-Amino-$β,β^1$-dicyanostyrene, 26
7-Amino-6-ethoxycarbonyl-1,3-dimethylpyrido [2,3-d]pyrimidine-2,4(1H, 3H)-dione, 185
3-Aminoisoquinoline, 20
Amino-methylamine exchange, 97
Ammonium bisulfite, 92
2-Anilino-3-aminoquinazolin-4(3H)-one, 110
2-Anilino-3-aminoquinazoline-4(3H)-hydrazone, 110
4-Anilino-5-bromopyrimidine, 4, 44
3-Aroylamino-5-methyl-1,2,4-oxadiazoles, 204
Aroylarylazo-1,2,4-triazolium salt, 220
1-Aryl-5-amino-1,2,3-triazole, 155
1-Aryl-5-amino-4-phenyl-1,2,3-triazoles, 156
3-Aryl-5-arylidene-2-methylthioimidazolin-4-one, 128
1-Aryl-4,6-dimethyl-2 (1H)-pyrimidinethiones, 108
4-Aryl-3,5-dioxo-2,3,4,5-tetrahydro-1,2,4-triazine, 120
1-Aryl-5-hydrazino-1,2,3-triazole, 162
3-Aryl-2-hydrazonothiazoline, 162
1-Aryl-4-nitroimidazole, 126
5-Aryl-1,2,4-oxadiazoles, 203
3-Arylamino-4-amino-5-oxo-4,5-dihydro-1,2,4-triazines, 120
2-Arylamino-4,6-dimethylpyrimidine, 109
2-Arylamino-4,6-dimethylpyrimidine N-oxides, 108

INDEX

5-Arylamino-4-phenyl-1,2,3-triazoles, 156
5-Arylamino-1,2,3-triazole, 155
5-Arylidene-3-amino-2-arylimino-
  imidazolidin-4-one, 128
1-Arylsulfonyl-4-nitroimidazoles, 127
6-π Assisted heterocyclization, 220
2-Aza-4-cyano-3-phenyl-1-piperidino-1,3-
  butadiene,12
5-Azacytosine, 148
5-Azauracil, 148
Azido-imine, 157
Azobenzene, 51, 67

B

Base-catalyzed fragmentation, 131
Base-induced fragmentation, 29
Benzamidine, 73, 137, 138
3-Benzoylamino-5-methyl-1,2,4-oxadiazole,
  4, 200
1-Benzylinosine, 116
1-Benzyl-N^6-methoxyadenine, 178
2-Benzyl-5-nitropyrimidine, 142
3-Benzyloxymethyl-1-ribosyl-5-cyanouracil,
  185
2-Benzylthio-3-phenylquinazolin-4(3H)-
  thione, 110
Betaine, 186
Bicyclic intermediates, formation
  between 3,5-dinitro-1-methylpyridin-
    2(1H)one and ketones/ammonia,
    132, 133, 134
  between 3,5-dinitropyridin-4(1H)one and
    diethyl sodio-3-oxopentanedioates,
    136, 137
  between 5-nitropyrimidine and amidines,
    140
  between 3-methyl-5-nitropyrimidin-
    4(3H)one and a ketone/ammonia,
    145, 146
  between 6-nitro-1,2,4-triazolo[1,5-
    a]pyrimidine and nitriles, 180, 181
Bifunctional nucleophiles, 130
Bipolar species, 159, 212
Bond breaking, 5, 7, 32, 65, 66, 100, 161,
  173, 204
π-Bonded S^{IV}, 214
Bond-switch process, 207, 209, 210, 213, 215
6-Bromo-5-deuterio-4-phenylpyrimidine, 11,
  28
4-Bromo-2,6-diphenylpyrimidine, 38

2-Bromo-4,6-diphenylpyrimidine, 38
3-Bromoisoquinoline, 19
5-Bromo-6-methylisocystosine,142
5-Bromo-4(N-methylanilino)pyrimidine, 11
6-Bromo-4-phenylpyrimidine, 2, 7, 12, 21,
  22, 26, 29
4-Bromo-6-phenylpyrimidine, 37
3-Bromopyridazine, 68
5-Bromo-4-t-butyl-6-deuteriopyrimidine, 41
5-Bromo-1,3,6-trimethyluracil, 142
6-Bromoquinazoline, 66
2-Bromoquinazoline, 66
2-Bromoquinoline, 19
Bucherer substitution, 92
1-t-Butylinosine, 116
4-t-Butyl-5-bromo-6-deuteriopyrimidine, 11
4-t-Butyl-5-bromo-5,6-dihydropyrimidine,
  42
4-t-Butyl-6-bromopyrimidine, 23
4-t-Butyl-6-chloropyrimidine, 23
3-t-Butyl-6-hydrazino-1,2,4,5-tetrazine, 82
2-t-Butyl-5-nitropyrimidine, 139
4-t-Butyl-5,6-pyrimidyne, 11, 41
4-t-Butylpyrimidine, 52
2-t-Butylpyrimidine, 137
1-Butylthymine, 119
5-t-Butyl-1,2,4-triazin-3-one, 75

C

Calculated frequencies, 207
Calculations
  ab initio calculations of the
    rearrangement of 5-amino-
    1,2,3,4-tetrazoles, 57, 192
  of charge densities on C-2,4,5,6 in
    4-phenylpyrimidine, 30, 31, 96, 99
  of π-electron stabilization energy
    (MNDO, PM3, AM1 method) of
    attack of ammonia on C-2,4,5,6 in
    N-methylpyrimidinium ion, 96, 30, 31
  of the N-amino-N-nitro transition state in
    1,4-dinitroimidazole, 125
  of the addition to the bridge head C-4 in
    2,3a,6a-triazaphenalenium salts, 186
Calf thymus DNA, 120
Cannizzaro reaction, 38
5-Carbamoyluracil, 135
Carbinolamine, 167
Carbon as Pivotal Atom, 218

Carbon-nitrogen skeleton rearrangement, 31
4-Carboxamido-5-(2-nitroanilino)-1H-1,2,3-triazole, 156
Charge-controlled reaction, 30, 31, 76, 96
Chichibabin amination, 23, 45, 49, 51, 58, 67, 76, 79
Chichibabin hydrazination, 81, 82
Chiral amino acid esters, 116
2-Chloroadenine, 61
3-Chloro-1,2,4-benzotriazine, 75
2-Chloro-4,6-dicyanopyrimidine, 38
2-Chloro-3,5-dinitropyridine, 17
2-Chloro-3,6-diphenylpyrazine, 67
2-Chloro-5,6-diphenylpyrazine, 67, 68
2-Chloro-4,6-diphenylpyrimidine, 56
6-Chloro-2,4-diphenylpyrimidine, 25
2-Chloro-4,6-diphenyl-pyrimidine, 63
4-Chloro-5-ethoxy-4-phenylpyrimidine, 32
3-Chloroisoquinoline, 20
4-Chloro-5-methoxy-2-phenyl[6-^{14}C]pyrimidine, 32
3-Chloro-6-methylpyridazine, 68
Chloromethyl sulfone anion, 96
2-Chloro-5-nitropyridine, 14, 59
3-Chlorophenanthro[9,10-e]1,2,4-triazine, 75
6-Chloro-5-[^{14}C-cyano]-4-phenylpyrimidine, 26
4-Chloro-2-phenylpyrimidine, 59
6-Chloro-4-phenylpyrimidine, 22
6-Chloro-4-phenylpyrimidine-3-oxide, 22
2-Chloro-6-phenylpurine, 61, 63
2-Chloro-4-phenylquinazoline, 56
2-Chloropyrazine, 65
2-Chloro-4,6,7-triphenylpteridine, 63
4-Cyanamino-5-carboxamido-1-methylimidazole, 58
3-Cyano-1,2-dimethyl-4-oxo-tetrahydroquinolinium salt, 197
2-Cyanoimidazole, 65
3-Cyano-5-nitropyridinium salt, 196
2-Cyano-4-phenylquinazoline, 56
4-Cyano-1-thia-2a,5a-diazaacenaphthene, 187
3-Cyanothiophenes, 183
Cine-substitution, 11, 41, 44, 123
Classification, 1
C–N donation, 138, 139, 141
Coalescence studies, 194, 203, 209
Conjugate base, 104, 106

Cornforth rearrangement, 190
Covalent anionic adducts, formation of
5-alkyl(aryl)amino-1,4-dinitroimidazole anion, 126, 127
5-(-N-aminomorpholine)-1,4-dinitroimidazole anion, 129
1-amino-3-bromo-1,2-dihydroisoquinolinide, 19
4-amino-2-bromo-1,4-dihydroquinolinide, 20
6-amino-5-bromo-4-t-butyl-1,6-dihydropyrimidinide, 36
6-amino-2-chloro-1,6-dihydro-5-nitropyridinide, 16
6-amino-4-chloro-1,6-dihydro-2-R-pyrimidinide, 32
4-amino-2-chloro-1,4-dihydro-3,5-dinitropyridinide, 18
6-amino-2-chloro-1,6-dihydro-3,5-dinitropyridinide, 18
2-amino-5-chloro-1,2-dihydro-3,6-diphenylpyrazinide, 67
6-amino-2-chloro-1,6-dihydro-4-phenylpyrimidinide, 36
4a-amino-3-chloro-dihydro-1,2,4-benzotriazinide, 76
4a-amino-3-chloro-dihydrophenanthro[9,10-de]-1,2,4-triazinide, 76
4-amino-2-chloro-dihydro-4,6-diphenyl-1,3,5-triazinide, 80
1-amino-1,2-dihydroisoquinolinide, 19
2-amino-1,2-dihydro-4-phenylpyrimidinide, 48
4-amino-1,4-dihydro-2-phenyl-1,3,5-triazinide, 76
2-chloro-1,6-dihydro-6-hydroxy-5-nitropyridinide, 15
2,3-dihydro-6-(chloro, bromo)-3-hydrazinopyridazinide, 68
Covalent hydration, 20
Covalent hydrazination, 107
Covalent neutral adducts, formation of
6-amidino-1,6-dihydro-1-methyl-4(5)-phenylpyrimidine, 130
6-amidino-1,6-dihydro-1,3,5-triazine, 148
6-alkylamino-5,6-dihydrothymine, 117
2-amino-1-alkyl-1,2-dihydropyridine, 91
2-amino-4,6-diethoxy-1,2-dihydro-1-ethylpyrimidine, 102

INDEX

4-amino-3,4-dihydropteridine, 62
6-amino-1,6-dihydro-1-methylpyrimidine, 95
6-amino-1,6-dihydro-4-phenylpyrimidine, 49
2-amino-1,2-dihydro-4-phenylpyrimidine, 49
6-amino-1,6-dihydro-6-hydrazino-1,2,4,5-tetrazine, 85
1-aryl-4,6-dimethyl-6-hydroxylamino-2(1H)pyrimidinethione, 109
6-cyanoalkyl-5-cyano-1,2-dihydrouracil, 185
3-cyano-1,2-dihydro-2-hydroxypyridine, 196
1,6-diamino-1,2-dihydropyridine, 163
6,7-diamino-5,6,7,8-tetrahydropteridine, 62
1,2-diamino-1,2-dihydro-2,4,6-trimethylpyrimidine, 105, 106
6-[di(aminocarbonyl)methyl]-5-nitrouracil, 135, 136
1,3-di(aminocarbonyl)-1,2-dihydro-2-hydroxypyridine, 197
1,3-di(aminocarbonyl)-1,6-dihydro-6-hydroxypyridine, 198
3,4-diaryl-3-ethoxy-2-methyl-5-phenylimino-1,2,4-triazole, 159
1,3-dimethyl-6-guanidinouracil, 142, 143
1,6-dihydro-6-(2,4-dinitroanilino)-1-R-pyridine, 87
1,6-dihydro-6-ethoxy-6-(ethylamin)pyrimidine, 103
1,4-dihydro-4-hydroxy-1-methyl-3-nitroquinoline, 99
1,6-dihydro-6-imino-1-isopropyl-4-phenylpyrimidine, 29
Crossover experiments, 194
Cyanamide, 130, 131
5-Cyano-1,2-dihydro-2-imino-1,4,6-trimethylpyrimidine, 188
5-Cyano-1,2-dihydro-2-imino-1-methylpyrimidine, 188
5-Cyano-4,6-dimethyl-2-(methylamino)pyrimidine, 188
Cyanogen bromide, 90
Cyanoguanidine, 143
Cyclization, 97
1,4-Cycloaddition, 141
Cycloaddition, 141, 142

Cyclopentapyrimidine, 145
Cytosine 3-oxide, 170
Cytosine, 168

D

Dealkylation, 98
Dealkoxylation, 98
Deamination, 104
Deethoxycarbonylation, 89
Deethylation, 100
Deethoxylation, 100
Degenerate conversion of
  (C–N$^+$) into (C–N), 130, 141
  (CCN) into (CCN), 132
  (CCC) into (CCC), 136, 137
  (NCC) into (NCC), 135, 136
  (NCC into (NCC), 144, 145 146,
  ($^+$NCN) into (NCN), 138, 139 148, 143, 148, 149
  $^+$N (aryl) into $^+$N (alkyl), 88, 89, 90
  $^+$N (cyano) into $^+$N (protein), 90
  $^+$N (alkyl) into N, 91, 95, 97, 101, 103
  $^+$N (aryl) into N, 91
  $^+$N (benzyl) into N, 98
  $^+$N (amino) into +N (oxyde), 107, 108
  $^+$N (aryl) into +N (amino), 110
  N (aryl) into N, 109, 129
  N (aryl) into N (amino), 111, 112
  N (nitro) into NH, 113, 114, 115
  N (nitro) into N (amino), 115, 129
  N (nitro) into into N (alkyl, aryl), 117, 124, 127
  N (alkyl$'$) into N(alkyl$''$), 118, 119
Degenerate rearrangements
  amidine rearrangements, 5, 158, 161, 162, 165, 166, 172
  Boulton–Katritzky rearrangement, 3, 155
  carbon-carbon interchange
    in 2-aminothiophenes, 183
    in 3-aminocarbonyl-1-methoxypyridinium salt, 198
    in 3-benzyloxymethyl-5-cyano-ribosyluracil, 185
    in 5-cyano-1,3-dimethyluracil, 184
    in 3-cyano-1,2-dimethyl-4-oxotetrahydroquinolinium salt, 197
    in 5-cyano-1-methylpyridinium salt, 196
    in 3-cyano-5-nitropyridinium salt, 196
    in 1,3-di(aminocarbonyl)pyridinium salt, 197

in 3,5-dicyano-1,2,6-
trimethylpyridinium salt, 196
in 1,2-dihydro-5-cyano-2-imino-1-
methylpyrimidine, 188
in 2,3-dimethyl-4-formy-1-
phenylpyrazolinone-5, 183
in 3-formamidinopyridinium salt, 198
in 4-hydroxymethylene-5-oxazolone,
183
Dimroth rearrangement, 3, 5, 7, 8, 29, 36,
71, 93, 101, 110, 131, 153, 157–159,
161, 163, 165, 167–175, 177, 178, 180,
188, 194
(1,2)-(4,5) *exo*annular rearrangement, 154
(1,2,3)-(5,6,7) *exo*annular rearrangement,
154
heteroatom–heteroatom interchange
in 2-aminopyridines, 163
in 4-aminopyrimidine 3-oxides, 170
in 2-aminothiophene, 183
in 2-amino(imino)quinazolines, 180
in amino-1,2,3,4-tetrazoles, 155, 157,
158
in 5-amino(hydrazino)-1,2,3-triazoles,
155–157
in 5-cyano-2-iminopyrimidines, 188
in 5-cyanouracils, 184, 185
in 4-cyano-1-thiadiazaacenaphthene,
187
in 5-cyanotriazaphenalenium chloride,
185
in cytosines, 168, 169
in 3,5-diarylimino-1,2,4-
thiadiazolidines, 159–161
in (di)imino-1,2,4-dithiazolidines, 161
in 1,3,4-dithiazolidines, 161
in 4-formylpyrazolinone, 183
in 2-hydrazinothiazolines, 162
in 1-hydroxy*iso*guanines, 176
in 4-hydroxymethyleneoxazolone, 183
in 5-imino-1,2,4-thiadiazolines, 159, 160
in iminoimidazolines, 162
in 2-iminopteridines, 175
in iminopyrimidines, 165, 167, 168
in 2-imino-1,3,5-triazines, 172
in 1-methyladenine(adenosine), 175,
177, 178
in 1-methylpurines, 176
in 1,2,4-triazolium-3-aminides, 158
Delocalized anion, 44

Demethylation, 92, 94, 95, 96, 97, 104
6-Deuterio-3,5-dinitro-1-methyl-2-pyridone,
132
4-Deuterio-3-nitro-5,6,7,8-
tetrahydroquinoline, 132
Deuterium-hydrogen exchange, 9, 11, 12, 28
Deuterium labeling, 11, 32
3-Dialkylaminomethyl-6H-imidazo[1,2-
c]quinazolin-5-one, 110
2,4-Diamino-1-aryl-6,6-mono (di)alkyl-1,6-
dihydro-1,3,5-triazine, 172
1,3-Di(aminocarbonyl)pyridinium salt, 197
Diamidino intermediate, 138
2,6-Diamino-4-methylpyrimidine, 52
2,4-Diamino-6-methylpyrimidine, 56
1,2-Diaminopyrimidinium salt, 110
2,6-Diaminopurine, 61
6,7-Diamino-5,6,7,8-tetrahydropteridine, 62
2,4-Diaminoquinazoline, 58
4,6-Dianilino-1,2-dihydro-2-imino-1-phenyl-
1,3,5-triazine, 172
Diaryloid, 201
Diaroyl effect, 201, 209
1,4-Diary-2-methylthiopyrimidinium iodide,
109
1,3-Diazacyclohepta-1,2,4,6-tetraene, 164
2,7-Diazatropylidene, 164
Diazo functionality, 157
Diazoimines, 156, 157, 191, 194
Diazo-imines, 157
5-Diazomethyl-4-alkoxycarbonyl-1,2,3-
thiadiazole, 213
5-Diazomethyl-4-methoxycarbonyl-1,2,3-
triazole, 220
Diazotation, 16
2,4-Di-*t*-butyl-6-chloropyrimidine, 25
Dicarbonylnitrile ylid, 190
Dicyanoazadiene, 27
3,5-Dicyano-1,2,6-trimethylpyridinium salts,
196
Descriptor frequence values, 206
3,5-Di(ethoxycarbonyl)pyridin-4-(1H)-one,
136
4,6-Diethoxy-1-ethylpyrimidinium
tetrafluoroborate, 101
3,4-Diformaldoxime furoxan, 204
Diformylamine, 146
Dihydroaminoazines, 16
1,4-Dihydro-6-ethoxy-1-ethyl-4-
iminopyrimidine hydrogen
tetrafluoroborate, 101

1,6-Dihydro-4-ethoxy-1-ethyl-6-
    iminopyrimidine hydrogen
    tetrafluoroborate, 101
1,2-Dihydro-3-formyl-2-imino-1-
    methylpyridine, 196
1,2-Dihydro-1-hydroxy-2,3,1-
    benzodiazaborine, 121
1,4-Dihydro-4-hydroxy-1-methyl-3-
    nitroquinoline, 99
1,2-Dihydro-2-hydroxy-1-methyl-3-
    nitroquinoline, 99
2,3-Dihydro-6-hydroxylamino-2-oxopurine,
    176
1,2-Dihydro-2-imino-1,4-
    dimethylpyrimidine, 168
1,2-Dihydro-2-imino-1-methylpyrimidine,
    165
2,3-Dihydro-2-imino-3-methylquinazoline,
    180
2,3-Dihydro-3-hydrazino-3-methyl-1,2,4,5-
    tetrazine, 85
2,3-Dihydro-3-hydrazino-3-R-1,2,4,5-
    tetrazine, 83
2,3-Dihydro-1,2,4,5-tetrazine, 84
1,6-Dihydro-1-methyl-6-thiopurine, 175
2,3-Dihydro-3-oxo-1,2,4-triazine, 71
1,4-Di(1-thymyl)butane, 119
1-Dimethylamino-5-cyano-2,3a,6a-
    triazaphenalenium chloride, 185
4-Dimethylamino-1,2-dihydro-2-imino-1-
    methylpyrimidine, 166
4-Dimethylamino-5-nitro-2-imino-1-
    methylpyrimidines, 166
Dimethylamino-6-nitro-1,2,4-triazolo[1,5-
    a]pyrimidine, 180
1,3-Dimethyl-5-azauracil, 148
1,3-Dimethyl-5-cyanouracil, 184
1,3-Dimethylpseudouridine, 144
1,2-Dimethylpyrimidinium iodide, 96
3,5-Dimethyl-1,2,4-triazole, 104, 105
1,3-Dimethyluracil, 142, 143
1,3-Dimethylurea, 143
4,6-Dimethoxypyrimidine, 44
1,1-Dimorpholinoethene, 141
1,3-Dinitroacetone, 136
1,4-Dinitroimidazole, 123
1,4-Dinitro-2-methylimidazole, 124, 126, 127
3,5-Dinitro-1-methylpyridin-2(1H)-one, 132
1-(2,4-Dinitrophenyl)pyridinium salts, 87
2,3-Diphenylquinoxaline, 66

3,5-Dinitropyridin-4(1H)-one, 136
4,6-Dioxo-6-H, 11H-11-methyl-3-
    phenyl[1,2,4]-triazino[3,4-
    b]quinazoline, 172
3,5-Diphenyl-4-acetylisoxazole, 190
4,6-Diphenyl-2-fluoropyrimidine, 72
6,7-Diphenyl-2-methylthiopteridine, 64
6,8-Diphenyl-2-methylthiopurine, 64, 65
3,5-Diphenyl-1,2,4-triazine, 71, 72, 74, 149
2,4-Diphenyl-1,3,5-triazine, 71
Dipolar species, 170
1,2-Di-R-4-(hydroxyamino),
    hexahydropyrimidines, 170
1,2-Di-R-4-amino-1,2,5,6-
    tetrahydropyrimidine 3-oxides, 170
1,4-Di(1,3,5-triazin-1-yl)benzene, 148
Diradical intermediate, 190
Dost's base, 160
Double ANRORC-type rearrangement, 29,
    173, 187
5-(β-D-Ribofuranosyl)isocytosine, 144
Dynamic NMR spectroscopic
    measurements, 203, 209

E

E- and Z-isomers, 157, 204, 206, 207, 215
Electrocyclic ring opening, 92, 93, 94
π-Electron deficiency, 17, 69, 71, 80
Electron density, 69, 76
Electron repulsion, 44, 143
Electron transfer, 51
4-Ethoxy-1-ethylpyrimidiniumm
    tetrafluoroborate, 98
4-Ethoxy-2-phenylpyrimidine, 100
3-Ethoxycarbonyl-6,7,8,9-tetrahydro-4(H)-
    pyrido[1,2-a]pyrimidin-4-one, 97
6-(Ethylamino)-4-phenylpyrimidine, 103
3-Ethyl-6-bromo-1,2,4,5-tetrazine, 85
1-Ethyl-3-cyanopyridinium iodide, 196
1-Ethyl-4-imino-2-phenylpyrimidine, 100
π-Excessive character, 123
1,3-Exoannular rearrangements, 153, 155,
    163, 165, 180, 182, 183

F

2-Fluoro-4-phenylpyrimidine, 72
2-Fluoro-4-phenylquinazoline, 55
3-Fluoro-5-phenyl-1,2,4-triazine, 72, 74
3-Formamidopyridinium salt, 198

3-Formyl-2-(methylamino)pyridine, 196
5-(Formylamino)-4-[(N-β-aminoethyl)formamidino]imidazole, 178
Formylcyanonitropropenide salt, 15
Fragmentation process, 73
Frontier orbital calculations, 30, 31, 99
Furoxan-furazan rearrangement, 8

G

Glutaconic dialdehyde, 92
G_0, G_1, G_2 and G_3 graph, 6, 7, 8
Guanidine, 142, 144, 148

H

3-Halogeno-5-phenyl-1,2,4-triazine, 149
6-Halogeno-4-phenylpyrimidines, 36
Halogenopyridazines, 67
6-Halogenopyrimidines, 21, 22
Halogeno-1,2,4,5-tetrazines, 67
3-Halopyridines, 9
Hammett relation, 126
Hammet's σ-values, 156
Heat of formation, 30, 31, 201
Hector's base, 159
Heterocyclization, 220
Hexahydropyrimidines, 170
Hexahydro-2,4,6-tris(imino)-1,3,5-trimethyl-1,3,5-triazine, 172
Homotetrazole aromaticity, 84, 85
Hydrazination, 68, 84, 120
Hydrazinodeamination, 85
Hydrazinodehalogenation, 67, 68, 85
Hydrazinodehydrogenation, 81, 82
Hydrazinodemethylthiolation, 110
Hydrazinolysis, 45
Hydrazinotetrazines, 81
6-Hydrazino-1,2,4,5-tetrazines, 85
Hydride transfer, 32
Hydroxydechlorination, 14, 15, 16, 59
5-Hydroxy-4-iminomethyl-1,2,3-triazole, 193
1-Hydroxyisoguanine, 176
4-Hydroxymethylene-5-oxazolone, 183
π-Hypervalent S^{IV}, 208

I

Imidazole, 65
6-Imino-8-azapurines, 176

2-Imino-3-β-carboxyethyl-2,3-dihydro-4-oxopteridine, 175
2-Imino-1-methylpyridine, 163
2-Iminobenzoyl-3-cyanamino-5,6-diphenylpyrazine, 63
Iminodeethoxylation, 98, 100
4-Iminomethyl-1-phenyl-1,2,3-triazoles, 191
ortho-Iminomethylenebenzylcyanide, 19
Iminopyrazolo[3,4-d]pyrimidines, 176
4-R-5-imino-1,2,4-thiadiazoline, 159
Immobilization, 89
Interchange. See Degenerate rearrangements
Interconversion. See Degenerate rearrangements
Isomerization. See Degenerate rearrangements
Z/E isomerizations, 157
Intermolecular cycloaddition, 117
Intramoleculair hydrogen bonding, 40, 85, 114, 192
Intermolecular migration, 57
Intermolecular transfragment reaction, 131, 135, 136, 137, 139, 142, 144
Inverse Diels–Alder cycloaddition, 139, 141, 142
3-Iodo-4-phenyl-1,2,4-triazine, 72
Iodopyrimidinyl radical anion, 23
Irradiation, 119, 120, 204
Isocyanates, 213
Isocytosine, 142, 143
Isomerization. See Degenerate rearrangements
Z/E isomerizations, 157
Isoselenocyanate, 212
Isothiocyanates, 213

K

Kinetic versus thermodynamic control, 18, 23, 31, 48, 93, 99
Kinetics of
  aminodenitration of 1,4-dinitroimidazole, 125, 126
  isomerization of 1-aryl-5-amino-4-phenyl-1,2,3-triazole, 156
  rearrangement of 1,2-dihydro-2-imino-1,4-dimethyl pyrimidine, 168
  rearrangement of 4-dimethylamino-2-imino-1-methyl-1,2-dihydropyrimidine, 166

rearrangement of 4-amino-5-nitro-6-iminopyrimidine, 166
rearrangement of 4-dimethylamino-5-nitro-2-imino-1-methylpyrimidine, 166
ring opening of 2-amino-3-methylpteridin 4(3H) one anion, 174, 175

L

Leaving group mobilities, 39
Lewis acids, 161
Liquid ammonia–potassium permanganate, 16

M

Meisenheimer adducts, 134
Meta telesubstitution, 33
Metal complexation, 31
4-Methoxy-6-aminopyrimidines, 44, 45
1-Methoxy-3-carbamoylpyridinium salts, 99
1-Methyladenine, 175
1-Methyladenosine, 178
2-Methylaminopyrimidine, 165
2-Methylaminopyridines, 163
6-Methylaminopurine, 175
3-Methyl-4-benzoylfuroxan oxime, 204
4-Methyl-6-chloropyrimidine, 24
Methyl α-cyano-β-(2-thienyl)acrylate, 218
3-Methylcytosine, 169
Methylguanidine, 143
7-Methylguanine, 58
5-Methylimino-4-phenyl-1,3,4-dithiazolidine 1-dioxide, 161
1-Methylinosine, 116
1-Methylisocytosine, 143
6-Methylisocytosine, 142
Methylisourea, 130
2-(Methylamino)pyrimidin-4(3H)-one, 143
1-Methylpyrimidinium methosulfate, 94
Michael addition, 97, 186
Mirror plane symmetry, 7
1-Morpholino-1-cyclohexene, 145

N

N^4-acetylcytosine, 169
N^6-alkoxy-7-alkyladenines, 178
N-alkyl-2-iminopyridines, 5
N-alkyl-3-R-pyridinium salts, 90
N-amino-4,6-diphenylpyrimidinium mesitylenesulfonate, 103
N^6(β-aminoethyl)adenosine, 177
N-aminomorpholine, 127
N-aminopyrimidinium mesitylenesulfonates, 107
N-amino-2,4,6-trimethylpyrimidinium mesitylenesulfonate, 104
N-benzyl-4-t-butylpyrimidinium salt, 98
N-benzyl-4,6-dimethylpyrimidinium salt, 98
1-N-butylthymine, 120
[N^4-β-carboxyethyl]cytosine, 169
N-cyano-N^1-(β-ethoxyvinyl)benzamidine, 32
N^+-cyanopyridinium polymer, 90
N-deethylation, 98
N^6-ethanoadenosine, 177
N-iminoylide, 104
Nitrene, 204
p-Nitroacetophenone, 146
3-(α-Nitroacetaldoxime)furazan, 204
Nitroamine, 113
Nitrodiformylmethane, 146
3-(α-Nitroethyl)-4-phenylfurazan, 204
Nitrogen as Pivotal Atom, 200
4-Nitroimidazole, 127
N-Nitroimidazoles, 116
N-Nitropyrimidones, 112
1-Nitroinosine, 114, 116
4-p-Nitrophenylpyrimidine, 146
5-Nitro-2-phenylpyrimidine, 138
5-Nitropyrimidine, 138, 139, 141, 142
3-Nitrouridine, 113, 116
5-Nitrouracil, 134
Nitrous oxide, 114, 116
6-Nitro[1,2,4-triazolo][1,5-a]pyrimidine, 179
Nitromethyl anion, 96
3-Nitro-4-methylpyridinium salt, 134
^{15}N labeled compounds
    5′-O-acetyl-2′,3′-O-isopropylidene-3-[^{15}N-amino]-[3-^{15}N]uridine,115
    2″,3′,5′-tri-O-acetyl-1-[^{15}N-amino]-[1-^{15}N]inosine, 116
    2-[^{15}N-amino]-4,6-diphenyl-1,3,5-triazine, 80
    3-[^{15}N-amino]isoquinoline, 19, 20
    3-amino[^{15}N]isoquinoline, 19
    2-amino-4-phenyl[1,3-^{15}N]pyrimidine, 36
    2-[^{15}N-amino]-4-phenylquinazoline, 55
    3-amino-5-phenyl[4-^{15}N]-1,2,4-triazine, 71
    2-[^{15}N-amino]pyridine, 163

3-[^{15}N-amino]-1,2,4-triazine, 69
3-benzyl[3-^{15}N]uridine, 113
5-bromo-4-t-butyl[1(3)-^{15}N]pyrimidine, 42
6-bromo-4-phenyl[1(3)-^{15}N]pyrimidine, 13
2-bromo[^{15}N]pyridine, 14
3-bromo[^{15}N]isoquinoline, 19
2-bromo-4-phenyl[1(3)-^{15}N]pyrimidine, 25
6-chloro-5-cyano-4-phenyl[1(3)-^{15}N]pyrimidine, 21
2-chloro[^{15}N]pyrazine, 65
2-chloro-4,6-diphenyl[(1),(3),(5)-^{15}N]-1,3,5-triazine, 80
2-chloro-4-phenyl[1,3-^{15}N]pyrimidine, 38
2-chloro-4-phenyl[1(3)-^{15}N]pyrimidine, 21
2-chloro-4-phenyl[3-^{15}N]quinazoline, 55
4-chloro[3-^{15}N]quinazoline, 53
1-chloro(bromo)-4-R-[2,3-^{15}N]phthalazine, 68
2,4-diphenyl-6-halogeno[1(3)-^{15}N]pyrimidines, 25
4,6-diphenyl-2-X-[1,3-^{15}N]pyrimidine, 38
2-fluoro-4-phenyl[1(3)-^{15}N]pyrimidine, 21
[^{15}N]hydrazinium hydrogen sulfate, 114, 116
2-imino-[3-^{15}N]imidazolidin-5-one, 163
[1-^{15}N]inosine, 114
2-iodo-4-phenyl[1(3)-^{15}N]pyrimidine, 21
2-methyl-4-nitro[1-^{15}N]imidazole, 125
1-methyl[1,3-^{15}N]pyrimidinium methyl sulfate, 95, 138
3-methylthio[4-^{15}N]-1,2,4-triazine, 69
[^{15}N]nitroamide, 125
3-nitro-[3-^{15}N]uridine, 113
4-phenyl-2-X-[1,3-^{15}N]pyrimidines, 34
5-phenyl[4-^{15}N]-1,2,4-triazin-3-one, 75
[3-^{15}N]pyrimidine nucleosides, 112
[3-^{15}N]purine nucleosides, 112
[3-^{15}N]quinazolone-4, 57
[1(3)-^{15}N]tetrazolo[1,5-*a*]pyridine, 164
[3-^{15}N]uridine, 113
N-methylpyrimidinium salts, 2, 49, 104, 137
4-(N-methylanilino)-5-brompyrimidine
N-methylformamidine, 147
N-methyl α-nitroacetamide, 132, 135, 136
N-methoxypyridinium salt, 92, 93
1-(*N*-Morpholino)-4-nitroimidazole, 127
NMR spectroscopic study of
 the covalent amination in
  N-alkylpyridinium salts, 91, 92

N-benzylpyrimidinium salts, 98
5-bromo-4-substituted pyrimidines, 41
2-chloro-3,6-diphenylpyrazine, 67
4,6-diethoxy-N-ethylpyrimidinium tetrafluoroborate, 101
3,5-dinitro-2-chloropyridine, 18
1,4-dinitroimidazole, 12
4,6-diphenylpyrimidine, 104
4,6-diphenyl-1,3,5-triazine, 79
6-ethoxy-4-oxopyrimidinium salts, 103
halogeno(amino)-1,2,4,5-tetrazines, 83–86
isoquinoline, 19
1-methylpyrimidinium salts, 94, 97
4-phenylpyrimidine, 48, 49
phenyl 1,3,5-triazine, 76
pyrimidine, 30
2-substituted-4-halogenopyrimidines, 31
4-substituted-2-halogenopyrimidines, 36
1,2,4-triazines, 69
bond switching in 5(β-amino-β-arylvinyl)isothiazoles, 214
bond switching equilibrium in pentacarbonyltungsten(0)-thioaldehyde-1,2-dithiol, 217
bond switching in 5-amidino-1,2,4-thiadiazoles, 207, 209
bond switching in 3-acetylamino-1,2,4-oxadiazole, 203
crossover experiments with benzotriazoles, 194
deprotonation of substiutuents in azines, 25, 44
the ring opening of N-methoxypyridinium salts, 93
the ring opening of1-methoxy-3-carbamoylpyridinium salts by liquid ammonia, 99
the structure of the intermediate from 5′-O-acetyl-2′,3′-O-isopropylidene-3-nitrouridine and [^{15}N]benzylamine, 113, 114
the structure of the intermediate from 2′,3,5′-tri-*O*-acetyl-1-nitroinosine and [^{15}N]benzylamine, 116
^{15}N NMR spectroscopy, 160, 210, 215
N–O bond fission, 201
N-phenyl-2-(ethoxycarbonyl)pyridinium salts, 89
N-phenyl-3-hydroxypyridinium chloride, 92

INDEX 251

^{15}N-Scrambling, 19, 20, 27, 42, 53, 71, 79, 163, 164
Nucleotide-TpdA, 119

O

O-methylisourea, 130, 131
1,2,4-Oxadiazoline, 171
9-Oxo-7-nitro-4,9-dihydrotriazolo[1,5a]pyrido[2,3-d]pyrimidine, 180

P

Pentacarbonyltungsten (0)-thioaldehyde-1,2-dithiol, 217
1-(Pentadeuteriophenyl)-3-aminoncarbonyl-4-deuteriopyridinium salt, 89
Phenylacetamidine, 142
1-Phenyl-5-acetamido-1,2,3,4-tetrazole, 158
Phenylation, 138
4-Phenyl-3-aroylimino-5-methylimino-1,2,4-dithiazolidine, 161
2-Phenylbenzimidazole, 66
1-Phenyl-2,3-dimethyl-4-formylpyrazolinone-5, 183
1-Phenyl-4-formyl-1,2,3-triazole, 192
3-Phenyl-5-methyl-4-(acetyl-d_3)isoxazole, 190
1-Phenyl-2-methyl-4-acetylpyrazolinone-5, 183
1-Phenyl-4-nitroimidazole, 127
2-Phenyl-5-nitropyrimidine, 142
5-Phenylimino-4-methyl-1,3,4-dithiazolidine 1 dioxide, 161
4-Phenyl-6-piperidinopyrimidine, 12
Phenylphosphorodiamidate (PPDA), 56, 75
2-Phenylpyrimidine, 137, 138
4-Phenyl-5,6-pyrimidyne, 12
4-Phenylpyrimidine, 47, 49, 50, 67, 145
5-Phenylpyrimidine, 51
5-Phenylpyrimidinium iodides, 131
Phenyl-1,3,5-triazine, 76
Phenyl-1,2,4-triazinediones, 130
Phenyl[1,2,4-triazino[4,3-a]quinazoline, 173
3-Phenylquinazolin-4-one, 130
Photoproducts, 116, 119
Photostimulation, 153
4-Piperidino-6-chloropyrimidine, 24
$ortho$-(Piperidinomethyleneamino)benzonitrile, 54

4-Piperidinoquinazoline, 54
Pivalamidine, 137, 138, 139, 141
1,3-Prototropic hydrogen shift, 157
6-(isoPropylamino)-4-phenylpyrimidine, 29
Pseudobase, 93, 97, 99
Pseudomonas pyrrocinia, 218
Purine, 62
Purinyl anion, 60, 61
2-Pyridylnitrene, 163
3,4-Pyridyne, 9, 10
2,3-Pyridyne, 9, 10, 11
2,3-Pyridine N-oxide, 11
Pyridine polymer, 90
5,6-Pyrimidyne, 12
Pyrimidine N-oxides, 104, 107
Pyrrolnitrin, 218, 220

R

Radical anion, 51
Radical scavengers, 23, 51, 67
Radioactive labeling, 27, 28 32, 65
RBR graphs, 4, 5, 7
1-R′-5-carbamoyluracil, 135
3-R-6-1,6-dihydro-hydrazino-1,2,4,5-tetrazine, 82
1-R-6-(2,4-dinitroanilino)-1,6-dihydropyridine, 87
1-R-4,4′-diphenyl-2-iminoimidazolidin-5-one, 162
Reactivity order, 56, 65, 74, 82
Rearrangements, 186
Regioselectivity, 96
Regiospecific chlorinations, 219
Retro–Michael reaction, 184
Reversed Dimroth reaction, 171
1-R-5-hydrazino-1H-1,2,3-triazole, 156
Ring-Bond-Redistribution Graphs, 4
Ring contraction
  of chloropyrazine into imidazoles, 66
  of 3-chloro-1,2,4-triazines into 1,2,4-triazoles, 71, 72
  of 4,6-diethoxypyrimidine into 3-methyl-1,2,4-triazole, 46
  of 2-methylthiopteridines into 2-methylthiopurines, 65

S

Second-order perturbation equation, 96
Side-chain participation, 3, 153

1,5-Sigmatropic shift of hydrogen, 216
$S_N(AE)$ mechanism, 11, 12, 14, 17, 18, 20, 24, 30, 31, 33, 40, 53, 63, 64, 72, 77, 80
$S_N(ANRORC)$ mechanism in the
  amination of 4-substituted pyrimidines, containing at C-2 a sulfur containing substituent, 39, 40
  aminodebromination in 3-bromoisoquinoline, 19, 20
  aminodebromination of 5-bromopyrimidines, 41, 42
  aminodechlorination of 6-chloro-5-cyano-4-phenyl pyrimidine, 26, 27
  aminodechlorination of 2-chloro-4,6-diphenyl-1,3,5-triazine, 80
  aminodechlorination of chlorobenzo-1,2,4-triazines, 75
  aminodechlorination of 2-chloropyrazines, 65
  aminodechlorination of 4-chloroquinazolines, 53, 54
  aminodehalogenation of 2-halogenoquinazolines, 55, 56
  aminodehalogenation of 2-halogeno-4-substituted pyrimidines, 34-38
  aminodehalogenation of 6-halogeno-4-substituted pyrimidines, 7, 13, 21, 24, 29
  aminodehalogenation of 4-halogeno-2-substituted pyrimidines, 31, 32
  aminodehalogenation of 2-halogenopteridines, 62, 63
  aminodehalogenation of halogenopurines, 58-61,
  aminodehalogenation of 3-halogeno-1,2,4-triazines, 71-73
  aminodehydrogenation of phenylpyrimidines, 46-51
  aminodehydrogenation of phenyl-1,3,5-triazines, 76, 77, 79
  aminodehydrogenation of quinazolines, 58
  aminodemethoxylation of 4,6-dimethoxypyrimidines, 45
  aminodemethylthiolation of 2-methylthiopteridines, 64
  aminodemethylthiolation of 3-methylthio-1,2,4-triazines, 69, 70
  aminodenitration in 1,4-dinitroimidazole, 123-127
  aminodenitration of 3-nitrouridines, 113, 114
  aminodenitration of 1-nitroinosines, 114, 116
  aminodeoxogenation of quinazolin-4-one, 56, 57
  conversion of the angular [1,2,4]triazino[4,3-*a*] quinazoline into the linear [1,2,4]triazino [3,2-*b*]quinazoline, 173, 174
  conversion of 1-phenyl-2,3-dimethyl-4-formylpyrazolinone-5 into 1-phenyl-2-methyl-4-acetylpyrazolinone-5, 183
  4-hydroxymethylene-5-oxazolone into oxazole-4-carboxylic acid, 183
  conversion of 6-amino-2-chloro-7-methylpurine into 7-methylguanine, 58
  conversion of N-alkyl-3-cyanopyridinium salt into 3-formyl-2-(methylamino)pyridine, 196
  conversion of 1,3-di(aminocarbonyl)pyridinium salt into, 3
  conversion of 3-formamidinopyridinium salt into 2-amino-3-formylpyridine, 198
  dealkyl(benzyl)ation of N-alkyl(benzyl)pyrimidinium salts, 94-100
  dealkylation of N-alkylpyridinium salts 91, 92
  deamination of N-aminopyrimidinium salts, 103-106
  detosylation of 2,3,1-benzodiazaborine, 121, 123
  hydrazinodeamination in amino1,2,4,5-tetrazines, 85, 86
  hydrazinodehalogenation of 1-chloro(bromo)phthalazines, 68
  hydrazinodehalogenation of halogeno-1,2,4,5-tetrazines, 85
  hydrazinodehydrogenation of 1,2,4,5-tetrazines, 81-84
  hydrazinedenitration in 1,4-dinitroimidazoles, 129, 130
  hydroxydechlorination of 2-chloro-5-nitropyridine, 15, 16
  interconversion of 1,2,4-triazolium-3-aminides, 159
Zincke exchange reactions, 87-90

S_NH *tele* amination, 43
S_Ni mechanism, 200
$S_{RN}1$ mechanism, 23
S-methyl*iso*ourea, 131
Sodium hydrazide, 81
Solvation, 50
Steric interference, 79
Steric strain, 112
Strained conformation, 84
Strand scission, 120
3-(Substituted amino)furazan 4-carboxamidoximes, 206
Sulfur-amino rearrangement, 144
Sulfur as Pivotal Atom, 207
Symmetrical intermediate, 203

T

Tautomeric equilibrium, 97
Tele amination, 59
Terephthalamidine, 148
3aλ^4-1,3,4,6-Tetraazapentalenium salt, 210
3aλ^4-1,3,4,6-Tetraazathiapentalene, 208
1,2,3,4-Tetrahydro-2,2-dialkyl-4-nitromethylenepyridine, 134
1,2,3,4-Tetrahydro-2,6-dioxopyrimido[1,2-*c*]pyrimidine, 169
1,2,4,6-Tetramethylpyrimidinium iodide, 96
1,2,3,4-Tetrahydro-4-imino-1-methyl-2-oxo-3-phenylquinazoline, 180
2,4,7,9-Tetramethyldipyrimido[1,2-*b'*,1,2'-*c*]hexahydrotetrazine, 106
Thermal induction, 153
Thermal interconversion, 190
Thermodynamic stability, 201
1,2,4-Thiadiazole[4,5-*a*]pyrimidine, 211, 212
Thiapentalene structure, 210
1,3-Thiazine, 144
2-Thio-4-iminotetrahydroquinazoline, 180
2-Thiopseudouridine, 144
2-Thiouracil, 143

Thiourea, 143
2-Thioxo-3-phenylquinazolin-4(1*H*, 3*H*)-one, 110
Thorpe carbon–carbon cyclization, 216
Thymidine, 118
(*p*-Toluenesulfonylamino)benzotriazole, 193
5-(*p*-Toluenesulfonylamino)-1,2,3,4-tetrazole, 158
1,3,5-Triazine, 148
1,2,4-Triazolium-3-aminides, 158
3-Trichloroacetylamino-5-methyl-1,2,4-oxadiazole, 201
Trigonelline, 91
1,3,7-Trimethylpyrido[2,3-*d*]pyrimidine-2,4(1*H*,3*H*) dione, 184
1,4,6-Trimethylpyrimidinium iodide, 96
1,3,6-Trimethyluracil, 142
2,4,6-Trimethylpyrimidine, 104
2,4,6-Trinitrochlorobenzene, 31

U

Uracil, 143
Urea, 143
4-Ureido-5-cyano-1-methylimidazole, 58

V

Vilsmeyer–Haak conditions, 150
Vilsmeyer–Haal reagent, 151

X

Xanthine-8-carboxylic acid, 65

Y

Ylid, 106, 107

Z

Zincke exchange reations, 87, 89

ISBN 0-12-020774-5